Fire Engineering
and Emergency Planning

Fire Engineering and Emergency Planning

Edited by

RONALD BARHAM

Department of Built Environment,
University of Central Lancashire,
UK

CRC Press
Taylor & Francis Group
Boca Raton London New York

CRC Press is an imprint of the
Taylor & Francis Group, an **informa** business

A TAYLOR & FRANCIS BOOK

CRC Press
Taylor & Francis Group
6000 Broken Sound Parkway NW, Suite 300
Boca Raton, FL 33487-2742

First issued in paperback 2019

© 1996 by Taylor & Francis Group, LLC
© Chapter 49: R.A Klein and CFRS Consulting Services, 1995
CRC Press is an imprint of Taylor & Francis Group, an Informa business

No claim to original U.S. Government works

ISBN-13: 978-0-419-20180-9 (hbk)
ISBN-13: 978-0-367-40145-0 (pbk)

Publisher's note   This book has been produced from camera-ready copy provided by the individual contributors.

A catalogue record for this book is available from the British Library

Visit the Taylor & Francis Web site at
http://www.taylorandfrancis.com

and the CRC Press Web site at
http://www.crcpress.com

Contents

Introductions

René Montet *President of CECAM*   *(unreported)*
Jean-Louis Mathieu *Director of CECAM*   *(unreported)*

Edith Cresson *European Commissioner for Science, Research and Development*

Benjamin J. M. Ale *Rijsinstituut voor Volksgezondheid en Mijeuhygiene (The Netherlands), Laboratorium voor Stralingsonderzoek (National Institute of Public Health and Environmental Protection, Laboratory of Radiation Research)*
Sven-Erik Magnusson *Professor, Lund University (Sweden), Institute of Science and Technology, Department of Fire Safety Engineering*
Bryan T. A. Collins *Her Majesty's Chief Inspector of Fire Services, The Fire Service Inspectorate of the The Home Office (U.K.)*

Chairs of sessions

Roland Borghi *Professor, LFMN-CORIA / Institut of Mechanics, Insa, Rouen (France)*
Marita Kersken-Bradley *Consulting Engineer, Ingenieur Büro Kersken+Partner, München (Germany)*
Sven-Erik Magnusson *Professor, Lund University (Sweden), Institute of Science and Technology, Department of Fire Safety Engineering*
Lindiwe Rubadiri *presently with University of Central Lancashire (England), Department of Built Environment; shortly to become Lecturer, School of Engineering, Gabarone, Botswana*
João L. Porto *Professor, Universidade do Porto (Portugal), Departamento de Engenharia Civil (University of Porto, Department of Civil Engineering) and Director Geral (Managing Director), Metro do Porto S.A. (Portugal)*
Denis T. Davis *Executive Chairman, Instituion of Fire Engineers; Chief Fire Officer, Cheshire County Fire Brigade (England)*
Ronald Barham *Visiting Professor of Construction Economics to Fachhochscule Augsburg (Germany), Fellow of Sheffield Univesity (England), Faculty of Law and Senior Lecturer at the University of Central Lancashire (England), Department of Built Environment*

List of contributors

Angell, E.L.   *Chartered Architect, Preston (England)*

Angell, J.C.   *Senior Lecturer, University of Central Lancashire (England), Department of Built Environment*

Areitio, J.   *Networks & Systems, Bilbao (Spain)*

Areitio, M.T.   *Networks & Systems, Bilbao (Spain)*

Baratov, A.N.   *Professor, Moscow State University of Building (Russia)*

Barham, R.   *Visiting Professor of Construction Economics to Fachhochschule Augsburg (Germany), Fellow of Sheffield Univesity (England), Faculty of Law and Senior Lecturer at the University of Central Lancashire (England), Department of Built Environment*

Borchiellini R.*Polytecnico di Torino (Italy), Departimento di Energetica*

Bordignon, L.   *Universitat Politecnica de Cataluya (Spain), Department of Chemical Engineering and Tesca Ricerca & Innovazione srl, Scanzorosciate (BG) (Italy)*

Bosley, K.   *Fire Research and Development Group, Fire and Emergency Planning Department, The Home Office (U.K.)*

Butler, G.W.   *Tyne and Wear Metropolitan Fire Brigade (England)*

Cafaro, E.   *Politecnico di Torino (Italy), Dipartamento di Energetica*

Cali M.   *Polytecnico di Torino (Italy), Departimento di Energetica*

Carver, S.   *University of Leeds (England), School of Geography*

Casal, J.   *Professor, Universitat Politècnica de Cataluya (Spain), Departament d'Enginyeria Quimica (Polytechnic University of Catalonia, Department of Chemical Engineering*

Chambers, S.M.   *Consultant, Get Science Write, Calbourne (England)*

Cheung, B.T.   *Deputy Senior Operations Officer, Cheshire County Fire Brigade (England)*

Cole, S.K.   *Atomic Energy Authority Technology Consultancy Services (SRD) (England)*

Davis, D.   *Chief Fire Officer, Cheshire County Fire Brigade (England)*

Donegan, H.A.   *Univestity of Ulster at Jordanstown (U.K.), Department of Mathematics*

Drouet, J-C.   *University Technology Institut of Aix-en-Provence (France), Health and Safety Department*

Dunlop, K.E.   *University of Ulster at Jordanstown (U.K.), Fire SERT Centre*

Evans, D.   *University of Central Lancashire (England), Department of Built Environment*

Fernández-Becerra, R.   *Jefe del Gabinete Técnico de Protección contra Incendios, Gerencia Municipal de Urbanismo, Sevilla, formerly Director del Programma de Prevenccion contra Incendios, Expo'92 (Spain)*

Flora, P.   *Lecturer, University of Central Lancashire (England), Faculty of Design and Technology*

Foster, R.   *University of Sunderland (England), School of Computing and Information Systems*

Gallina, G.   *CNR (Italian National Research Council) - ICITE - Sesto Ulteriano (MI) (Italy)*

Goldstein, R. *Professor and Head of Laboratory, Institute for Problems in Mechanics of the Russian Academy of Sciences, Laboratory on Mechanics of Strength and Fracture Materials and Structures*

Graham, T. *University of Central Lancashire (England), Department of Built Environment*

Green, C.A. *University of Sunderland (England), School of Computing and Information Systems*

Gremyachkin, V. M. *Institute for Problems in Mechanics of the Russian Academy of Sciences*

Hassani, S.K.S. *University of Ulster at Jordanstown (U.K.), Fire SERT Centre*

Hiorns, N. *Atomic Energy Authority Technology Consultancy Services (SRD) (England)*

Humpage, P. *West Midlands Fire Service (England)*

Irving, P.J. *University of Sunderland (England)*

Kendal, S.L. *University of Sunderland (England)*

Ketchell, N. *Atomic Energy Authority Technology Consultancy Services (SRD) (England)*

Klein, R. A. *Institüt fur Physiologische Chemie der Universität Bonn (Germany) and Principal Scientific Adviser to Cambridgeshire Fire and Rescue Service (England)*

Lancia, A. *Professor, Universitat Politecnica de Cataluya (Spain), Department of Chemical Engineering and Tesca Ricerca & Innovazione srl, Scanzorosciate (BG) (Italy)*

Leathley, B.A. *Risk Management Consultants, Four Elements Ltd., London (England)*

Leisenheimer, B. *Universität Karlsruhe (Germany), Engler-Bunte-Institut, Lehrstuhl und Bereich Feuerungstechnik*

Leukel, W. *Universität Karlsruhe (Germany), Engler-Bunte-Institut, Lehrstuhl und Bereich Feuerungstechnik*

Makhviladze, G.M. *Professor of Fire Engineering, University of Central Lancashire (England), Department of Built Environment and Head of Research Centre for Fire and Explosion Studies*

Morris, E. *Media and Education Officer, Cheshire County Fire Brigade (England)*

Mott, B. C. *Senior Lecturer, Bournmouth University (England), Department of Product Design and Manufacture*

Mutani, G. *Polytecnico di Torino (Italy), Departimento di Energetica*

Myers, M. *Tyne and Wear Emergency Planning Unit (England), Fire and Civil Defence Authority*

Ndumu, A.N. *University of Central Lancashire (England), Department of Built Environment*

Ndumu, D.T. *University of Central Lancashire (England), Department of Built Environment*

Nielson, M.T. *Tyne and Wear Fire Brigade (England)*

Ody, K. *Risk Management Consultants, Four Elements Ltd., London (England)*

Onishcenko, D.A. *All Russia Research and Design Institute for Offshore Oil and Gas Recovery (VNIPIMORNEFTEGAS), Institute for Problems in Mechanics of the Russian Academy of Sciences*

Parsons, K. *Lecturer, University of Central Lancashire (England), Department of Built Environment*

Payne, C.F.　*Emergency Planning Consultant, Christopher Payne & Associates, (England)*

Piccinini, N.　*Professor, Polytcnico di Turino (Italy), Dipartimento di Scienza del Materiali e Ingegneria Chimica*

Planas, E.　*Universitat Politecnica de Cataluya (Spain), Department of Chemical Engineering*

Platten, A.　*University of Central Lancashire (England), Department of Built Environment*

Pollock, A.J.　*Univestity of Ulster at Jordanstown (U.K.), Department of Mathematics*

Ranaboldo, L.　*Politecnico di Torino (Italy), Dipartamento di Energetica*

Reynolds, C.　*Fire Research and Development Group, Fire and Emergency Planning Department, The Home Office (U.K.)*

Roberts, J.P.　*Professor and Head of Department, University of Central Lancashire (England), Department of Built Environment*

Roman, J.L.　*ITSAMAP FUEGO, Madrid (Spain)*

Rubadiri, L.　*presently with University of Central Lancashire (England), Department of Built Environment; shortly to become Lecturer, School of Engineering, Gabarone, Botswana*

Saluzzi, A.　*Politecnico di Torino (Italy), Dipartamento di Energetica*

Sheilds, T. J.　*Professor and Director of Fire SERT Centre, University of Ulster at Jordanstown (U.K.)*

Silcock, G.W.H.　*University of Ulster at Jordanstown (U.K.), Fire SERT Centre*

Sime, J.　*Research Consultant, Jonathan Sime Associates (England)*

Simms, W.I.　*Univestity of Ulster at Jordanstown (U.K.), School of the Built Environment*

Sini, M.　*Tesca Ricerca & Innovazione srl, Scanzorosciate (BG) (Italy)*

Smith, P.　*University of Sunderland (England), School of Computing and Information Systems*

Smith-Hansen, L.　*Risø National Laboratory (Denmark), Systems Analysis Department*

Stealey, J.R.　*Emergency Planning Consultant, John Stealey Associates, (England)*

Stephens, P.J.　*Atomic Energy Authority Technology Consultancy Services (SRD) (England)*

Volpert, A.I.　*Technicon (Israel), Department of Mathematics*

Volpert, V. A.　*Université-Lyon-1 (France), Laboratiore d'Analyse Numerique*

Warren, P.　*Deputy Director, The Fire Research Station, Building Research Establishment (England)*

Webber, D.M.　*Atomic Energy Authority Technology Consultancy Services (SRD) (England)*

Whitehead, K.　*Greater Manchester Fire Brigade (England)*

Zabegayev, A.V.　*Professor and Pro-Rector, Moscow State University of Building (Russia)*

Preface

Throughout Europe there is a considerable number of fires reported each year, resulting in the loss of many lives. The amount of damage to environment and to property is colossal, with re-instatement work on non-domestic property often costing hundreds of millions of ecu. Add to this the cost of business interruption and increased insurance premiums and the imperative of lessening the risk of fire, and for good emergency planning, through education and research becomes obvious.

This book represents the proceedings of EuroFire'95, the first European Symposium on Research and Applications in Fire Engineering and Emergency Planning, held at the Centre Européen de la Chambre des Artisans et des Métiers (CECAM), Nîmes, France, from the 25th to the 27th of March, 1995. The Symposium was organised by the Department of Built Environment (Centre for Research in Fire and Explosion Studies) of the University of Central Lancashire, England, with the assistance of the European Commission (Human Capital and Mobility Programme) and the Institution of Fire Engineers. The intended purpose is to make a start on "bridging the gap" between the various sectors of the European fire community, to promulgate state of the art applications and to flag up new areas and trends in research and practice.

Themes
The Symposium took, as its themes for this inaugural meeting:
- Fundamental Research
- Applied Research
- Education and Operations

These themes are reflected in the organisation and arrangement of this book, through which the strategic objective of establishing a comprehensive and rigorous review of the situation and trends in each of these areas, within the member states of the European Union and within the world fire community, generally, is being pursued.

This strategic object has also been addressed by each if the contributing authors. Many of the contributions also range across the boundaries between the main themes, clearly indicating that there has been, in recent developments, an integration of education, research and applications across the boundaries of the formerly separate and specific domains of fire engineering and emergency planning. This also demonstrates that science and engineering is now being considered in association with law and economics in pursuance of a common goal: the protection of humanity and its environment.

Secondary aims
Over the last five years, or so, there has been an acceleration in the quest for more and better tools for use in the provision of protection and the maintenance of safety.

Knowledge transfer facilitated by more widely-attended conferences, refereed journals and other relevant publications, is vital, particularly in an emergent area of scientific endeavour where there is an identified lack of communication between small cadres of established and knowledgeable specialist researchers, an ever growing body of younger researchers and an ever more inquisitive and interested body of "users" of the applied science.

The Inaugural Symposium was convened, therefore, with the additional intention that it should provide:

- a vehicle for young researches involved in fire engineering and emergency planning to meet and to participate in a significant high-level meeting and to benefit from contact with established authorities with a high level of experience and expertise in the subject area
- a specifically European forum for the presentation of research findings and for the publication of the results
- a facility where, by exchange of results, knowledge and views, a significant step could be taken towards the effective harmonisation of aspects of fire engineering safety, operations and management in the member states of the European Union to assist with the addressing of proposals to extend E.C. Directives and Codes of Practice affecting fire safety and emergency planning

The reasoning behind this last proviso was an intention to establish a comprehensive and rigorous research and knowledge base for E.U. wide common codes in fire safety which required community wide collaboration and activity in the fundamental areas of fire safety research. Action at a European level in a co-ordinated and structured manner, encouraging the participation of the future leaders in these areas, is seen as a pre-requisite of this and makes the organisation of international colloquia, symposia and conferences important mechanisms for its achievement.

The future
The intended continuation of this initiative in the form of a series of biennial conferences/symposia will make a substantial contribution to the goal of European harmonisation and provide a facility for the exchange of information in the fields of fire engineering and safety management.

This series of conferences, of which the contents of this publication record the first, will continue to have direct relevance to improving health and safety standards by increasing the necessary understanding of the scientific and engineering fundamentals required for fire safe design of environments by increasing the quality and quantity of both pure and applied research in the relevant areas. This first conference has identified several areas where research and technological development is most urgently needed and it is anticipated that future conferences will continue to facilitate advances in the understanding of fire safety engineering.

And finally ...

It would be remiss of me not to pay tribute to those individuals and organisations, too many to mention separately, whose enthusiastic and unstinting assistance has made it possible to bring this project to fruition. To them I say a heartfelt thank-you, merci, gracias, obrigado, danke, dank u, kiltos, grazie, tak, takk, tack, hvala, ... *etcetera.*

Dr. Ronald Barham
 Editor   *Centre for Research in Fire and Explosion Studies*
Department of Built Environment
University of Central Lancashire
Preston
Lancashire
England
PR1 2HE

EUROFIRE '95 INTRODUCTORY REMARKS

EDITH CRESSON
European Commissioner for Science, Research and Development

A message forwarded by letter from the Cabinet of the Commissioner:

Mme Cresson regrets that she is unable to be with you today, due to the development of an increasingly intensive workload but has asked that you be informed that she shares your views that the subjects of fire safety, fire engineering, emergency planning and risk management are important topics which have an impact on the quality of life of all European citizens. Accordingly she asks that we bring to your attention the following matters which are extracted from her recent presentation to the European Parliament:

"One of the priorities of my portfolio as Commissioner for Science, Research and Development is that of employment. Economic recovery is necessary for Europe and for the creation of new jobs - but it cannot take place on its own. Major trans-European research, which I spearheaded in 1989 under the French Presidency, will, in my opinion, save the day. It is necessary, also, to come to a long-standing international agreement to develop the competitiveness of our enterprises, notably in the areas vital to this future - namely: innovation and the development of industrial strategy. To this end, the "White Book" on growth, competition and employment, introduced by M. Jaques Delors, put the accent on two elements: research and education.

"It is pleasing to note that these two domains of Community action, leading towards the future, come under the auspices, now, of a single Commissioner. It is significant that it will now be the same in most of the European member states. For example, in the recent re-organisation of the German Government the combination of these two responsibilities carries the name 'Minister for the Future' - Zukunfs Ministerium. My primary concern will be, therefore, to strengthen the links between these two politics. In this respect, the theme of Framework IV, to which the European Parliament attaches much importance, is vital, i.e. the theme of applied socio-economic research .... which, to some extent, covers the area that you are here to listen to today. .... I appreciate that the joint efforts of Professor Ruberti, my predecessor, and the outgoing President, M. de Sana, were subjected only to short delays with the less important financing schemes. As a result, by this year, the execution of some twenty scientific programmes have, or will be, commenced. Fundamental research will continue to be favoured but I attach a certain importance to the visibility of action on the remaining priorities, in particular on biotechnologies, transport, information technology and, most importantly, environment.

"For your information it will be desirable that the remaining projects present, firstly, a really European dimension and, secondly, take account of the needs of the citizens and the market. It is equally important to ensure the diffusion of the results of research and that they are explained and are readily available to all European citizens. The need for research effort is fundamental. Therefore, it is necessary for the laboratories, the research centres and the enterprises that the E.C. decisions on projects are made quickly and that delays in payment of research funds are reduced. We must not forget that the European institutions exist to conduct a European politic and, in the area of research and its applications, we must look beyond our national practices to the position Europe faces in the politics of research and innovation and from our principal competitors, the United States and Japan. In the same spirit, I have to follow the open view of Professor Ruberti for whom the Joint Research Centre made much of the research in Europe understandable and to say that the European Parliament should pay attention to the scientific community and to the European Commission on the necessity of putting research at the service of the citizens of Europe in its efforts to encourage development."

"I wish you every success for this important event."

PART ONE

FUNDAMENTAL ISSUES

1 INTRODUCTION

B. J. M. ALE
Rijsinstituut voor Volksgezondheid en Mijeuhygiene
(The Netherlands), Laboratorium voor Stralingsonderzoek
(National Institute of Public Health and Environmental
Protection, Laboratory of Radiation Research)

The organisers have asked me to set the conference off by giving you an overview of what you might expect and saying something about why you are here.

The fire research and fire fighting community is about the same as in other techniques like chemicals. You have two main groups: at one extreme are the senior academics - who know it all and do very fine scientific research; at the other extreme, the fire-fighters - who put the fires out and are supposed to have knowledge of what these researchers have done, but unfortunately do not speak to them. In between these two extremes you have the young, upcoming academics - who might have access to the senior academics (if the senior academics have time) but are unlikely to see more of a fire than the fire drill in the laboratory.

Nevertheless, it should be noted that fires become more complex. Not necessarily by scale, because we used to have whole cities burn to the ground. In terms of fire technology that is relatively easy to deal with because they were mainly wood fires - but now fires are more complex: we now have fires in chemical industries; we have vapour cloud explosions; and we frequently have to deal with complicated chemical reactions. For example, you might have a fire on your hands in a nuclear power station but, equally, a 747 might crash into an apartment building (as happened in the Netherlands three years ago) and you might find yourself in the situation that we were - where you need all the equipment that you can possibly lay your hands on and you need all the knowledge that you possibly can lay your hands on - and within about half an hour, CNN has broadcast it all over the world. Members of the families of those potentially involved in the accident started to phone in from all over the world. This caused a breakdown of the telephone system and, eventually, the radio communication system. At this point, they started to use overseas short wave communication of their own, thus wrecking the radio system as well and so, as you can see, things can get nasty.

Also, in such situations, your emergency planning and your evacuation planning is subjected to severe complications. First of all people might no longer believe what you tell them to do; and, secondly, you might end up with enormous traffic jams. In the Boeing 747 disaster, we needed traffic policemen to actually direct the traffic of the ambulances and the fire trucks, themselves. We also needed police assistance in keeping out bystanders who wanted to make video tapes of the disaster for the home video. With all this complexity, it is necessary to understand not just the behaviour of complex fires but also to understand the behaviour of a complex built environment and to understand the behaviour of people within it.

These are the matters that we are going to discuss during this conference. The first day will mainly be devoted to fundamental research, the second day will then slide

Fire Engineering and Emergency Planning. Edited by R. Barham.
Published in 1996 by E & FN Spon. ISBN 0 419 20180 7.

towards applications in concrete situations and emergency planning, and the last day will mainly be spent looking at education, operations and emergency preparedness.

Other people will set the pace for the next days; I will shortly give you the pace for today. First, we will cover flame behaviour, fire behaviour, combustion processes, and flashover - looking at how a small fire can become a big one in a very short time; and then the behaviour of all sorts of new building materials that might create new problems. In the second session, we then look at applications of those series into actual buildings. We look at situations such as where you might have critical or vulnerable or very expensive equipment (such as computer equipment) in a building which you need to protect but still you want to take the fire out and, then, in the third session we look at emergency planning and crowd control, the logistics of how to be there on time and how to organise your preparedness plan.

These various matters should nicely fill the day - starting off with theory and ending with practice. However, this day should not only be a listening day. As I said at the beginning, one of the drawbacks of having a well-developed and very interesting science is that the practical applications get lost because the academics get bogged down in solving differential equations while others are trying to build more fire trucks. So if you have time, and I hope you will take the time, speak to each other. Figure out what the academics can mean to you. And to the academics, I say don't think that you know it all - because these fire-fighters have taken the heat of the actual fire and know, in all probability, what they are facing in a live fire situation.

With that I wish you a very nice day.

FIRE ENGINEERING

2 RESEARCH IN FIRE ENGINEERING: COMBUSTION AND EXPLOSION SCIENCE AND INDUSTRIAL PROBLEMS

G.M. MAKHVILADZE
Department of Built Environment,
University of Central Lancashire, UK

Abstract
The state of the art in fire engineering is crucially dependant on achievements in combustion and explosion science and problems originating from within industry. In this connection the following two basic issues are considered: modern trends in combustion and explosion science and modern industrial problems which are important for fire engineering. Examples of recent studies are given.
Keywords: Fire engineering, research, combustion, explosion, industry.

1 Introduction

To begin a discussion on research in fire engineering the mutual links between Fire Engineering, Combustion and Explosion Science and Industry (see Fig. 1) should be considered.

Fire Engineering (FE) is essentially based on Combustion and Explosion Science (CES) which has been providing the necessary background, solutions of principle, new qualitative information, methods and techniques. Also, it puts forward new concepts for FE and revises former approaches.

From this perspective, it is important to follow the modern trends in CES that seem promising for FE and could be incorporated into FE research in the near future. This is the first main question arising when evaluating the research in FE along with the link CES------> FE (see sketch in Fig. 1):

* What are the most important modern studies in CES from the perspective of FE?

In turn, FE tends to adopt, use and apply those issues originated with CES. Eventually, this stimulates new engineering solutions, implementation and improvement of codes, regulations and standards, development of proper preventive anti fire and anti-explosion measures,

Fire Engineering and Emergency Planning. Edited by R. Barham.
Published in 1996 by E & FN Spon. ISBN 0 419 20180 7.

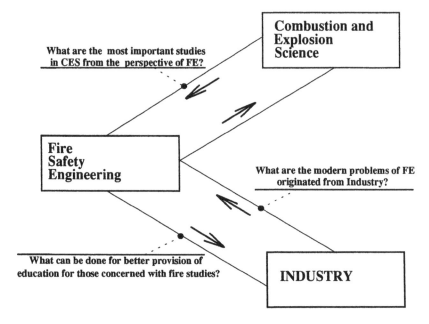

Fig. 1. Mutual links

provision of required fire design, implementation of adequate testing facilities and equipment, and further development of legislation. All this allows FE to meet the needs of modern industry in fire protection including building, petro-chemical industry, transportation, and energetics. Obviously, a high enough level of education must be achieved. That is why, along with the link FE -----> Industry the second important question arises :

• What can be done for better provision of education for those concerned with fire studies?

This question will be discussed in a different session of the conference and is not pursued here.

From another perspective, FE is under permanent pressure from industry which poses new technical problems and requirements. Along with the link Industry -----> FE the third question arises that needs to be answered :

• What are the modern problems of FE originating from Industry?

This cycle of reciprocal links is closed: in its turn, FE with its efforts to resolve the problems originated with industry has an influence on CES, demanding new methods, methodologies and approaches to be incorporated in research in fire and explosion studies.

Hopefully, the questions posed above will be a catalyst for further discussion. Proper answers to these questions and related discussions would give a chance to describe the state-of-the-art of research in FE. Having said that, from here on the author is going to answer, certainly partially, two of the basic questions posed above. These answers can not be but affected by the personal interests of the author. Also, the field is too broad for any detail review in one short talk. Nevertheless, this short introduction to our session should help to formulate our tasks for further discussion.

2 Basic tendencies of modern CES

Flames, deflagrations, explosions, detonation used to be and still remain the main areas of CES. But, like any modern science CES includes and makes use of results from many related sciences, including heat and mass transfer, natural convection, gas dynamics of reacting flows, chemical kinetics and thermodynamics, turbulence studies, computational fluid dynamics (CFD), material science, environmental studies, many areas of applied mathematics and experimental physics. This combination of disciplines results in new developments in the field of CES. Consider some of those issues that will be important in the future of FE.

2.1 New investigative techniques and methods. Laser spectroscopy
Over the last years new experimental delicate methods have been developed and incorporated into combustion studies [1], [2].

Progress in the development of non-intrusive laser spectroscopic diagnostic techniques and instrumentation now enables comprehensive investigations of reacting flows and the fine structure of chemical reaction fronts [2], [3]. In particular, the application of Laser Induced Fluorescence (LIF) provides images of the instantaneous spatial distributions of species and temperatures in short duration, high speed flow fields, distribution of the unburned fuel in non premixed turbulent flames. The Laser Spark emission technique (LASS) is being developed as an *in situ,* real time monitor for the elemental composition of individual coal and fly-ash particles and coal-ash deposits. The Infrared Computed Tomography (ICT) method has been developed to measure two-dimensional temperature distributions. During the last decade Coherent Anti-Stokes Raman Scattering (CARS) has proven to be very attractive for the diagnostics of various combustion processes. Laser light scattering is the most promising method for the characterisation of droplets and aerosols, in future it will be tested for spray combustion of hydrocarbon mixtures.

Laser spectroscopy as an analytical technique will be applied more and more in FE to study the fine structure of fires, flames and mixing zones, to analyse the processes relevant to properties and formation of the smoke particles and coal-ash deposits, to find spatio-temporal distributions of smoke, to study the processes of multi-phase releases resulting from a loss of containment..

2.2 New investigative techniques and methods. Computational modelling
Already, numerical modelling has widely spread in FE. During the last ten years, techniques of Computational Fluid Dynamics (CFD) have been widely developed for predicting the consequences of building fires [3], [4]. Based on the proliferation of modern numerical methods, computer hardware and software into many areas of CES, one could easily foresee the same tendency in FE.

On this stage, numerical modelling in CES is characterised by the adoption of turbulence theory models and relevant computational field models for combustion problems (see [1] to [3], [5], [6]). Obviously, the novel modelling being developed in CES for turbulent flows with nonpremixed [7] and premixed ([8] to [10]) flames, spray combustion [11] and two-phase reacting systems flows [12] will continue to be applied to fire and explosion problems. This as well as efforts at further improving the credibility of field models and submodels is expected to

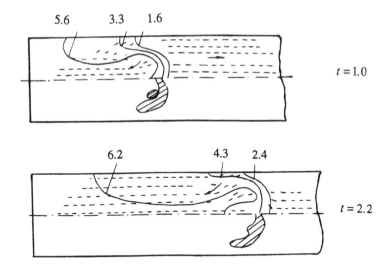

Fig. 2. Cellular structure of flame, propagating in plane channel.

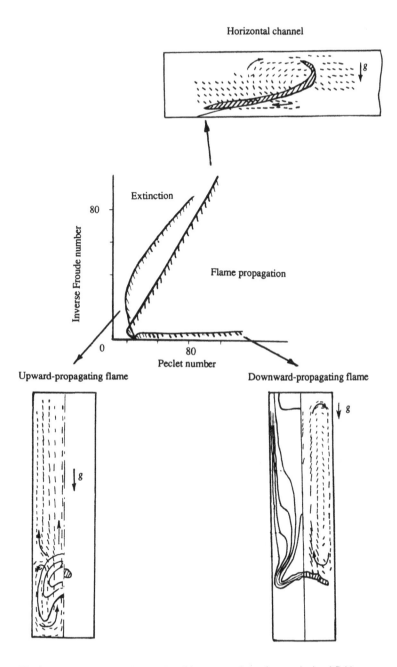

Fig. 3. Flammability limits in a combustible gaseous mixture in a gravitational field.

be very intensive and in many connections   will be responsible for the future level of theoretical and numerical research in FE.

2.3 Transition to the spatial description of combustion processes

Over the last decade combustion studies having solved the main 1D problems and based on the results obtained has turned to the spatial 2D and 3D description of combustion and explosion phenomena.   This has allowed consideration of more complicated and more realistic gas dynamic situations. Such a development resulted in two important issues in CES.

1. Renewal and revision of classical results (see e.g. [13]) obtained in the different areas of CES (thermal explosion, self-ignition, ignition by heated surfaces, flame and detonation propagation).   Non-one-dimensional processes, in particular natural convective flows, change significantly the critical conditions. When applied to FE this results in new criteria for ignition, self-ignition and flammability limits and more generally, a need to rethink matters.

For example, the whimsical cellular structure of a flame in a tube, arising due to the spatial thermal-diffusional instability of a laminar flame front (see fig. 2 from [14]; a flame propagating from left to right is shown at two successive moments; in the upper parts: isotherms and flow fields, in lower parts; flame front; all the distributions are symmetrical about the horizontal middle axis) results in the opposite dependence of flame propagation velocity on Lewis number in comparison with classical formula (see [13]). This is a significant factor in the estimation of flame velocities and quenching diameters.

It should be noted, that the value of the flammability limits and the flame propagation velocity (fire spread velocity) depend on the relation between the characteristic time of the chemical reaction in the flame, t_r, and the characteristic time for heat losses, t_{cool}. For near-limit mixtures, close to flammability limits, these two specific times are of the same order of magnitude, $t_r \sim t_{cool}$.

For gases, the heat losses time is actually a characteristic hydrodynamics time, t_{hd}, because the intensity of heat and mass transfer processes depends on fluid motions.   Different combustible systems reveal different flow patterns and different characteristic hydrodynamics times. Accordingly, the flammability limits and flame velocity will be different.

This is illustrated clearly in Fig. 3, where summary results on numerical modelling of flame propagation in a channel under a gravity force field, in a given mixture [15] are presented. In the middle of the figure the flammability limits on the plane (Peclet number, inverse Froude number) are plotted for the three different orientations of the channel relative to the gravity force. Each curve separates the area of flame propagation (below the curve) from the area of flame extinction (above the curve). In all the cases the flames cool due to natural convective motions. Because of different flow patterns in each case, the mechanism of flame extinction is also different. In a horizontal channel (upper picture; flow field and chemical reaction zone are shown) the main mechanism for heat losses is convective heat transfer through the upper wall. In a vertical channel and for lower ignition (left: isotherms and flow field, right: chemical reaction zone) the flame is cooled due to convective heat losses from lateral flame surfaces. For downward-propagating flames (left: isotherms, right: flow field and chemical reaction zone), extinction is due to the formation of longitudinal convective vortexes in the combustion products; these vortices promote the heat transfer from the flame through lateral cold walls.

This example shows the difficulty which is known to occur in FE, that results of tests vary, sometimes significantly, depending on the testing facilities used in different institutions. Even slight changes in geometry or in some conditions change the characteristics of heat and mass transfer have considerable influence on combustion characteristics under near-critical conditions.

2. Investigation (mostly by means of numerical methods) of large scale combustion and related phenomena that can not be treated, in principle as 1D problems. Many have became available for research in the last few years. Especially, one could note, chemical reaction front propagation and combustion product movements in both, confined volumes (closed and vented vessels, tubes, channels, enclosures) and in the open atmosphere. These studies being applied to FE give a possibility to analyse on a new level such problems as fire spread and smoke movement in enclosures (see section 3.1), dynamics of big open fires and accompanying convective flows (see [1] to [3]) and the behaviour of hazardous releases (dense vapour and particulate clouds, plumes, jets, fireballs, pool fires).

2.4 Dynamics and combustion of multi-phase mixtures

This area is being developed intensively in CES, especially in connection with spray and dust combustion [1], [2]. Different experimental and theoretical approaches have been developed and applied, including those based on the dynamics of heterogeneous media [16]. These studies are promising for FE because of their applicability for loss prevention in process industries, in particular, for the study of phenomena relevant to the loss containment (rupture of pressurised vessels, pipelines, etc.).

2.5 Deflagration-detonation transition, large scale detonation

Very often, fires begin with or are followed by explosions. For this reason such traditional areas of CES as deflagration-detonation transition, flame propagation in obstacle congested volumes, explosion scaling methodology, fast deflagrations, are of interest for FE. For example, these phenomena are of concern to the petroleum installations. In recent years significant attention has been given to numerical modelling (see, e.g. the review[17]) and large scale laboratory and field experiments ([18], [19]).

The other very fast developing area is the study of reasons for the onset of spontaneous accidental explosions (see e.g. [20], [21] and references there in). In particular, the concept of spontaneous explosions generated in non-uniformities of auto-ignition time delay has been recently treated by several teams. The concept could provide insight into the onset of accidental explosions (for example, in vented vessels for pressure relief systems [22]).

3 Modern problems of FE originated within industry

The broad spectrum of these problems reflects the industrial needs. It includes risk assessment, fire resistance, toxicity assessment, early diagnostics of accidents, big fires and ecology, testing, codes of practices, standards, regulations, etc.

We consider here only two problems taking into account their importance with regard to human losses and world-wide spreading.

3.1 Fire spread and smoke movement in enclosures

Research on fires in multi-room housing and industrial buildings, atriums, off- and onshore platforms is still in its infancy. But it is expected to develop very rapidly in the future. The study should include multi-purpose targets: prediction of fire behaviour and smoke movement inside compartments and buildings in whole, criteria of safety, visibility and toxicity of the smoke, heat fluxes, effective fire and smoke control and preventive measures (detection and fire protection systems, safety of escape routes, safe havens, ventilation, suppression systems).

Fire modelling is growing rapidly, and much progress is being made in different places [3]. There are four principal ways [23] to model enclosure fires: experimental, stochastic, field and zone models.

Some advantages of zone modelling are not yet completely exploited. One can remember an effective treatment of flashover phenomenon [24], [25] on the basis of classical thermal explosion well studied in CES. Recent development of this approach has been made [26] where modern non-linear dynamics was applied with the same aim. At this conference, in line with classical explosion theory, critical conditions for flashover are reported [27].

But the most attention in this area is given to the field modelling (see e.g. [3], [4]). Considerable achievements connected with the UK Fire Research Station's activity have resulted in the development of mathematical computer models and codes. Nevertheless, fire safety of building needs further development of models describing fire dynamics in enclosures, the study of gas dynamics, and the physical processes that are important in many fire situations, such as two-phase flows accompanying fire in enclosures.

The dispersed phase (soot, ash, aerosols, droplets) in the fire process can be formed from 'external' sources such as interaction of sprinkler jets and extinguisher powders with combustion products, from ruptured tubes and so on; and 'internal' sources, for example large enough smoke particles which are formed in fire and cannot be considered as gaseous components, production of aerosols in chemical reactions and droplets from plastics melting in fires. Fire-flow field interaction, models of fire growth, soot formation process, analysis extension for complex enclosures, incorporation of some novel computational methods (multi-grid methods, parallel processors, various differencing schemes) still remain as areas in which further activity is essential.

3.2. Large open fires

These fires arise on industrial sites as a result of loss containment followed by fuel releases and, potentially, result in severe accidents. Industry poses the main tasks for specialists in CES concerned with this sort of accidents: what are possible typical "scenarios" of the accidents, including worst-case' scenarios, effects on the surroundings and how to estimate the fuel mass (energy) released.

The 1989 railway catastrophe in Russia, near the city of Ufa (capital of Bashkiria) in the Eastern part of the Ural mountains is an extreme example of how dangerous the releases of liquefied hydrocarbons into the atmosphere can be. The Ufa-catastrophe, the World's largest disaster, followed such a the release into the atmosphere resulting in a huge fire which killed or injured more than 1000 people.

The cause of the disaster was the rupture of a pipeline for pumping liquefied fuel (propane, butane and heavier fractions of hydrocarbons) under a pressure of 28 atmospheres. As a result of hydrocarbon release a dense vapour cloud covered a large forest area of 2.5 km and railway tracks. Passenger trains travelling in opposite directions came into the cloud. When they passed each other, a series of explosions initiated a huge fire and fireball, heat radiation from which was the major killer of people.

Fire and the accompanying hydrodynamics consequence specified in [28], i.e. "hydrodynamics stage" of accident following by multi-explosions, were the key phenomena for the exploration of possible accident scenario and the estimation of the involved fuel mass. This approach has proven to be useful. Analysis in [28] shows that formation and rising of a buoyant fireball provides on explanation for the brief hurricane that followed. Further study could lead to the creation of an independent method for the assessment of vapour cloud hazards, based on relations between the mass of released fuel and characteristics of the flow field induced by the fire.

4 Conclusion

This brief consideration shows clearly that the future development of FSE is dependant on flexible and close reciprocal links with CES allowing for effective solutions of modern problems coming from industry. Also, better provision of education for those concerned with fire studies and fire protective communities should be born in mind..

The motto of the EuroFire'95 is "Bridging the Gap". Being applied to research in FE, this motto would denote necessary unification of efforts of specialists from these two fields "to go forward together".

5 Acknowledgement

The authors thanks British Engineering and Physical Sciences Research Council for supporting his work in fire modelling, Grant Ref. GR/J85035, and fuel cloud behaviour, Grant Ref. GR/K 13486.

6 References

1. *Twenty-fifth Symposium (International) on Combustion.* (1994) The Combustion Institute, Pittsburgh, Pennsylvania.
2. *Twenty-fourth Symposium (International) on Combustion.* (1992) The Combustion Institute, Pittsburgh, Pennsylvania.
3. *Fire Safety Science - Proceedings of the Forth International Symposium.* (1994). International Association for Fire Safety Science, Gaithersburg, Maryland.
4. Cox, G. (1992) Some recent progress in the field modelling of fire, in *Science and Technology* (ed. F.Weicheng, F.Zhuman), International Academic Publishers, Hefei, pp. 50-9.]
5. Bradley, D., Gaskell, P.H. and Gu, X.J. (1994) Application of a Reynolds stress, stretched flamelet, mathematical model to computations of turbulent burning velocities and comparison with experiments. *Combustion and Flame*, Vol. 96, No. 3, pp. 221-48.
6. Achurst, Wm.T. (1994) Combustion modeling and turbulent structure, in *Twenty-fifth Symposium (International) on Combustion,* The Combustion Institute, Pittsburgh, Pennsylvania.
7. Bray, K.N.S. (1990) Studies of the turbulent burning velocity. *Proceedings of the Royal Society London*, Vol. 431, pp. 315-35.

8.   Cant, R.S., Bray, K.N.C., Kostiuk, L.W. and Rogg, B. (1994) Flow divergence effects in strained laminar flamelets for premixed turbulent combustion. *Combustion Science and Technology*, Vol. 95, No. 1-6, pp. 261-76.

9.   Duclos, J.M., Veynante, D. and Poinsot, T.A. (1993) A comparison of flamelet models for premixed turbulent combustion. *Combustion and Flames*, Vol. 95, No. 5, pp. 101-17.

10.   Mantel, T. and Borghi, R. (1994) A new model of premixed wrinkled flame propagation based on a scalar dissipation equation. *Combustion and Flame*, Vol. 96, No. 4, pp. 443-57.

11.   Borghi, R. and Loison, S. (1992) Studies of Dense-Spray Combustion by Numerical Simulation with a Cellular Automation, in *Twenty-fourth Symposium (International) on Combustion*, The Combustion Institute, Pittsburgh, Pennsylvania, pp. 1541-7.

12.   Silverman, I. and Sirignano, W.A. (1994) Multi-droplet interaction effects in dense sprays. *International Journal of Multiphase Flow*, Vol. 20, No. 1, pp. 99-116.

13.   Zeldovich, Ya.B., Barenblatt, G.I., Librovich, V.B. and Makhviladze, G.M. (1985) *Mathematical theory of combustion and explosions*, Plenum Publishing Corporation, New York.

14.   Zulinyan, G.A., Makhviladze, G.M. and Melikhov, V.I. (1992) Effect of the Lewis number on the flame propagation mechanism. *Combustion, Explosion and Shock Waves*, Vol. 28, No. 6, pp. 614-9.

15.   Makhviladze, G.M. and Melikhov, V.I. (1991) Flame propagation in a closed channel with cold side walls. *Combustion, Explosion and Shock Waves*, Vol. 27, No. 2, pp. 173-83.

16.   Nigmatulin R.I. (1991) *Dynamics of multiphase media*, 2 volumes Academic Press, Washington.

17.   Hjertager, B.H. (1993) Computer modelling of turbulent gas explosions in complex 2D and 3D geometries. Review *Journal of Hazardous Materials*, Vol. 34, part 2, p.173-97.

18.   Lenoir, E.M. and Davenport, J.A. (1993) A survey of vapour cloud explosions: second update. *Process Safety Progress*, Vol. 12, No. 1, pp. 12-33.

19.   Bull, D.C. (1992) Review of large-scale explosion experiments. *Plant/Operation Progress*, Vol. 11, No. 1, p. 33-40.

20.   Makhviladze,G.M. and Rogatykh, D.I. (1991) Nonuniformities in initial temperature and concentration as a cause of explosive chemical reactions in combustible gases. *Combustion and Flame*, Vol. 87, No. 3-4, pp. 347-56.

21.   Proceedings of the Zel'dovich memorial. (1994) *Vol 2: Combustion, detonation, shock waves*. Russian Section of the Combustion Institute, Moscow

22.   Frolov, S.M., Gelfand, B.E.and Tsyganov, S.A. (1990) A possible mechanism for the onset of pressure oscillation during venting. *Journal of Loss Prevention in the process industries*, Vol. 3, No. 1, pp. 64-7.

23.   SFPE Handbook of Fire Protection Engineering (1988) National Fire Protection Association, USA.

24.   Thomas, P.H. (1980/1) Fires and flashover in rooms - a simplified theory. *Fire Safety Journal*, Vol. 3, pp. 67-76.

25.   Thomas, P.H., Bullen, M.L., Quintiere, J.C. and McCaffrey, B.J. (1980) Flashover and instabilities in fire behaviour. *Combustion and Flame*, Vol. 38, pp. 159-71.

26.   Bishop, S.R., Holborn, P.G., Beard, A.N. and Drysdale, D.D. (1993) Nonlinear dynamics of flashover in compartment fires. *Fire Safety Journal*, Vol. 21, pp. 11-25.

27.   Graham, T., Makhviladze, G. and Roberts, J. (1995) Critical conditions for flashover in enclosed ventilated fires, in this book.

28. Gelfand, B.E., Makhviladze, G.M., Novozhilov, V.B., Taubkin, I.S. and Tsyganov, S.A. (1992) Estimating the characteristics of an accidental explosion of a surface vapour air cloud *Combustion, Explosion and Shock Waves*, Vol. 28, No. 2, pp. 179-84.

3 CRITICAL CONDITIONS FOR FLASHOVER IN ENCLOSED VENTILATED FIRES

T. GRAHAM, G. MAKHVILADZE and J. ROBERTS
Department of Built Environment,
University of Central Lancashire, UK

Abstract
This paper examines the nature, role, and critical relationship between non dimensional parameters for flashover. A two-zone model for the development of a fire in a ventilated enclosure is proposed and used. The treatment of the model performed with the aid of non-dimensional variables results in the formulation of critical conditions for flashover; these critical conditions are obtained for small and high thermal inertia of the walls.
Keywords: Compartment fire, enclosure, zone model, critical condition, flashover

1 Fire development and flashover in a compartment

Typical fire development in an enclosure is characterised by a temperature history such as that sketched in Fig. 1.

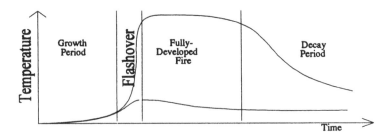

Fig. 1. Stages in a compartment fire temperature history.

Fire Engineering and Emergency Planning. Edited by R. Barham.
Published in 1996 by E & FN Spon. ISBN 0 419 20180 7.

Following ignition there is a slow growth period, which is limited by the pyrolysis rate of the fuel supply (fuel controlled combustion). Early in the fire development the combustion products are usually segregated in a well-stirred ceiling layer whose properties are roughly homogeneous. The hot smoke and warming room boundaries radiate heat back to the fire. Radiative feedback can enhance the reaction rate so that the fire accelerates towards the fully developed stage.

The transition (which is usually very rapid) to a fully developed fire is called 'flashover'. Flashover fires are disastrous [1], and an increasing problem now [2] because of the use of modern materials in buildings. It is a major objective of research to understand and to predict this process so that precautions can be made during building design. Apart from flashover, the regime of quasi-steady low-intensity fire is possible (lower curve) with small heating of the upper layer.

Thomas [3] and Thomas et al [4] presented a theoretical treatment of the flashover phenomena using the fundamentals of classical thermal explosion theory. Further development of this approach was made by in [5,6] with the application of modern nonlinear dynamics. In this treatment the flashover point is represented by a fold catastrophe.

In this paper we suggest the two-zone model for the compartment fire and, in line with the classical explosion theory approach we deduce the characteristics for the flashover phenomenon, namely the analytical expression for the critical temperature and critical conditions in terms of determining parameters.

2 The zone model

Zone models are known to assume that the compartment is divisible into two homogeneous regions: a hot/smoke zone and a cool/lower zone. Each zone is represented by average thermodynamic/gas properties. In this approach we lose local detail in return for simplicity, but gain the ability to interpret complicated bulk phenomenon with physical clarity. Another advantage is that one could specify a small number of the most important (determining) parameters and obtain quite simple relationships, describing the main features of the system.

Main assumptions:

- the compartment can be divided into two zones which may be represented by average temperatures,
- the fluid motion is very slow in comparison to the velocity of sound,
- flashover takes place during the early development of the fire, within the fuel controlled combustion regime; during early fire development the density of the lower zone may be assumed to be its initial value $\left(\rho_L \approx \rho_0 \right)$,
- the wall surfaces surrounding the zones can be described by two temperatures, the lower zone and wall surfaces below the thermal discontinuity are at the initial temperature.

The equation of energy conservation takes the form of a heat balance:

$$mc_p \frac{dT}{dt} = G - L \tag{1}$$

The left hand side being the change in internal energy of the hot layer, t is the time, T is the smoke/hot zone temperature, m is the total mass in the hot layer, c_p is the specific heat capacity (at constant pressure). On the right hand side G is the net heat gains and L is the net heat losses from the hot zone. The initial condition is $T(t=0) = T_0$, where T_0 is the reference (ambient room) temperature.

We treat the case of a developing fire and reasonably assume that flashover occurs during the early development of the fire, and do not account here for factors such as fuel/air exhaustion. As a consequence of the assumption $\rho_L \approx \rho_0$, at the beginning of flashover the neutral plane and the thermal discontinuity plane coincide.

The heat gains for the smoke layer are given by:

$$G = \chi \Delta h_c \dot{m}_f, \qquad \dot{m}_f = \frac{A_f}{\Delta h_{vap}} \left[\dot{q}''(T_0) + \alpha_u(T)\sigma\left(T^4 - T_0^4\right) \right] \tag{2}$$

where χ is the efficiency of the combustion process (the fraction of the theoretical heat that would reach the smoke layer), Δh_c is an effective heat of combustion, Δh_{vap} is the effective heat of vaporisation of the solid fuel, $\dot{m}_f$ is the mass burning rate of the fuel, A_f is the pyrolysing area (surface area of fire), $\dot{q}''(T_0)$ is the incident heat flux to the fuel surface from the fire, $\alpha_u(T)$ is the radiation feedback coefficient from the hot layer at temperature T, σ is the Stefan-Boltzman constant. The right hand side of the equation contains two terms; the first being the heat gain to the smoke layer from a free-burning fire, and the second term describes the radiation feedback to the fire from the upper zone [3] and compartment walls.

The heat losses from the smoke layer are given by:

$$L = \dot{m}_{out} c_p A_V \left(T - T_0\right)\left(1 - D\right) + \left[A_U - \left(1 - D\right)A_V\right]h_c\left(T - T_W\right) + \left(1 - D\right)A_V h_V\left(T - T_0\right)$$
$$+ \alpha_g \sigma\left(A_U - \{1 - D\}A_V\right)\left(T^4 - T_w^4\right) + \alpha_g \sigma\left(A_L + \{1 - D\}A_V - A_f\right)\left(T^4 - T_0^4\right) \tag{3}$$
$$+ \alpha_g \sigma A_f\left(T^4 - T_f^4\right)$$

Here $\dot{m}_{out}$ is the total mass flow of smoke/gas out of all vents and doorways, A_v is the total area of the vent, A_u is the surface area of wall surrounding the smoke layer (including any windows), D is the fractional height of the thermal discontinuity plane ($D = Z_D / H_V$, Z_D is the height of the thermal discontinuity plane above the bottom of the vent, H_V is the height of the vent), h_c is the convective heat transfer coefficient for the hot wall surfaces, T_w is the surface temperature of the walls surrounding the hot zone, h_V is a convective heat transfer constant for the vent, α_g is the emissivity of the gas layer, σ is the Stefan-Boltzmann constant, A_L is the surface area of walls surrounding the cool/lower zone (including referred parts of any vents), T_f is the surface temperature of the fuel bed.

The right hand side of (3) contains six terms. The first is an enthalpy flow out of the vent. The second and third terms are convective heat losses to the walls surrounding the hot layer, and out of the vent respectively. The last three terms are the radiative heat transfer from the hot zone to the hot wall surfaces, the cool zone and the vent, and the fuel bed areas respectively.

The parameter D varies from 0 to 1 and can in principle be found from the mass balance equation $\dot{m}_f \approx \dot{m}_{out}$ (in line with the assumption $\rho_L = \rho_0$ we ignore the mass influx into the room through the lower part of the vent under the neutral plane). However, this makes further calculation algebraically difficult, and is not necessary at this stage in the study. We shall assume that D is a constant (does not depend on T) and we will give results for different values of D.

Henceforth we shall use dimensionless time introduced by the following formula $\tau = t / t_*$, where $t_* = mc_p T_0 / E_S$ is the characteristic time of heating of the upper layer by heat from a freely burning fire, the value $E_s = \chi A_f \dot{q}''(T_0) \Delta h_c / \Delta h_{vap}$ is the characteristic heat flux (per unit time) to the smoke layer for a freely burning fire (that is to a smoke layer at the ambient temperature), and dimensionless temperature, $\theta = T / T_0$. With these dimensionless variables equation (1) with the initial condition takes the form:

$$\frac{d\theta}{d\tau} = 1 + \left(\varepsilon_k - \varepsilon_{R,L} \right)\left(\theta^4 - 1 \right) - \varepsilon_{C,H}\left(\theta - \theta_w \right) - \varepsilon_{out}\left(\theta - 1 \right) - \varepsilon_{C,L}\left(\theta - 1 \right)$$
$$- \varepsilon_{R,W}\left(\theta^4 - \theta_w^4 \right) - \varepsilon_{R,f}\left(\theta^4 - \theta_f^4 \right), \quad \theta(0) = 1, \tag{4}$$

where

$$\varepsilon_k = \left(\chi \frac{\Delta h_c}{\Delta h_{vap}} \right) \alpha_u \sigma A_f T_0^4 / E_s, \qquad \varepsilon_{R,L} = \alpha_g \sigma \left(A_L + \{1 - D\} A_V - A_f \right) T_0^4 / E_s,$$

$$\varepsilon_{out} = \dot{m}_{out} A_V \left(1 - D \right)\left(\theta - 1 \right) c_p T_0 / E_s, \qquad \varepsilon_{C,H} = \left(A_U - \{1 - D\} A_V \right) h_c T_0 / E_s,$$

$$\varepsilon_{C,L} = \{1 - D\} A_V h_V T_0 / E_s, \qquad \varepsilon_{R,W} = \alpha_g \sigma \left(A_U - \{1 - D\} A_V \right) T_0^4 / E_s,$$

$$\varepsilon_{R,f} = \alpha_g \sigma A_f T_0^4 / E_s.$$

Here ε_{out} is a dimensionless scale for enthalpy flow out of the vents of the compartment, $\varepsilon_{R,j}$ (j=w,L,f) is the dimensionless scale for radiative heat transfer from the hot zone to the hot walls, to the lower zone and to the fire bed, $\varepsilon_{C,k}$ (k = H,L) is a dimensionless scale for heat convected from the hot layer to the wall surfaces and the vent surfaces. Here on we shall ignore the term $\varepsilon_{out}(\theta-1)$ in comparison with the heat gains due to heat flux from the fire, because $\dot{m}_{out} \sim \dot{m}_f$, but $\Delta h_c >> c_p(T - T_0)$. Assuming also that $\varepsilon_{R,f}\left(\theta^4 - \theta_f^4 \right) \approx \varepsilon_{R,f}\left(\theta^4 - 1 \right)$, and determining the wall temperature T_W as in [5] ($\theta_w = 1 + \beta(\theta - 1)$, $0 \le \beta \le 1$) we have:

$$\frac{d\theta}{d\tau} = 1 + a_1\left(\theta^4 - 1 \right) - a_2\left(\theta - 1 \right) + a_3 \theta^3 + a_4 \theta^2 \tag{5}$$

where

$$d\tau' = a_0 d\tau, \; a_0 = 1 + 2\beta^2(1-\beta)(\beta-3)\varepsilon_{R,W}, \; a_1 = \left[\varepsilon_K - \varepsilon_{R,L} - \varepsilon_{R,f} - (1-\beta^4)\varepsilon_{R,W}\right]/a_0,$$

$$a_2 = \left[\varepsilon_{C,H}(1-\beta) + \varepsilon_{C,L} - 4\beta(1-\beta)^3\varepsilon_{R,W}\right]/a_0, \; a_3 = 4\beta^3(1-\beta)\varepsilon_{R,W}/a_0, \qquad (6)$$

$$a_4 = 6\beta^2(1-\beta)^2\varepsilon_{R,W}/a_0.$$

The simplest model is described by the four determining parameters a_1, a_2, a_3, and a_4. Solutions to (5) in the non-dimensional variables, also the initial condition and critical condition are given by some function of these parameters and time:

$$\theta(\tau) = f(\tau', a_1, a_2, a_3, a_4), \; \theta(0) = 1, \; g(a_1, a_2, a_3, a_4) = 0 \qquad (7)$$

3 Critical conditions

An intersection of the curves of gain function G and loss function L represents quasi-steady behaviour. Generally speaking, there are at most three solutions to the balance condition (see Fig. 2, which is called Semenov's diagram in classical thermal explosion theory) but the intermediate solution is unstable and not observed in practice. This is because any small perturbations will result in a large change in temperature. The number of intersections may change.

During fire development the losses and gain function curves move relative to each other. With the approximations used here the gain curve does not actually move and Fig. 2 shows just a variation in losses.

Consider the behaviour of heat losses as the walls are heated. The heat conducted away is proportional to the temperature difference; so losses begin to fall. A quasi-steady state in the fuel controlled regime (point P) moves towards higher temperatures (Fig. 2). In the case of flashover this movement of the point continues until a tangency of the curves occurs at the point P_* followed by a rapid increase in temperature.

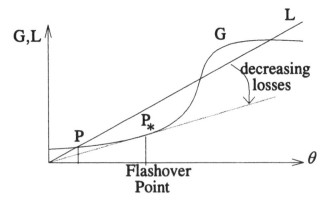

Fig. 2. Semenov's diagram, critical points in development.

Small perturbations beyond the critical point lead to a catastrophic jump to the only remaining quasi-steady state. This jump or bifurcation to a higher temperature represents flashover. Extinction is represented by a jump to a lower temperature (for example when fuel is depleted and heat gains decrease there would be such an extinction). The phenomenon is well known in thermal explosion theory [7].

Differentiating the heat conservation equations gives the tangent to the curves. The critical points are found where the tangent is zero. If the tangent is positive then gains exceed losses and the fire develops, conversely a negative tangent represents a decaying fire. Our aim is to study the lower critical point.

3.1 Walls of large thermal inertia
Consider different cases of β (which describes the thermal response of the walls). In the case of a large thermal inertia there is a delay for the wall surface temperature to rise before flashover and $\beta=0$.

The coefficients become:

$$a_0 = 1, \ a_1 = \varepsilon_K - \varepsilon_{R,L} - \varepsilon_{R,f} - \varepsilon_{R,W}, \ a_2 = \varepsilon_{C,H} + \varepsilon_{C,L}, \ a_3 = a_4 = 0 \tag{8}$$

and we have only two determining parameters. Equation (5) takes the form:

$$\frac{d\theta}{d\tau} = 1 + a_1\left(\theta^4 - 1\right) - a_2\left(\theta - 1\right) \tag{9}$$

which gives the critical conditions:

$$\begin{cases} a_1\left(\theta_*^4 - 1\right) - a_2\left(\theta_* - 1\right) + 1 = 0 \\ 4a_1\theta_*^3 - a_2 = 0, \ \ or \ \theta_* = \sqrt[3]{a_2 / 4a_1} \end{cases} \tag{10}$$

Hence the critical temperature is

$$\theta_* = \left(\frac{\left(A_U h_C + (1-D)A_V \left(h_v - h_c\right)\right)T_0}{4\alpha_g \sigma A T_0^4 \left[\chi \frac{\Delta h_C}{\Delta h_{vap}} \frac{\alpha_U}{\alpha_g} \frac{A_f}{A} - 1 \right]} \right)^{1/3} = \left(\frac{b_1}{b_2 \frac{A_f}{A} - 1} \right)^{1/3} \tag{11}$$

where $b_1 = \dfrac{A_U h_C T_0 \left(1 + (1-D)\dfrac{A_V}{A_U}\dfrac{h_v - h_c}{h_C}\right)}{4\alpha_g \sigma A T_0^4}$, $b_2 = \chi \dfrac{\Delta h_C}{\Delta h_{vap}} \dfrac{\alpha_U}{\alpha_g}$, $A = A_U + A_L$ is the total inner surface area of the compartment. Clearly if $\theta > \theta_*$ there is no stationary solution, gains exceed losses and flashover takes place. Equation (11) is a simple expression in terms of the ratio of convection heat losses from the smoke layer, in comparison to the net radiation heat gains.

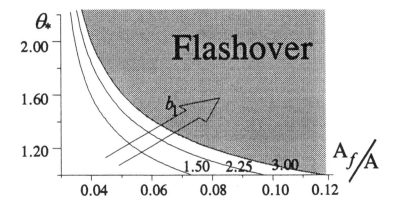

Fig. 3. Critical temperature as a function of dimensionless fire area.

The dependence of the critical temperature θ_* as a function of the non-dimensional fire area is presented in Fig. 3 for the value $b_2 = 34.25$ (from the data in [5]).

The region of flashover is above these curves. The greater the fire area, the easier for the fire to flashover. The greater the value of b_1 (the more intensive the convective heat losses, or the greater the vent area), or the greater the value of b_2 (the less intensive the heat release of the fire), the greater the critical temperature and the greater the difficulty for the fire to achieve flashover. All these considerations are physically reasonable.

Noting that usually $A_v \ll A_u$, $h_v \approx h_c$, we have from (11)

$$\theta_* \approx \left(\frac{A_u h_c T_0}{4\alpha_g \sigma A T_0^4 \left[\frac{A_f}{A} \left(\chi \frac{\Delta h_c}{\Delta h_{vap}} \frac{\alpha_u}{\alpha_g} - 1 \right) \right]} \right)^{\frac{1}{3}} \tag{12}$$

In this limit, the critical temperature θ_* does not depend upon the vent area.

It is physically clear that $\theta_* \geq 1$ ($a_2 \geq 4a_1$), i.e.

$$\frac{A_f}{A} < \frac{\Delta h_{vap}}{\chi \Delta h_c} \frac{\alpha_g}{\alpha_u} \left[\frac{A_u h_c T_0}{4\alpha_g \sigma A T_0^4} \left(1 + (1-D) \frac{A_v}{A_u} \frac{h_v - h_c}{h_c} \right) + 1 \right] \approx \frac{\Delta h_{vap}}{\chi \Delta h_c} \frac{\alpha_g}{\alpha_u} \left[\frac{A_u h_c T_0}{4\alpha_g \sigma A T_0^4} + 1 \right] \tag{13}$$

This shows how the non-dimensional fire area is limited by the ratio of convective heat losses to radiative heat gains for a quasi-stationary regime of fire to exist. This is a necessary but insufficient condition for the existence of the quasi-steady state.

Substituting θ_* from the second equation in (10) back into the first, gives the critical relationship between the non-dimensional parameters a_1 and a_2 which can be presented in the parametric form:

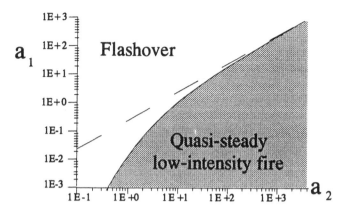

Fig 4 Regions of quasi-steady (low intensity) fire and flashover.

$$a_1 = \frac{1}{3\theta_*^4 - 4\theta_*^3 + 1}, \qquad a_2 = \frac{4\theta_*^3}{3\theta_*^4 - 4\theta_*^3 + 1} \tag{14}$$

This curve is shown in Fig. 4. The dotted straight line $a_2 = 4a_1$ is the asymptote of (14) at $\theta_* \to 1$ and also gives the range of parameters $a_2 > 4a_1$, where $\theta_* > 1$. Thus, the quasi-steady regime of low-intensity fire with little heating of the upper layer exists for the shaded area of parameters lying under the solid curve. Outside this area flashover occurs.

3.2 Walls of low thermal inertia

If the walls have time to attain the temperature of the smoke layer, then $\beta=1$, $\theta_w=\theta$ and again the process is described by equation (9) with the two coefficients being:

$$a_1 = \varepsilon_K - \varepsilon_{R,L} - \varepsilon_{R,f}, \qquad a_2 = \varepsilon_{C,L} \tag{15}$$

In comparison with (8) these simply omit the radiative and heat losses $\varepsilon_{R,W}$ and $\varepsilon_{C,H}$ because temperatures in the upper zone and surrounding walls are equal to each other. The analysis of these equations is similar to that in section 3.1. The equations (9), (10), (14) and Fig. 4 are the same. The formula for the critical temperature takes the form:

$$\theta_* = \left(\frac{h_V T_0}{4\alpha_g \sigma T_0^4} \frac{\{1-D\}A_V}{\left(A_L + \{1-D\}A_V\right)\left[\chi \frac{\Delta h_c}{\Delta h_{vap}} \frac{\alpha_U}{\alpha_g} \frac{A_f}{A_L + \{1-D\}A_V} - 1 \right]} \right)^{1/3} \tag{16}$$

The greater the vent area and the smaller the fire area, and its intensity, the higher is the critical temperature; as should be expected. In comparison with the case of walls of high thermal inertia, the critical temperature depends significantly upon the vent area.

4 Conclusions

A reasonable assumption that heating of the lower zone is small during the earlier stage of fire in a compartment, resulted in the formulation of a quite simple two-zone model that allows the determination of the main characteristics of the system. As a result, the behaviour of fire is described by two non-dimensional determining parameters a_1 and a_2 for both small and large values of thermal inertia of the walls. In the case of intermediate values of thermal inertia in the compartment walls, the parameter β should be taken into account.

Simple explicit formulae for the critical temperature of flashover were obtained. Flashover was considered to be the absence of a quasi-steady-state solution to the basic heat balance equation. The relevant area of the parameters was also found.

References

1. SFPE Handbook of Fire Protection Engineering (1988) National Fire Protection Association, USA.
2. Rasbash, D.J. (1991) Major disasters involving flashover. *Fire Safety Journal*, Vol. 17, pp. 85-93.
3. Thomas, P.H., Bullen, M.L., Quintiere, J.G. and McCaffrey, B.J. (1980) Flashover and instabilities in fire behaviour. *Combustion and Flame*, Vol. 38, pp. 159-71.
4. Thomas, P.H. (1980/1) Fires and flashover in rooms - a simplified theory. *Fire Safety Journal*, Vol. 3, pp. 67-76
5. Holborn, P.G., Bishop, S.R., Beard, A.N. and Drysdale, D.D. (1993) Nonlinear dynamics of flashover in compartment fires. *Fire Safety Journal*, Vol. 21, pp. 11-45.
6. Drysdale, D.D., Holborn, P.G., Bishop, S.R. and Beard, A.N. (1993) Experimental and theoretical models of flashover. *Fire Safety Journal*, Vol. 21, pp. 257-66.
7. Frank-Kamenetskii, F.K. (1969) *Diffusion and heat transfer in chemical kinetics*, Plenum Publishing Corporation, New York.

4 THE THEORETICAL MODEL OF CARBONIZED ORGANIC SOLID FUELS COMBUSTION

V.M. GREMYACHKIN
Institute for Problems in Mechanics,
Russian Academy of Sciences, Moscow, Russia

Modern buildings are faced inside by wood and plastic plates which are very dangerous in case of buildings firing. There are two stages in wood or plastic plates combustion. First stage is volatile materials burning out and the second stage is the combustion of carbonized materials which are formed after volatile components burned out. The second stage of the face plates combustion is the most dangerous because of combustion temperature is highest in this stage. Besides, the carbonized materials at high temperature can interact with carbon dioxide and steam but not only with oxygen. In this case the gaseous fuels (mixture of carbon oxide and hydrogen) may be formed what is very dangerous for fire development. These circumstances made the investigations of carbonized materials combustion very important for the simulation of fire development in modern buildings.

The most investigations deal with char particles combustion [1-4] for what the kinetics of carbon interaction with different reactances is determined. Here we shall consider the carbonized material plates combustion what is more important for buildings fire consideration.

Let's have a plate of porous char and the gas layer over char surface having the thickness δ. On the external boundary of the gas layer the temperature and the composition of the gas phase are known: $T = T_0, z_j = z_j^0 (j = 1 - CO_2, 2 - CO, 3 - H_2O, 4 - H_2, 5 - -CH_4, 6 - O_2)$.

Let us consider the diffusive equations in form of carbon, oxygen and hydrogen atoms conservation and the heat transfer equation in form of the full enthalpy conservation

$$\frac{d}{dx} \sum m_j I_j / \mu_j = \Phi_c / \mu_c \qquad (1)$$

$$\frac{d}{dx} \sum n_j I_j / \mu_j = 0 \qquad (2)$$

$$\frac{d}{dx} \sum l_j I_j / \mu_j = 0 \qquad (3)$$

$$\frac{d}{dx} (\sum I_j h_j + I_h) = \Phi_c h_c - I_R \delta(x) \qquad (4)$$

where m_j, and n_j, and l_j are the numbers of carbon, oxygen and hydrogen atoms in molecules $C_{m_j} O_{n_j} H_{l_j}$, and $I_R = \sigma(T_s^4 - T_w^4)$ is the elimination flux of heat from char surface, and

$$I_j = mz_j - \rho D \frac{dz_j}{dx}; \qquad I_h = mcT - \lambda \frac{dT}{dx}$$

are the flows of substances and heat.

Also the next equations are considered

Fire Engineering and Emergency Planning. Edited by R. Barham.
Published in 1996 by E & FN Spon. ISBN 0 419 20180 7.

$$\sum z_j + z_i = 1 \tag{5}$$

$$\frac{dm}{dx} = \Phi_c \tag{6}$$

$$m = -\frac{K}{\mu}\rho\frac{dP}{dx} \tag{7}$$

taking into consideration that the sum of relative mass concentrations is unit; the alteration of mass flow is associated with rate of carbon consumption; the existing of gas flow in porous char media is associated with pressure gradient by Darcy's law.

It is necessary also to consider the equations of chemical reactions kinetics of char interaction with reactances.

It can be assumed that all chemical reactions proceed in diffusive regime [5-6] on the char surface and in the thin flame over the char surface. This diffusive model demands nothing the kinetic data for utilization. Such model rather well describes the combustion of liquid hydrocarbon fuels droplets, when only the homogeneous chemical reactions take place. However, this model is not able to describe the complex regularities of char combustion.

If the equations of all chemical reactions kinetics are considered [7-8], the most common theoretical model takes place. However, the equations in this model are nonlinear and very complex for solution. The big volume of information received in a result of these equations numerical solution are very hard to analyze. The lack of real kinetic data is also considerable hard for utilization of this model.

The estimations [9] show that in case of carbon combustion in air, as in case of liquid droplets combustion the homogeneous reactions transfer from diffusive to kinetic regimes for carbon particles size about a few microns only. However, the heterogeneous chemical reactions proceed in kinetic regime. Thus it can be assumed that the heterogeneous chemical reactions proceed in kinetic regime but the homogeneous chemical reactions proceed in diffusive regime in process of char combustion and gasification.

It is necessary to note that the temperature of liquid droplets is limited by boiling temperature of liquid fuels which is rater low. Because the homogeneous reactions in case of liquid droplets combustion can proceed only in the thin flame where the temperature is maximum. The temperature of char particles may be very high. In this case the homogeneous chemical reactions can take place in gas phase everywhere but not only in the thin flame front.

If the homogeneous reactions proceed in diffusive regime, when the chemical reaction rate is significantly more reactances flows, the chemical equilibrium in gas phase must take place as the process of chemical reaction proceedng is the process of chemical equilibrium establishment. The conditions of chemical equilibrium give the additional equations [10]

$$p_j = p_c^{m_j} p_o^{n_j} p_h^{l_j}/K_j \tag{8}$$

where K_j are the equilibrium constants of substances formation from elements.

If the heterogeneous reactions proceed in diffusive regime the equilibrium between solid and gas phases must take place also. In this case the partial pressure of carbon vapor must be equal to partial pressure of satiation carbon vapor inside the char. If the heterogeneous reactions proceed in kinetic regime it is necessary to get the kinetic equation for rate of carbon consumption

$$\Phi_c = \rho S[\gamma_1(z_1 - z_{e1})^{q_1} + \gamma_2(z_3 - z_{e3})^{q_2} + \gamma_3(z_4 - z_{e4})^{q_3} + \gamma_4(z_6 - z_{e6})^{q_4}] \qquad (9)$$

where z_{ej} are the equilibrium concentrations of reactances at the char surface and γ_j are the rate constants of the corresponding heterogeneous chemical reactions.

The boundary conditions for case of char plate combustion and gasification may be written in form

$$x = \infty, I_j = 0, I_h = 0, m = 0; \quad x = \delta, T = T_0, z_j = z_j^0, P = P_a.$$

Besides, it is necessary to determine the conditions on the char surface also. Such conditions must be uninterruption of the temperature, the gas components concentrations and the gas flow velocity.

It is necessary to mark that the theoretical model includes consideration of char porous structure from which the penetration K and internal char surface S are depended.

The considering theoretical model of char plate combustion and gasification in a mixture of gaseous reactances arbitrary composition and temperature is able to determine:

(1) the char burning out rate in dependence on environment gas temeperature and composition;

(2) the distributions of the temperature, pressure and reactances concentrations in gas phase and inside porous char;

(3) the surface temperature of char and elimination flux from the char surface;

(4) the distribution of the carbon consumption rate inside char which depends on temperature, pressure and reactances concentrations;

(5) the rates of the individual reactances formation (consumption) including hydrogen and carbon oxide formation.

REFERENCES

1.Laurendau N.M.,Progress in Energy and Combustion Sciences 4:221-270 (1978)

2.Dutta S.,Wen C.Y., Belt R.J.,Ind. Eng. Chem. Process Des.Dev. 16:20-28 (1977)

3.Chin G., Kimura S., Tone S., Otake T., Ind. Chem. Eng. 23:105-120 (1983)

4.Smith I.W., Combust. Flame 14:237-248 (1979)

5.Gloving A.M., Pesochin V.R., Tolmachev I.Y., Physica goreniya i vzriva 18(No 2): 28-35 (1982)

6.Libby P.A., Blake T.R., Combust. Flame 36: 139-169 (1979)

7.Hitrin L.N., Physics of Combustion and Explosion, MGU, Moscow, 1957,p.337-411.

8.Srinevas B., Amundson R.N., AIChE Journal 26:487-496 (1980)

9.Gremyachkin V.M., Schiborin F.B., Physica goreniya i vzriva 27(No 5): 67-73 (1991)

10.Thermodynamic properties of individual substances (Ref., V.P.Glushko,Ed.), Nauka, Moscow, 1978.

11.Lee S., Angus J.C., Edwards R.V., Gardner N.C., AIChE Journal 30: 583-593 (1984)

12.Gremyachkin V.M., Roschina L.M., Physica goreniya i vzriva 26(No5):63-67 (1990)

5 SELF-GENERATED TURBULENCE OF LAMINAR AND TURBULENT TRANSIENT FLAME FRONTS INSIDE A CLOSED SPHERICAL VESSEL

B. LEISENHEIMER and W. LEUCKEL
Universität Karlsruhe, Engler-Bunte-Institut,
Lehrstuhl und Bereich Feuerungstechnik, Germany

Introduction

Optical measuring technique with very high resolution allows transient records of flame front positions and thus supply new possibilities of evaluating laminar and turbulent burning velocity (Palm-Leis et al., 1969, Dowdy et al., 1990). Therefore effort and good progress has been made in investigating laminar deflagrations in closed vessels especially in the early stage of explosion after ignition. Subject of this paper is the investigation of laminar and turbulent flame front behaviour in the later stage of an explosion process, where experimental data of burning velocity and knowledge about the effect of self-generated turbulence in a spherical propagating flame front are still limited.

It will be shown that laminar explosion experiments led to a radial range of the propagating flame front in which the influence of stretch due to the overall spherical shape as well as compression of the unburnt gas can be neglected, so that the burning velocity S is mainly influenced by self-generated acceleration processes. In this particular regime of flame front radius also turbulent flame fronts are accelerated by additional self-generated turbulence, and the flame front propagation is based on the effective turbulence intensity $(u'/S_{L,p})_{eff}$ resulting from superposition of fan-generated $(u'/S"d.$ and flame-generated $(u'/S_{L,p})_{gen}$ turbulence intensity. Finally, a linear increase of reduced turbulent burning velocity $S_T/S_{L,p}$ with increasing effective turbulence intensity $(u'/S_{L,p})_{eff}$ can be deduced.

Experimental

The pressure/time histories of premixed lean and rich fuelgas/air-mixture flame fronts ($p_0 = 1.013$ bar, $T_0 = 298$ K) ignited by spark discharge ($E = 50$ mJ) in the centre of a closed spherical explosion bomb (volume $V = 1.16$ m^3, radius $R = 0.651$ m) were recorded by piezoelectric pressure transducers (Piezotronics) in order to evaluate laminar and turbulent burning velocity as function of radius (Fig. 1). By using different fuel gases (CH$_4$, H$_2$,

Fire Engineering and Emergency Planning. Edited by R. Barham.
Published in 1996 by E & FN Spon. ISBN 0 419 20180 7.

CH_4/H_2, C_2H_2/N_2), the laminar burning velocity of a theoretical one-dimensional planar flame front $S_{L,p}$ was varied in a wide range between $0.2 < S_{L,p} < 3.4$ m/s. In order to generate a homogeneous mixture prior to ignition as well as to induce different scales of turbulence, 2,4 and 8 small (d = 0.25 m, 500 < n < 2500 rpm) or 8 large (250 < n < 1300 rpm) ventilation fans were mounted symmetrically inside the explosion vessel, and were run in wide ranges of the rotational speed n.

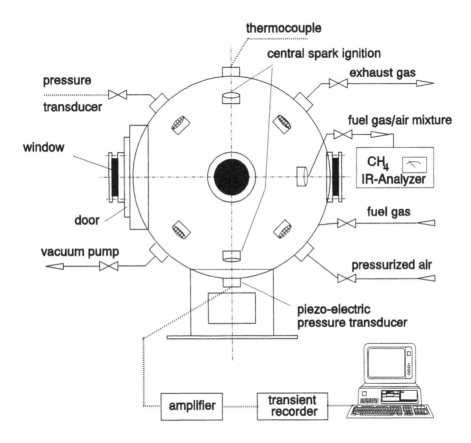

Fig. 1:   Experimental set-up

To determine the influence of turbulent velocity fluctuation amplitude and integral length scale on flame shape, binarized images of laminar and turbulent propagating flame fronts were recorded by a CCD-camera (Nanocam Proxitronic) in the early stage of combustion, using laser light sheet illumination from a 4.5 W Argon/Ion laser (Innova 70).

An advanced method of measuring turbulence characteristics in the isothermal system is introduced combining a 5-hole Pitot-probe sonde providing the $\bar{u}$ values ($\bar{u}$ = time averaged mean value and spatial direction of velocity vector) with hot-wire anemometry providing the u' values (u' = RMS-value of velocity fluctuation). The single hot wire probe was orientated towards negative direction with respect to the mean velocity vector $\bar{u}$, which leeds to the representative turbulence intensity u'. The integral length scale L_T was obtained calculating the cross-correlation function (inverse Fourier Transformation of cross power density spectra) of the signal of two hot-wire probes using a 16 bit signal analyzer (Tectronix).

Calculation of burning velocity

Based on theoretical considerations using the base principles of mass and energy conservation, and introducing several justified simplifications such as ideal gas law, spherical flame propagation, isentropic compression of fresh and burnt mixture (isentropic exponent κ = 1.4) and no heat loss to the vessel wall, it can be deduced (Lewis et al., 1934) that pressure p increases proportionally to the increasing mass fraction of burnt mixture M in the vessel ($p \sim M_b$), finally reaching the theoretical maximum pressure p_e at the end of explosion, corresponding to the final state of isochoric combustion. Dissociation processes in the burnt gas were taken into account.

Although the experimental maximum pressure at the end of the combustion process (inflection point) is actually lower than the theoretical one due to heat loss and reaction quenching near to the wall, the simplifying considerations made above are justified in the early stage of explosion. In addition to the geometrical assumption of spherical flame propagation and neglect of the flame front volume compared to the fresh and burnt gas volume, one obtains an analytic relation between motion of the flame dx relative to the fuelgas/air-mixture and the pressure rise dp

$$\frac{dx}{dp} = \frac{1}{3}\frac{r}{p_e - p_0}\left(\frac{p}{p_0}\right)^{-\frac{1}{\kappa}}\left[1 - \frac{p_e - p}{p_e - p_0}\left(\frac{p}{p_0}\right)^{-\frac{1}{\kappa}}\right]^{-\frac{2}{3}} \tag{1}$$

Hence, introducing the dimensionless pressure p'=(p/p$_0$-1) and the reduced time t'=(p$_e^{1/3}$/R)t and replacing the flame motion dx by dx = S dt one arrives at:

$$\int_0^{p^{\wedge 1/3}} \frac{d(p^{\wedge 1/3})}{(p'+1)^{1/3\kappa}\left[1 + p_e'\frac{(p'+1)^{1/\kappa}-1}{p'}\right]\cdot S_{(p')}} = t'_{theor.} \tag{2}$$

In order to get an expression of $S_{(p')}$, it is suitable to plot the pressure/time record $p'^{1/3}$ as a function of t' in order to avoid a mathematical singularity at t=0 (dp=0) for numerical integration. Transforming the first derivative of this function, which represents the integrand of eq. 2, the burning velocity S can be evaluated as function of (p') in the range p > 1.02 bar respectively r > 0.1 m.

Results and Discussion

In order to define a radial regime during a deflagration process in which the flame front is predominantly influenced by acceleration, laminar stoichiometric methane/air explosions were investigated from the point of ignition up to the end of combustion. Taking into account the experimental data of flame velocity in the very early stage of explosions in a closed vessel (Bradley *et al.*, 1993) the typical behaviour of burning velocity S, (Fig. 2) is represented by three main zones, where S_L is predominantly influenced by different effects: *Stretch, Acceleration and Compression.*

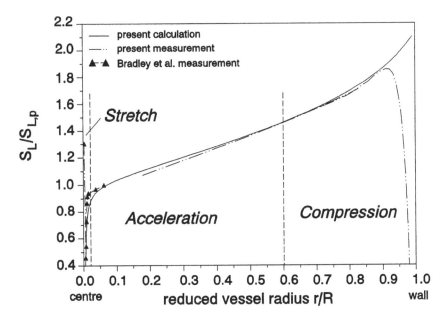

Fig. 2:   Reduced laminar burning velocity $S_L/S_{L,p}$ as a function of reduced vessel radius r/R

For spherical flame propagation with increasing radius the influence of flame front stretch (flame curvature and straining, hydrodynamic effects) reduces the one-dimensional planar laminar burning velocity $S_{L,p}$ linearly with the stretch factor k (Markstein, 1951, Clavin, 1985): $S_L = S_{L,p} -Lk$, where $k=(1/A)(dA/dt)=(2/r)(dr/dt)$ and L = Markstein lenght scale *(Stretch)*.

Above a certain radius $r/R \approx 0.02$ of a spherically propagating flame front, stretch effect due to the overall spherical shape of the flame front can be neglected compared with the instantaneous local flame curvature due to flame instabilities or turbulence motion. In this regime of *Acceleration* we determined an approximately linear increase of S_L with vessel radius r: $S_L/S_{L,p} \sim r/R$. During laminar deflagration, hydrodynamic flame instabilities due to thermal expansion of the gas and transport effects implies an initial wrinkling of the flame front (Sivashinsky, 1977). Under those conditions of diffusional-thermal instability, Sivashinsky (1977) emphasized that the flame front spontaneously becomes turbulent characterized by a constant increase in its mean propagation velocity, which is in good aggreement with experimental results (laser light sheet images, burning velocity SL) of spherical laminar flame propagation (Fig. 2).

The following exponential increase of S_L *(Compression)* can be described by a function $S_L \sim (p/p_0)^z$, where z represents pressure ($S_L \sim p^{-0.47}$) and temperature ($S_L \sim T^2$) dependency of the laminar burning velocity S_L of stoichiometric methane/air flames (Andrews *et al.*, 1972) which leads to z = 0.1 and is also in good agreement with the experimental behavior of S_L (Fig. 2). In the last stage of combustion the flame front is close to the wall, and the quenching of combustion process due to heat losses causes a sharp decrease of S_L.

Introducing the commonly used Markstein number Ma = L/δ_L (L = 0.15 mm; Bradley *et al.* 1993) and Karlovitz number Ka$=k\delta_L/S_L$ (δ_L = laminar flame thickness), the typical behaviour of S_L for laminar deflagration of stoichiometric methane/air-mixtures in closed vessels can be described in a wide radial range (r/R < 0.9) by superposition of stretch, acceleration and compression:

$$\frac{S_L}{S_{L,p}} = \frac{1}{1+MaKa} \cdot \left[1+0.7\frac{r}{R}\right] \cdot \left[\frac{p}{p_0}\right]^{0.1} \tag{3}$$

Concluding these results there is a region of the propagating flame front between 0.02 < r/R < 0.6, in which the influence of stretch due to the overall spherical shape as well as compression of the unburnt gas can be neglected, hence the burning velocity is mainly influenced by acceleration processes and further evaluation in this direction is promising.

In order to study the acceleration of various fuel gases during turbulent explosion experiments, turbulence parameters of the fan-generated turbulent flow field were measured. In the region of evaluation of turbulent burning velocity S_T, turbulent macro length scale is fairly constant $L_T \approx 24$ mm and independent of the number, size and speed of the fans. Turbulence intensity u' increases approximately linearly from 1.4 m/s (r/R = 0. 1) up to the maximum value 2.4 m/s (r/R 0.5) and followed by a weak decrease to 2.3 m/s around r/R = 0.65.

Moreover, variation of the orientation of the hot wire probe in the explosion bomb has shown, that the fan induced turbulent flow field in this inner region of the vessel can be regarded as isentropic but not homogeneous. These destributions are linear functions between the centre of the explosion bomb and half of radial distance to the wall, which represents the regime of detailed evaluation.

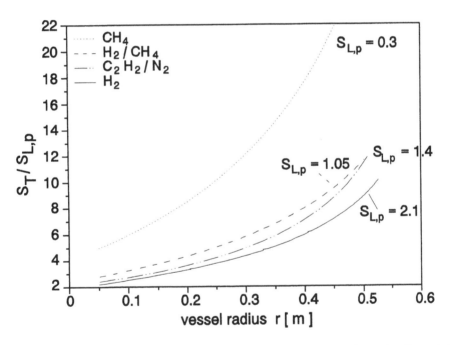

Fig. 3:   Reduced turbulent burning velocity $S_T/S_{L,p}$ as a function of vessel radius r [m]

As a consequence of these turbulence conditions a strong increase of turbulent burning velocity S_T during turbulent flame front propagation in the vessel was obtained (Fig. 3). Hence, assuming the preservation of the isotropic turbulent flow field during each explosion experiment, S_T can be correlated with turbulence intensity u' at each respective radial position, and it became obvious that S_T increases progressively.

On the other hand, under moderate turbulence conditions like in the present case, linear increase of S_T with increasing u' must be assumed (Abdel-Gayed *et al.*, 1984, Fansler *et al.*, 1990, Leuckel et *al.*, 1990).

Therfore, as could be expected from the laminar experiments described above, it can be deduced that the turbulent flame front is accelerated by self-generated additional to the fan-generated turbulence, i.e. the propagation of the flame front is based on the effective turbulence intensity $(u'/S_{L,p})_{eff}$ resulting from superposition of fan-generated $(u'/S_{L,p})_{eff}$ and self-generated $(u'/S_{L,p})_{meas}$ turbulence. This generated part of turbulence increases in direct proportion to the increasing overall flame front surface $(u'/S_{L,p})_{gen} \sim (r/R)^2$.

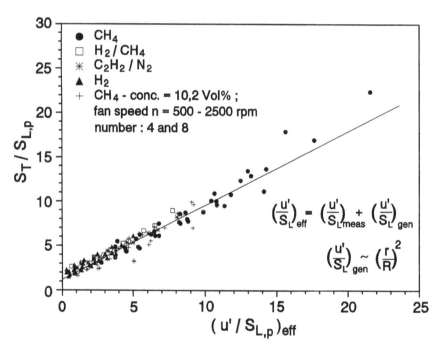

Fig. 4:   Reduced turbulent burning velocity $S_T/S_{L,p}$ as a function of effective turbulence intensity $(u'/S_{L,p})_{eff}$

Taking into account additional self-generated turbulence for all fuelgas/air mixtures and turbulence conditions investigated, the reduced turbulent burning velocity $S_T/S_{L,p}$ plotted against effective turbulence intensity $(u'/S_{L,p})_{eff}$ demonstrates the expected linear behaviour (Fig. 4).

Conclusions

Based on the evaluation of laminar burning velocity S_L as well as on fan-generated turbulent burning velocity S_T of premixed transient fuelgas/air explosions inside a constant volume spherical vessel as a function of various turbulence characteristcs of flow field - namely turbulence intensity u' and macro length scale L_T - the following results can be concluded:

- Fan-generated turbulence intensity u' increases approximately linearly with increasing vessel radius and fan speed.

- Turbulent macro lenght $L_T = 24 \pm 2$ mm is constant along the vessel radius and independent of number, diameter and speed of the fans generating turbulence.

- Self-generated turbulence causes linear increase of $S_L/S_{L,p} \sim (r/R)^2$ during the laminar explosion process in good agreement with theorical predictions.

- Self-generated turbulence causes a progressive increase of $S_T/S_{L,p} \sim (r/R)^2$ during an explosion process with initial fan-generated turbulence.

- With respect to additional self-generated turbulence the reduced turbulent burning velocity $S_T/S_{L,p}$ increases linearly with increasing effective turbulence intensity $(u'/S_{L,p})_{eff}$.

References

Abdel-Gayed, R. G., Al-Khishali, K. J. and Bradley, D. (1984). Turbulent Burning Velocities and Flame Straining in Explosions. *Proc. R. Soc. Lond.* A391, p. 393.

Andrews, G. E. and Bradley, D. (1972). Ihe Burning Velocity of Methane-Air Mixtures. *Comb. Flame 19,* p. 275.

Bradley, D., Ali, Y., Lawes, M. and Mushi, E. M. J. (1993). Problems of the Measurement of Markstein Lengths with Explosion Flames. *Joint Meeting of the British and German Sections of the Combustion Institute*, Queens College, Cambridge, p. 404.

Clavin, P. (1985). Dynamic Behaviour of Premixed Flame Fronts in Laminar and Turbulent Flows. *Prog. Energy Combust. Sci. 1* 1, p. 1.

Dowdy, D. R., Smith, D. B. and Taylor, S. C. (1990). The Use of Expanding Spherical Flames to Determine Burning Velocities and Stretch Effects in Hydrogen/Air Mixtures. *Twenty-third Symposium (International) on Combustion,* p. 325, the Combustion Institute.

Fansler, T. D. and Groff, E. G. (1990). Turbulence Characteristics of a Fan-Stiffed Combustion Vessel. *Comb.Flame* 80, p. 350.

Leuckel, W., Nastoll, W. and Zarzalis, N. (1990). Experimental Investigation of the Influence of Turbulence on the Transient Premixed Flame Propagation Inside Closed Vessels. *Twenty-third Symposium (International) on Combustion,* p.729, the Combustion Institute.

Lewis, B. and Von Elbe, G. (1934). Determination of the Speed of Flames and the Temperature Distribution in a Spherical Bomb from Time-Pressure Explosion Records. *Jour. Chem. Phys.* 2, p. 283.

Markstein, G. H. (1951). Experimental and Theoretical Studies of Flame-Front Stability. *J. Aero. Sci. 18,* p. 199.

Palm-Leis, A. and Strehlow, R. A. (1969). On the Propagation of Turbulent Flames. *Comb. Flame* 13, p. 111.

Sivashinsky, G.I. (1977). Nonlinear Analysis of Hydrodynamic Instability in Laminar Flames-I. Derivation of Basic Equations. *Acta Astronautica* 4, p. 1177.

6 ALGORITHMS FOR THE CALCULATION OF EGRESS COMPLEXITY

A.J. POLLOCK and H.A. DONEGAN
School of Computing and Mathematics, University of Ulster at Jordanstown, Newtownabbey, UK

Abstract
Any set of building compartments expressed as a network of nodes and arcs can be viewed as a system of egress. Information measures, based on Shannon entropy, can be derived from such a system to reflect its complexity.

Egress complexity measures can be used to compare buildings under normal or adverse working conditions and they can assist in determining the best position to place an additional exit to an existing network. This paper provides the algorithms necessary to calculate these measures and illustrates their application with examples.
Keywords: Building, complexity, egress, entropy, fire safety, graph, network, algorithm.

1 Introduction

Suppose that a naive occupant is positioned at a random location within a building. It is a fundamental question to ask - how much information does the occupant need to infer before successful egress occurs? Proposed measures for such information were recently introduced by Donegan et al[1]. However, the measures were restricted to network representations of buildings without circuits. This paper resolves this restriction and proposes enhanced algorithms. The algorithms are based on a pseudo information measure (1) referred to as the *nodal information*, adapted from Quinlan's ID3 algorithm[2].

$$I_i = n_i^+ \log_2 \frac{n_i^+ + n_i^-}{n_i^+} + n_i^- \log_2 \frac{n_i^+ + n_i^-}{n_i^-} \qquad 1 \leq i \leq k \tag{1}$$

Fire Engineering and Emergency Planning. Edited by R. Barham.
Published in 1996 by E & FN Spon. ISBN 0 419 20180 7.

The values for n_i^+ and n_i^- represent positive and negative information steps respectively and k is the total number of non-exit nodes. A positive information step is a traversal between two compartments via a previously untravelled path. When a path is visited a second time then backtrack occurs and this is called a negative information step.

2   Calculating Complexity Measures

It is possible to calculate an egress complexity measure $\Sigma_k\, I_i$ referred to as the exit complexity, envisaged as the resistance of that exit. Algorithm **A1** characterises the procedure for calculating this value for a given floor network **G**.

A1 -   Algorithm to calculate the exit complexity for a floor network having a single exit, no circuits and k non-exit nodes

$S_{1.1}$ Initialise $i = 1$;
$S_{1.2}$ $n^+ = k$;
do {

$\quad S_{1.3}$ Select non-exit node N_i in the floor network **G**;
$\quad S_{1.4}$ Denote the number of arcs on the shortest route to the exit as x;
$\quad S_{1.5}$ $n^- = n^+ - x$;
$\quad S_{1.6}$ Substitute the values for n^+ and n^- into equation (1) to obtain the nodal
$\quad\quad\quad$ information value I_i for N_i;
$\quad S_{1.7}$ $i = i + 1$;
} while ($i \le k$)
$S_{1.8}$ Calculate $\Sigma_k\, I_i$ to obtain the exit complexity value for **G**.

When an additional exit is inserted into a floor network the new egress complexity value should be decreased ie the chance of egress is improved. Therefore **A1** requires modification to reflect such phenomenon. In a multiple exit situation this can be modelled as follows, see[3]:

$$\frac{1}{r} = \frac{1}{c_1} + \frac{1}{c_2} + \dots + \frac{1}{c_m} \quad \text{for } 1 \le j \le m \tag{2}$$

where c_j represents the exit complexity for the jth exit and r is the *global complexity*. An algorithm to calculate r follows.

A2 -   Algorithm to calculate the global complexity for a floor network with m exits and no circuits

$S_{2.1}$ Initialise $j = 1$;
do { $S_{2.2}$ Select exit node E_j in **G**;
$\quad\quad S_{2.3}$ Remove exit nodes $E_1..E_{j-1}$, $E_{j+1}..E_m$ and afortiori connecting paths from
$\quad\quad\quad\quad$ **G** to obtain **G'**;
$\quad\quad S_{2.4}$ Apply **A1** to **G'** to calculate the exit complexity value E_j for E_j;

$S_{2.5}$ $j = j + 1$;
} while ($j \leq m$)
$S_{2.6}$ Calculate $[1/E_1 + ... + 1/E_m]^{-1}$ = the global complexity r.

These algorithms are unable to cope with a floor network **G** containing circuits. To overcome this situation the following method based on graphical spanning trees is proposed. Each spanning tree for **G** must include a single exit and all the non-exit nodes in **G**. Such spanning trees are classified as *admissible spanning trees*. The spanning tree with the maximal (or worst case) exit complexity is then used to represent the global complexity value for **G**. Consider the floor network N_1 in Fig. 1:

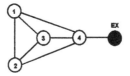

Fig. 1. Floor Network N_1

The first step towards generating an exit complexity value for N_1 is to examine all its admissible spanning trees. Graph theory permits 16 admissible spanning trees which are shown in Fig. 2 with their resultant exit complexity values outlined in Table 1.

Table 1. Spanning Tree Complexity Values

Spanning Tree	Exit Complexity	Spanning Tree	Exit Complexity
S1	23.43	S9	21.53
S2	21.53	S10	19.63
S3	21.53	S11	16.02
S4	21.53	S12	21.53
S5	16.02	S13	16.02
S6	21.53	S14	16.02
S7	19.63	S15	19.63
S8	16.02	S16	16.02

Clearly, the trees 5,8,11,13,14,16 are isomorphic ie they have the same structure, hence obtaining the same value of 16.02 for exit complexity. This is also true for trees 2,3,4,6,9,12 having the measure 21.53 and trees 7,10,15 with 19.63. The global complexity for N_1 is 23.43, the largest (or worst case) value. An algorithm to formalise the above procedure is outlined in **A3**.

A3 - Algorithm to calculate the global complexity for a floor network with circuits and a single exit

$S_{3.1}$ Generate a set *T* of admissible spanning trees for **G**;
$S_{3.2}$ Denote the total number of admissible spanning trees as t;

$S_{3.3}$ Initialise h = 1;
do {
    $S_{3.4}$ Select admissible spanning tree T_h;
    $S_{3.5}$ Apply **A1** to T_h to calculate the exit complexity E_h for T_h;
    $S_{3.6}$ h = h +1;
} while (h ≤ t)
$S_{3.7}$ The global complexity = max$\{E_1...E_t\}$.

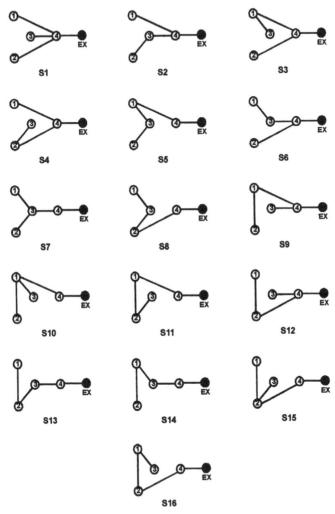

Fig. 2. Admissible Spanning Trees

The algorithm **A3** is a primer for **A4** which characterises floor networks with multiple exits.

A4 - Algorithm to calculate the global complexity of a floor network with circuits
and m exits

$S_{4.1}$ Generate a set T of admissible spanning trees for G;
$S_{4.2}$ Initialise j = 1;
do {
    $S_{4.3}$ Select an exit node E_j in G;
    $S_{4.4}$ Select the subset T_j of admissible spanning trees from T containing exit E_j;
    $S_{4.5}$ Denote the total number of spanning trees in T_j as t;
    $S_{4.6}$ Initialise h = 1;
    do {
        $S_{4.7}$ Select admissible spanning tree T_h from T_j;
        $S_{4.8}$ Apply **A1** to T_h to calculate the exit complexity E_h for T_h;
        $S_{4.9}$ h = h +1;
    } while (h ≤ t)
    $S_{4.10}$ The exit complexity C_j for $E_j = \max\{E_1...E_t\}$;
    $S_{4.11}$ j = j +1;
while (j ≤ m)
$S_{4.12}$ Calculate $[1/C_1 +...+ 1/C_m]^{-1}$ = the global complexity r.

3   Egress Comparison using Complexity Measures

Consider a four-storey building A represented by networks illustrated in figure 3.
Suppose that B is another building also represented by the networks in figure 3, but
with the following exits and stairwells removed: EX2, SW1.2, SW2.2 and SW3.2. The
results are shown in table 2.

Table 2. Buildings Complexity Results

A	Ground Floor	Result	First Floor	Result	Second Floor	Result	Third Floor	Result
	EX 1	196.34	SW 1.1	131.78	SW 2.1	130.53	SW 3.1	127.73
	EX 2	188.14	SW 1.2	139.75	SW 2.2	138.49	SW 3.2	141.00
	Global	96.08	Global	67.82	Global	67.20	Global	67.02

B	Ground Floor	Result	First Floor	Result	Second Floor	Result	Third Floor	Result
	EX1	196.34	SW 1.1	131.78	SW 2.1	130.53	SW 3.1	127.73
	Global	196.34	Global	131.78	Global	130.53	Global	127.73

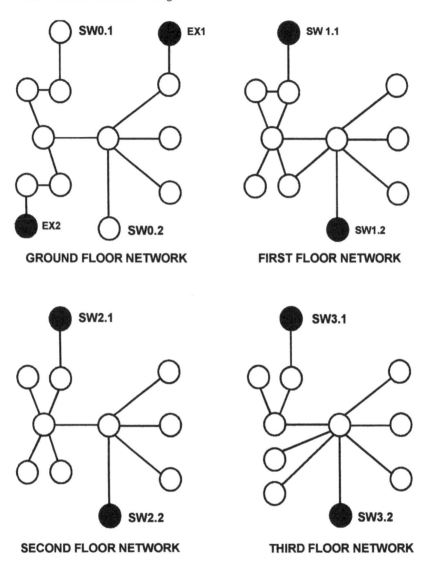

Fig. 3. Four Storey Network

3.1   Some Observations

Clearly because of the increased number of exits in A the global complexity values for B are correspondingly larger than for A.

Since the ground floor in A has two additional non-exit nodes it has a larger global complexity than its counterparts on the remaining floors. This is also the case for building B.

Since there are circuits on the first floor of building A, both of the exits yield higher exit complexity values than the corresponding exits on the second floor.

A key observation is illustrated by comparing the exit complexities for SW2.2 and SW3.2. The latter has the higher value as it connects to a node with a greater number of connecting paths.

4   Positioning Exits

Optimal positions for additional exits are based on the relative maximal amount of backtrack associated with each position. The following algorithm determines appropriate backtrack values.

A5 - Algorithm to calculate backtrack for a floor network with k non-exit nodes and m exits

$S_{5.1}$ Generate a set T of admissible spanning trees for **G**;
$S_{5.2}$ Initialise i = 1;
do {

    $S_{5.3}$ Select non-exit node N_i in **G**;
    $S_{5.4}$ Initialise j = 1;
    do {

        $S_{5.5}$ Select an exit node E_j in **G**;
        $S_{5.6}$ Select the subset T_j of admissible spanning trees from T containing exit E_j;
        $S_{5.7}$ Denote the total number of admissible spanning trees in T_j as t;
        $S_{5.8}$ Initialise h = 1;
        do {

            $S_{5.9}$ Select admissible spanning tree T_h from T_j;
            $S_{5.10}$ Denote the number of arcs on the shortest route to exit E_j as x;
            $S_{5.11}$ The backtrack $B_{1.h}$ for $T_h = k - x$;
            $S_{5.12}$ h = h + 1;
        } while (h ≤ t)
        $S_{5.13}$ The backtrack $B_{2.j}$ for E_j = max$\{B_{1.1}...B_{1.t}\}$;
        $S_{5.14}$ j = j + 1;
    } while (j ≤ m)
    $S_{5.15}$ The backtrack $B_{3.i}$ for N_i = max$\{B_{2.1}...B_{2.m}\}$;
    $S_{5.16}$ i = i + 1;
} while (i ≤ k)
$S_{5.17}$ The backtrack for **G** is ($B_{3.1}$ + ... + $B_{3.k}$).

4.1   Example of Backtrack Usage

Consider figure 4 which illustrates a floor plan with a single exit and suppose that a second exit is to be added at any of 9 possible locations.

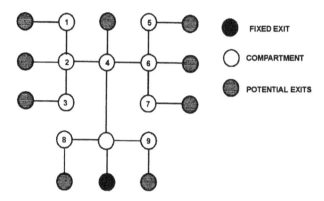

Fig. 4. Potential Exits

The respective backtrack values are shown in table 3.

Table 3. Backtrack Results

2nd Exit Position	1	2	3	4	5	6	7	8	9
Backtrack Value	76	77	76	78	76	77	76	72	72

From these results the optimal position for the second exit is at node 4. Algorithm **A5** uses the notion of association[4] to determine optimal locations. It places an additional exit at a position with least association to other exits on the floor and simultaneously maximises association with non-exit nodes.

5   References

1.   Donegan, H.A., Pollock A.J. and Taylor I.R. (1994) Egress Complexity of a Building, Proc. 4th ISFSS, Ottawa Canada, pp.601-612.
2.   Quinlan, J.R. (1982) Semi-autonomous Acquisition of Pattern-based Knowledge, Introductory Readings in Expert Systems, Gordon and Breach, New York, pp.192-207.
3.   Donegan, H.A. and Pollock, A.J. (1995) A Mathematical Basis for Egress Complexity, (submitted for publication).
4.   Kernohan, G., Rankin, G.D., Wallace, G.D. and Walters, R.J. (1973) PHASE: An Interactive Appraisal Package for Whole Hospital Design, Computer Aided Design, Vol 5., pp.81-89.

7 MODELIZATION OF COMBUSTION WITH COMPLEX KINETICS

V.A. VOLPERT
Laboratoire d'analyse numérique, Université Lyon, France
A.I. VOLPERT
Department of Mathematics, Technion, Haifa, Israel

Abstract
This paper is devoted to a new approach to study combustion waves and branching chain flames with complex kinetics. We develop a mathematical theory of travelling wave solutions for some classes of parabolic systems, and show that reaction-diffusion systems desribing chemical waves under some conditions can be reduced to these classes of systems. We obtain conditions of uniqueness and stability of combustion waves and cold flames, and find their velocity.
Key words: combustion, cold flames, complex kinetics.

1. Introduction

Development of the modern combustion theory is determined to the large extent by the works of Zeldovich and Frank-Kamenetskii on the thermal propagation of flames and by the works of Semenov on branching chain flames. In the first case the chemical reaction is strongly activated, and the propagation of the flame occurs basically due to the heat diffusion. The simplest example can be given by the one-step chemical reaction of the first order,

$$A \longrightarrow B.$$

The reaction rate under the mass action law in this case has the form

$$\frac{dB}{dt} = ke^{-E/RT}A,$$

where A and B denotes also the concentration of the corresponding species, T is the temperature, E the activation energy, R the gas constant, k is the pre-

Fire Engineering and Emergency Planning. Edited by R. Barham.
Published in 1996 by E & FN Spon. ISBN 0 419 20180 7.

exponential factor. If the activation energy E is sufficiently large, then the reaction takes place in a narrow temperature interval near the adiabatic temperature, and the reaction zone is localized in space. It allows to apply the infinitely narrow reaction zone method developed by Zeldovich and Frank-Kamenetskii [28], [29] and to find the wave velocity and the condition of its stability.

The simplest reaction describing branching chain flames is the following one-step autocatalytical reaction:

$$A + B \longrightarrow 2B.$$

The propagation of the flame in this case occurs due to the diffusion of the active center B. If we suppose that the diffusion coefficients of the species A and B equal to each other, then the model describing the propagation of the cold flame in this case is the same as considered by Kolomogorov, Petrovskii, and Piskunov in [9]. We know that the waves exist for all values of velocities greater or equal to the minimal velocity, they are stable in a certain sense, and the value of the minimal velocity is found explicitly.

It appears that even these very simple examples describe the basic properties of flames. However the aim to describe a more realistic complex chemistry seems very attractive, and there are many works devoted to this problem.

The method of infinitely narrow reaction zone can be generalized for more complicated chemical reactions [6] - [8], [13], [25]. The best studied examples are independent reactions:

$$A \longrightarrow B, \quad C \longrightarrow D,$$

consecutive reactions:

$$A \longrightarrow B \longrightarrow C,$$

and parallel or competitive reactions:

$$A \longrightarrow B, \quad A \longrightarrow B.$$

If each reaction occurs in its own narrow reaction zone and the interaction between them is only due to the temperature field, then the infinitely narrow reaction zone method is applicable. There are some other examples and generalizations [1], [26], [27]. Sometimes physical arguments help to simplify a problem and to reduce it usually to one of the examples above [14], [15].

There are also some other approaches to study more mathematical questions of existence of solutions [2] - [5], [10], [19] and to study some properties of attractors of the corresponding dynamical systems [11].

In our works [18] - [24] we develop another approach. It is based on the following two ideas:

- We develop a mathematical theory of travelling wave solutions for the monotone parabolic systems:

$$\frac{\partial u}{\partial t} = a \Delta u + F(u), \tag{1}$$

where $u = (u_1, ..., u_n)$, $F = (F_1, ..., F_n)$, a is a diagonal matrix, and the function F satisfies the following condition

$$\frac{\partial F_i}{\partial u_j} \geq 0, \quad i \neq j. \tag{2}$$

We find, in particular, the wave velocity and show its stability,

- We show that under some conditions the models describing combustion waves or branching chain flames with complex kinetics can be reduced to the systems of equations from this class.

Thus we can apply the developed mathematical theory to study chemical waves.

2 Reaction-diffusion systems

We consider a chemical reaction of the general form

$$\sum_{j=1}^{m} \alpha_{ij} A_j \rightarrow \sum_{j=1}^{m} \beta_{ij} A_j, \quad i = 1, ..., n. \tag{3}$$

Here $A_1, ..., A_m$ are concentrations of the reactants, α_{ij}, β_{ij} the stochometric coefficients. Under the usual approximation of constant density the distribution of the temperature and the concentrations can be described by the reaction-diffusion system

$$\frac{\partial T}{\partial t} = \kappa \frac{\partial^2 T}{\partial x^2} + \sum_{i=1}^{n} q_i W_i, \tag{4}$$

$$\frac{\partial A_j}{\partial t} = d \frac{\partial^2 A_j}{\partial x^2} + \sum_{i=1}^{n} \gamma_{ij} W_i, \quad j = 1, ..., m. \tag{5}$$

where T is the temperature, κ and d are the coefficients of the heat and mass diffusion, respectively, q_i is the adiabatic heat release of the i-th reaction, $\gamma_{ij} = \beta_{ij} - \alpha_{ij}$, W_i is the rate of the i-th reaction,

$$W_i = K_i(T) A_1^{\alpha_{i1}} \times ... \times A_1^{\alpha_{im}}.$$

The functions $K_i(T)$ determine the temperature dependence of the reaction rate and usually have the form of the Arrhenius exponent,

$$K_i(T) = k_i^0 e^{-E_i/RT},$$

where k_i^0 is a constant, E_i is the activation energy of the i-th reaction, R is the gas constant.

We consider the case where the diffusion coefficients of all species are equal to each other and to the coefficient of thermal diffusivity κ. It means physically that we consider a reaction in the gaseous phase, and the molecular weights of gases are close to each other. Some of the results, namely those on the wave existence, can be obtained without this condition. However in this work we discuss basically the questions of wave stability, unicity and velocity. It is well known for the simple one-step kinetics that conditions of stability depend on the relation between κ and d. If $\kappa \neq d$ and the dimensionless parameter called Zeldovich number, $Z = qE/RT_b^2$ is sufficiently large, then the flame is unstable [28]. (Here T_b is the adiabatic temperature.) If $\kappa = d$, then it is always stable.

We can put the question whether the flame with complex kinetics is stable in the case where $\kappa = d$. The answer is known. For the parallel reactions (see the examples in Section 1) the wave can be nonunique [7], [8]. Under some conditions on parameters there are three waves propagating with different velocities. Two of them are stable and one is unstable.

So we can say that there are two types of instabilities of chemical waves: the thermo-diffusional instability which can appear if $\kappa \neq d$ even for a simple kinetics, and the kinetic instability which can appear for a complex kinetics even if $\kappa = d$. We discuss here conditions of the kinetic stability. Using the results for monotone systems [19]-[23], we can say that the wave is unique and stable if the reaction-diffusion system can be reduced to the monotone system. Thus the conditions of reducibility to the monotone systems give the conditions of stability and uniqueness of the wave.

3 Conditions of reducibility

We consider the kinetic system of equations

$$\frac{dA}{dt} = \Gamma W, \tag{6}$$

where A is the vector of concentrations, W is the vector of reaction rates, Γ the matrix of stoichometric coefficients,

$$\Gamma = \begin{pmatrix} \gamma_{11} & \cdots & \gamma_{n1} \\ \cdot & \cdot & \cdot \\ \gamma_{1m} & \cdots & \gamma_{nm} \end{pmatrix}, \qquad \gamma_{ij} = \beta_{ij} - \alpha_{ij}.$$

Denote by r the rank of the matrix Γ. Without loss of generality we can suppose that the first r columns of this matrix are linearly independent. We introduce the new functions u_i, $i = 1, ..., n$ by the equalities

$$A_j = \sum_{i=1}^{r} \gamma_{ij} u_i + A_j^+, \tag{7}$$

where A_j^+ are positive constants which determine the balance polyhedron. Since the columns $r+1, ..., n$ of the matrix Γ are linearly dependent of the first r columns, then we have

$$\gamma_{kj} = \sum_{1}^{r} \lambda_{ki} \gamma_{ij}, \quad k = r+1, ..., n, \tag{8}$$

where λ_{ki} are some numbers. Substituting these expressions into the system of equations (6), we obtain after some transformations

$$\frac{du_i}{dt} = W_i + \sum_{i=1}^{r} \lambda_{ki} W_k, \quad i = 1, ..., r. \tag{9}$$

If the reactions are not isothermic, we need also the expression of the temperature through the new variables:

$$T = \sum_{i=1}^{r} q_i u_i + T^+, \tag{10}$$

where T_+ determines the value of the adiabatic temperature, q_i is the heat release of the i-th reaction.

After making the indicated substitutions, we denote the right hand-side of (9) by $F_i(u)$:

$$F_i(u) = W_i + \sum_{i=1}^{r} \lambda_{ki} W_k, \quad i = 1, ..., r. \tag{11}$$

We find now the conditions of monotonicity, i.e. the conditions of validity of the inequality (2). We assume that all functions $K_i(T)$ are nondecreasing. This condition is satisfied, in particuluar, for the Arrhenius temperature dependence of the reaction rate.

The direct computations give:

$$\begin{aligned}
\frac{\partial F_i}{\partial u_j} &= W_i \left[\sum_{k=1}^{m} \frac{\nu_{ik}\gamma_{jk}}{A_k} + \frac{K_i'(T)}{K_i(T)} q_j \right] \\
&+ \sum_{l=r+1}^{n} \lambda_{li} W_i \left[\sum_{k=1}^{m} \frac{\nu_{lk}\gamma_{jk}}{A_k} + \frac{K_i'(T)}{K_i(T)} q_j \right].
\end{aligned}$$

Sufficient conditions of monotonicity take the form

$$\nu_{ik}\gamma_{jk} \geq 0, \quad \lambda_{li}\nu_{lk}\gamma_{jk} \geq 0, \quad q_j \geq 0, \quad \lambda_{li} q_j \geq 0$$

for all $i, j = 1, ..., r$, $i \neq j$, $l = r+1, ..., n$, $k = 1, ..., m$.

These conditions are not necessary. For example we can consider the heat releases q_j of alternating signs. Moreover these conditions are not convenient for application. They can be given in a different form and simplified [19]. However here we restrict ourselves to some more simple particular cases and examples.

An important particular case is that of *linearly independent* reactions. It means that the rank of the matrix Γ is n and its columns are linearly independent. In this case the conditions of monotonicity become very simple:

If each species is consumed in no more than one reaction and all reactions are exothermic, then the condition of monotonicity is satisfied.

Thus we obtain the following result. If the reaction does not contain parallel stages and all elementary reaction are exothermic, then the flame in kinetically stable and unique. Its velocity admits the minimax representation.

We recall the result of Khaikin and Khudyaev which we mentioned in the previous section. For an example of two parallel reactions they show that the combustion wave can be nonunique [7], [8]. Thus if the conditions of unicity and stability are not satisfied, the solution can really be nonunique and unstable.

Here are some examples of reaction without parallel stages. Independent reaction:

$$A_1 + A_2 \to ..., \; A_3 + A_4 \to ..., \; A_5 + A_6 \to ...,$$

sequantial reactions:

$$A_1 \to A_2 \to A_3 \to ...$$

or

$$A_1 + A_2 \to A_3, \; A_3 + A_4 \to A_5, \; A_5 + A_6 \to A_7,$$

nonbranching chain reactions

$$A_2 + B \to AB + A, \; B_2 + A \to AB + B,$$

We note finally that we obtain the conditions of monotonicity for reactions with parallel stages, with heat releases of alternating signs, for linearly dependent reaction, for reversible reaction, and for other classes of reactions [19] - [23].

References

[1] Berestycki H., Nicolaenko B., Scheurer B. (1985) Travelling wave solutions to combustion models and their singular limits, *SIAM Journal of Mathematical Analysis,* **16**, No. 6, pp. 1207-1242.

[2] Bonnet A. (1992) Travelling waves for plannar flames with complex chemistry reaction network. *Communications on Pure and Applied Mathathematics,* **45**, No. 12.

[3] Heinze S. (1987) Traveling waves in combustion processes with complex chemical networks. *Transactions of American Mathematical Society,* **304**, No. 1, pp. 405-416.

[4] Kanel' Ya.I. (1963) The stationary solution of the set of equations in the theory of combustion, *Doklady Physical Chemistry,* **149**, No. 1-6, pp. 241-243.

[5] Kanel' Ya.I. (1990) Existence of a traveling-wave solution of a Belousov-Zhabotinskii system, *Differential Equations,* **26**, No. 4, pp. 478-485.

[6] Khaikin B.I., Filonenko A.K., Khudyaev S.I. (1968) Flame propagation in the presence of two successive gas-phase reactions, *Combustion, Explosion and Shock Waves,* **4**, No. 4, pp. 343-349.

[7] Khaikin B.I., Khudyaev S.I. (1978) On nonuniqueness of stationary propagating reaction zones with competing reactions, Preprint of the Institute of Chemical Physics, Chernogolovka, 13 p. (in Russian).

[8] Khaikin B.I., Khudyaev S.I. (1979) Nonuniqueness of combustion temperature and rate when competing reactions take place, Doklady *Physical Chemistry*, **245**, No. 1-3, pp. 225-228.

[9] Kolmogorov A.N., Petrovsky I.G., Piskunov N.S. (1937) A study of the equation of diffusion with increase in the quantity of matter, with application to a biological problem, *Bulletin of Moscow Univiversity, Mathematics and Mechanics*, **1**, pp. 1-26 (in Russian).

[10] Manley O., Marion M., Temam R. (1987) Equations of combustion in the presence of complex chemistry, *Indiana University Mathematical Journal*, **42**, No. 3, pp. 941-967.

[11] Marion M. (1991) Attractors and turbulence for some combustion models, *IMA Volumes in Mathematics and Applications*, Vol. 35.

[12] Matkowsky B.J., Sivashinsky G.I. (1979) An asymptotic derivation of two models in flame theroy associated with the constant density approximation, *SIAM Journal of Applied Mathematics*, **37**, 686-699.

[13] Merzhanov A.G., Rumanov E.N., Khaikin B.I. (1972) Multizone burning of condensed systems, *Zhurnal Prikladnoi Mekhaniki i Tekhnicheskoi Fiziki*, No. 6, pp. 99-105 (in Russian).

[14] Novozhilov B.V., Posvyansky V.S. (1973) Propagation velocity of a cold flame, *Combustion, Explosion and Shock Waves*, **9**, No. 2, pp. 191-194.

[15] Novozhilov B.V., Posvyansky V.S. (1974) Verification of the method of quasistationary concentrations in the problem of cold-flame propagation, *Combustion, Explosion and Shock Waves*, **10**, No. 1, pp. 81-86.

[16] Tang M.M., Fife P.C. (1980) Propagating fronts for competing species equations with diffusion, *Arch. Ration. Mech. and Anal.*, **73**, pp. 69-78.

[17] Terman D. (1988) Traveling wave solutions arising from a two - step combustion model, SIAM Jornal of Mathematical Analysis, **19**, No. 5.

[18] Volpert A.I., Volpert V.A. (1990) Application of the rotation theory of vector fields to the study of wave solutions of parabolic equations. Transactions of Moscow Mathematical Society, Vol. 52, pp. 59-108.

[19] A.I.Volpert, V.A.Volpert, V.A.Volpert (1994) Travelling wave solutions of parabolic systems. *American Math. Society, Providence.*

[20] Volpert V.A., Volpert A.I. (1989) Existence and stability of traveling waves in chemical kinetics. *In: Dynamics of Chemical and Biological Systems*, Nauka, Novosibirsk, pp. 56-131 (in Russian).

[21] Volpert V.A., Volpert A.I. (1989) Waves of chemical transformation having complex kinetics. *Doklady Physical Chemistry* **309**, No. 1-3, pp. 877-879.

[22] Volpert V.A., Volpert A.I. (1990) Existence and stability of waves in chemical kinetics. *Khimicheskaya Fizika*, **9**, No. 2, pp. 238-245 (in Russian).

[23] Volpert V.A., Volpert A.I. (1990) Some mathematical problems of wave propagation in chemical active media. *Khimicheskaya Fizika* , **9**, No. 8, pp. 1118-1127 (in Russian).

[24] Volpert V.A., Volpert A.I. Wave trains described by monotone parabolic systems. To apppear in Nonlinear World.

[25] Volpert V.A., Khaikin B.I., Khudyaev S.I. (1981) Combustion waves with independent reactions, *In: Problems of Technological Combustion* Vol. 1, Chernogolovka, pp. 110-113 (in Russian).

[26] Volpert V.A., Krishenik P.M. (1986) Stability in two-stage combustion wave propagation under controlled conditions, *Combustion, Explosion and Shock Waves*, **22**, No. 2, pp. 148-156.

[27] Volpert V.A., Krishenik P.M. (1986) Nonsteady propagation of combustion waves in a system of successive reactions with endothermal stages, *Combustion, Explosion and Shock Waves* **22**, No. 3, pp. 285-292.

[28] Zeldovich Ya.B., Barenblatt G.I., Librovich V.B., Makhviladze G.M. (1985) The Mathematical Theory of Combustion and Explosions, *Consultants Bureau*, New York.

[29] Zeldovich Ya.B., Frank-Kamenetsky D.A. (1938) A theory of thermal propagation of flame, *Acta Physicochimia U.S.S.R.*, **9**, pp. 341-350.

PROTECTION OF BUILT ENVIRONMENT

8 RESULTS COMPARISON OF SMOKE MOVEMENT ANALYSIS IN BUILDINGS USING STEADY-STATE AND TRANSIENT MODELS

R. BORCHIELLINI, M. CALI and G. MUTANI
Dipartimento di Energetica, Politecnico di Torino, Italy

Abstract

In this paper, two mathematical models are compared regarding their predictive ability to perform fire hazard analysis connected with smoke and toxic gases diffusion. After a classification of the computer programs used to estimate fire and smoke behaviour into buildings, the attention is turned on differentiation between two types of models: the "transient" models which are able to describe in detail fire behaviour and the "steady state" models, less detailed but easier to use. For instance, the "transient" models provide the temperature profile in the fire origin room, while the "steady state" models need this profile as an input.

In this work a comparison of the results obtained with three computer programs which simulate smoke movement represent the building with a uni-dimensional fluid network is exposed. The parameters varied in this study are the wind effect on the external facades of the building, the thermal gradients between the nodes of the network and finally the type of building examined.

The "steady state" models are expected to be used more frequently owing to their simplicity while the "transient" ones analyse in particular the event. This paper focuses on application limits of both models.

1. Introduction

In these last years fire simulation with computer programs is gaining an important part in fire safety analysis. Actually, using these tools it is possible to estimate the diffusion of fire into a building with low expense of time and money.

The development of a fire is a really complex phenomenon and computer programs use different keys to identify the problem. There are two main ways to classify models describing the fire course into a building; those classifications depend on the way the

Fire Engineering and Emergency Planning. Edited by R. Barham.
Published in 1996 by E & FN Spon. ISBN 0 419 20180 7.

building and the fire behaviour are considered. The first classification is based on building representation in which the phenomenon develops. It can be used a uni- or bi-directional network with nodes, representing zones or compartments, connected by branches that are windows or doors. Along this network there is fire and smoke spread; those models, very flexible and easy to use for big structures with high number of compartments and connections are called "zone models". Besides, "field models" work on a tri-dimensional dominion and are used to describe fire spread with much more detail into a few compartments.

Independently from first classification, a second one stands, which differently selects the models: the ones that describe the event giving time dependent results and others that give a steady fire analysis; the first are called "transient" and the second are steady state models. The comparison between these two last types of models could give the right conditions in which a transient model is useful and whereas should be used a steady state one. The utilisation of these two types of models involves different costs in time and money, so, for a detailed analysis it is sometimes required to use a transient model while in other cases the steady-state one is sufficient. The aim of this work consists in the determination of the application limits of these two types of models and for this reason there were selected three different but still comparable, computer programs.

2. The examined programs for the comparison

The comparison is made between three programs: CFAST 2.0 [1], ASCOS [2] and CONTAM [3]. These programs were used to estimate, the first the complete fire course, the second and the third the smoke diffusion and airflow into confined spaces. Those three programs utilise a "zone model" representation; CFAST 2.0 uses a "transient", while ASCOS with CONTAM a "steady -state" approach to study the event. This work compares the smoke net flow across the building which results from the three programs changing the data that mainly influence the smoke spread into a building: wind velocities thermal gradients and types of building.

2.1. CFAST 2.0: simulation of fire behaviour with a transient model

CFAST 2.0 has been developed to improve in FAST [4] (Fire And Smoke Transport) the fire characteristics and to introduce the vertical connections between compartments. CFAST 2.O is a zone model with two homogeneous volumes in every zone: in the lower there is mainly air, in the upper the hot smoke and gases coming out from the burning objects, and in the burn room there is a fire plume convecting mass and energy between the zones.

The solution is obtained by solving the following set of equations over small increments of time: conservation of enthalpy, mass, and the Bernoulli equation for momentum coupled with the ideal gas law.

This program calculates the evolving distribution of the time-dependant characteristics that vary with the fire behaviour.

2.2. ASCOS: estimate of smoke propagation with a steady-state model

ASCOS (Analysis of Smoke Control Systems) is a program for steady air flow analysis of smoke control systems along a network. The purpose of this program is to verify how the ventilation system can be utilised to evacuate smoke which is due to fire or how it limits the smoke movement into the building.

ASCOS is a uni-directional network model that connects several nodes (every one with homogeneous characteristics) through vents located at reference heights for every zone level.

The solution is obtained by solving the following set of equations: conservation of mass, and the Bernoulli equation for momentum coupled with ideal gas law. The output gives the steady state pressures and net flows along the examined network.

2.3. CONTAM (version 94): a program for the airflow analysis

CONTAM represents the building as a mono-dimensional network in which every node is an enclosed region (as a room), with uniform air, temperature and contaminant concentration. CONTAM was developed to have a contaminant diffusion and airflow analysis program utilising a mouse-driven graphic interface. CONTAM combines most of the capabilities of many previous programs in which there is ASCOS.

The airflows calculation may be made in three simulation modes: steady, transient (up to 24 hours) and cyclic (24-hours steady-periodic). In this paper the airflows analysis is made in the steady state mode.

3. Tests cases

The correspondence among CFAST 2.0, ASCOS and CONTAM has been found by testing three cases significant to understand which are the main variables causing and influencing the smoke flow.

The selection about the test cases was made considering at first a simple case where there are only two zones connected by a vertical vent. Then, in the second case, the network becomes more complex with addition of two levels and a shaft. These two first buildings are paper cases because of compartments and vents dimensions which allow an easy description of the network with all the programs. Finally the last case takes place into an existing apartment.

In this comparison the first step is to sketch a network representing the building and the path along which the fire develops with a zone model. Then, the resulting smoke flow from CFAST 2.0, ASCOS and CONTAM is analysed by changing the following variables:

1. the wind velocity working on the outside facades of the building; the considered values are: 0, 1, 3 and 9m/s. In CFAST and ASCOS it must be described as a wind power law; the values needed to describe this law are: an exponent of 0.2 and a reference height of 10 m for wind speed, instead CONTAM requires a wind pressure profile for every opening connected with outside;

2. the fire power used in CFAST is as follows:

type A: the material heat of combustion is 34,3 MJ/kg, the fire area is 3 m² and the heat release is 1,6 MW (after 15 minutes from the ignition both are constant) and the mass ratio of hydrogen to carbon produced in pyrolysis is 0,333;

type B: the material heat of combustion is 17 kJ/kg, the fire area is 3 m² and the heat release is 784 W (after 15 minutes from the ignition both are constant) and the mass ratio of hydrogen to carbon produced in pyrolysis is 0,333.

With these two types of fire CFAST 2.0 gives for every test case a different temperatures distribution; the first with higher and the second with lower temperature gradients.

The outside ambient conditions are always the following: temperature 21°C, barometer absolute pressure 101325 Pa, relative humidity 60% and altitude meteo station 0 m.

To compare the three programs the relation of the temperature of every zone is the following:

$$T_{\text{ASCOS or CONTAM}} = \left[\frac{T_{\text{Upper layer}} * V_{\text{Upper layer}} + T_{\text{Lower layer}} * V_{\text{Lower layer}}}{V_{\text{Upper layer}} + V_{\text{Lower layer}}} \right]_{CFAST}$$

(1)

where: T is the temperature and V is the volume (because CFAST divides the zone in two layers, while ASCOS and CONTAM have only one zone with homogeneous characteristics).

The comparison of the resulting mass flows (kg/s) is made calculating the relative difference in percent by the term δ calculated as follows:

$$\delta = \frac{\left| G_{\text{programA}} - G_{\text{programB}} \right|}{G_{\text{programA}}} * 100$$

(2)

3.1. Case 1: two connected zones on different levels

This first case is a paper test case very simple and analytically easy to solve. With these characteristics it is possible to control how the Codes treat the data.

The building with the relative network used by the computer programs in Case 1 is shown in Figure 1. The particular shape is made to consider the stack-effect having a simple network. The dimension of the two nodes are: width 3 m, depth 5 m and height 3 m. All the vents are opened having a flow area of 1 m² at a relative height of 1,5 m from the floor of the considered node. The fire is into the node 1.

3.2. Case 2: four zones on different levels connected with a shaft

This second case is a paper test case with two more levels and a shaft. The building with the relative network used by the computer programs in Case 2 is shown in Figure 2. The dimensions of the four nodes are: width 3 m, depth 5 m and height 3 m and the horizontal area of the shaft has the same dimensions of the area four zones. All the vents connected with the outside are partially closed with a flow area of 0,01 m² at a relative height of 1,5 m from the floor of the considered node; the inside vents are opened doors with an area of 2 m². The fire is into the node 1.

3.3. Case 3: house test floor

This last case is a real apartment located at the ground floor of a house. This floor with the relative network, the dimensions of the six nodes and the relative connections are

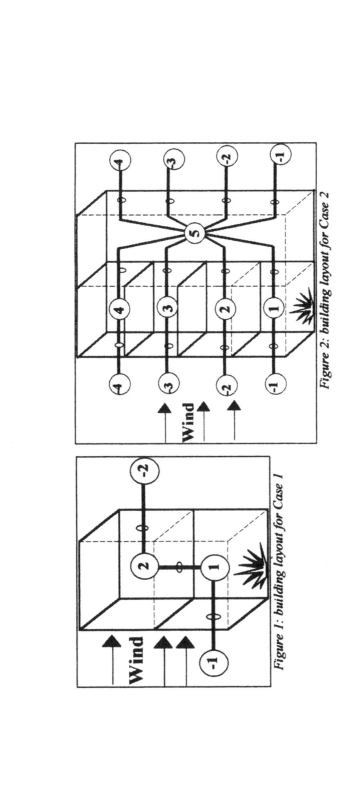

Figure 1: building layout for Case 1

Figure 2: building layout for Case 2

shown in Figure 3. All the vents connected with the outside are partially closed with a flow area that is 1% of the effective area; the inside vents are opened doors with a flow area of 1,68 m². The fire is in the entrance (node 3 in the network).

4. Results

In each case the total airflow passing through the building is compared for given external and internal conditions. The equation 2 gives the comparison index of the three programs.

4.1. Results for Case 1

The geometrical characteristics of the building given as input are all the same for the programs tested. The results for Case 1, changing the wind speed and the type of fire (and so the temperature distribution do the building), are shown in Figures *4* and *5*. The considerable mass flows at high wind velocities is due to the three vents completely opened (1 m² of flow area).

From the two Figures 4 and 5 it is very clear that the results have almost the same behaviour. The transient and the steady state models have similar mass flows even with distinct mathematical models; the δ between these two models is always under the 10%.

From the same figures, the programs have a more increased slope with Fire B where here are law thermal gradients; this means that the results are more influenced by the values due to the wind velocity effect. The δ between the three programs reaches only once about the 50% (comparison between the steady state programs and CFAST with fire B and no wind); this last time the net mass flows that pass the building are really low and so the δ is not a good comparison value.

4.2. Results for Case 2

The geometrical characteristics of the building are simplified for steady-state programs (example: the vents areas are at a fixed reference heights for every level), while for the transient program are given with detail as in a building project .This time the network is more complicated with more connected levels and a shaft which represents a stairwell. This is a typical network for a building with apartments in every floor.

The results for Case 2, changing the wind speed and the temperature distribution into the building, are shown in Figures 6 and 7. Figure 6 gives an almost independent total mass flow from the outside conditions. This means that changing the wind pressure on the building facades, the mass flows are due only by the strong thermal gradients that there are between the inside zones (it must be reminded that this time the vents flow area is only 0.01 m²). In Figure 7 there are the results with low thermal gradients (Fire B); this time the outside conditions differently influence the resulting mass flows comparing transient and steady-state models.

Another thing to notice from Figures 6 and 7 is the big difference, not only in the behaviour, but in the values of the resulting mass flows using the transient, CFAST, and the steady-state models ASCOS and CONTAM. In Figure 8 there is a bar chart which shows the δ index to compare the three programs for the case with low thermal gradient. In the legend there are the ways the programs are compared; for example: Contam/Ascos means that these two programs are compared by the term δ with the Ascos mass flow in

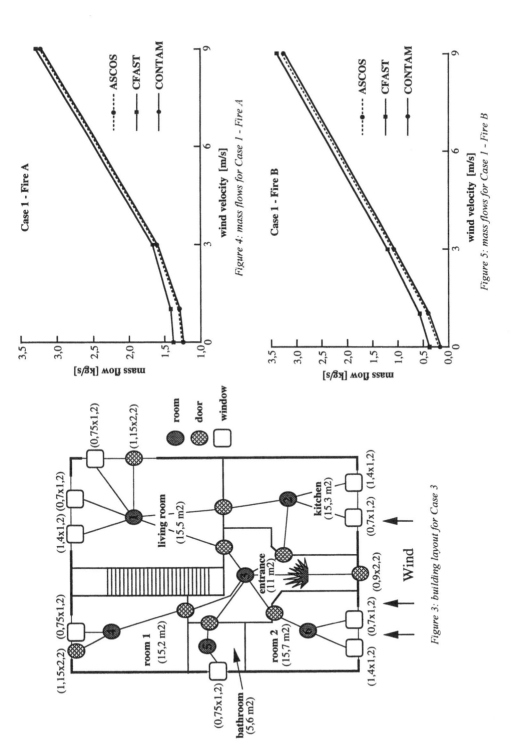

Figure 4: mass flows for Case 1 - Fire A

Figure 5: mass flows for Case 1 - Fire B

Figure 3: building layout for Case 3

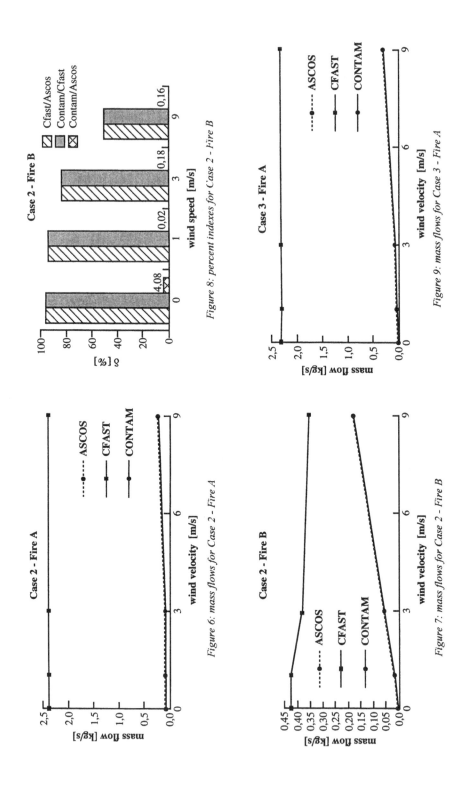

Figure 6: mass flows for Case 2 - Fire A

Figure 7: mass flows for Case 2 - Fire B

Figure 8: percent indexes for Case 2 - Fire B

Figure 9: mass flows for Case 3 - Fire A

the denominator of equation 2. Between CFAST and ASCOS or CONTAM the δ is far from admissible values (with an average value of 88%). Instead between the two steady state models the δ has an average value of 0.7%; these last models are made with the aim to solve complicated networks and so should be used in these cases.

4.3. Results for Case 3

This last test case is the most important one because it takes place in an existing apartment. The geometrical characteristics of the building are given in detail by the transient program, while they are approximately given for the steady-state programs.

The results for Case 3, changing the wind speed and the temperature distribution into the building are shown in Figures 9 and 10. These figures show that, between transient and steady-state models resulting smoke flows there is no correlation. From the bar chart in Figure 11 there is the calculated δ in case of high thermal gradients (Fire A). Between ASCOS and CONTAM the δ arises 31% (fire A and no wind). Certainly with such low mass flows the results can be considered equal. From Figure 12 it is possible to understand how low are the differences between the resulting mass flows between CONTAM and ASCOS for the second and the third case with fire A: when zero is a permitted value, percentage differences can be very misleading.

With a complex network the mathematical model used makes the difference.

5. Conclusions

In this work is exposed a comparison of the results obtained with three computer programs simulating the smoke movement in building. These programs are different even if they all use a "zone" model describing the smoke spread along a network: one is a "transient" and the others are "steady-state" programs.

Even if the smoke movement is not a steady-state phenomenon to compare these codes, it was necessary to find steady conditions with the "transient" program (fire with characteristics no more time dependent). Then transient model was useful to determine the temperature profiles produced by a fire. Simulations with this type of programs are long and with a detailed introductions of the data. With the "steady-state" programs everything gets easier, but those last ones are just valid for people with good technical abilities in fire safety and especially, just satisfying for some applications: one is the study of the ventilation system to evacuate the smoke due to a fire. In this case, the ventilation system induces homogeneous zones into the building (zone with fixed characteristics) and every zone could be represented with a homogeneous node of the network These programs are the most flexible and easy to use for big buildings with many levels and connections obtaining results with certain degree of approximation which has to be controlled. These kind of conclusions could be given from the results of Case 2 where the complexity of the network has a substantial importance on the solution. In these cases the use of steady-state models gives sufficient degree of approximation.

To study the smoke movement for fire security the transient models, as CFAST, seamed to be the more detailed. With this kind of program it is possible to know, not only the smoke net flow that passes through a building, but even how the smoke is distributed in a room and how the people in the rooms are involved with fire effects.

From this work it is possible to stress that in fire security analysis the best thing to do is

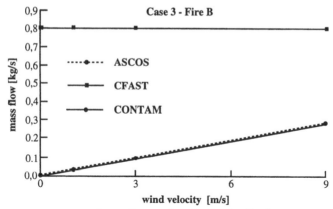

Figure 10: mass flows for Case 3 - Fire B

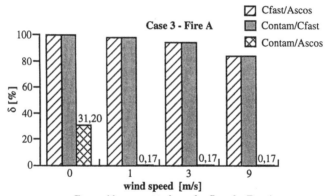

Figure 11: percent indexes for Case 3 - Fire A

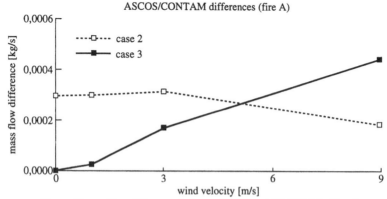

Figure 12: resulting differences between ASCOS and CONTAM for Fire A

to verify analyses with at least two models. Today, for the fire safety project, still remains the use of H_2O spray sprinklers in buildings; sprinklers eliminate fire and smoke with 95% reliability.

6. Acknowledgements
The authors want to express gratefulness to **Scot Deal** and **George Walton** (NIST, Gaithersburg, MD, USA) for the precious collaboration and the friendly support in getting at first the computer programs and then the right results.

7. References
[1] Portier R. W., Reneke P. A, Jones W. W., Peacock R. D. (December 1992): "A User's Guide for CFAST Version 1.6", BFRL, NIST, Gaithersburg, MD.
[2] Klote J. H., Milke J. A (1992): "DESIGN OF SMOKE MANAGEMENT SYSTEM, ASHRAE Special Publications, American Society of Heating, Refrigerating and Air-conditioning Engineers, Inc., Society of Fire Protection Engineers, Atlanta, GA.
[3] Walton George N. (March 1994): "CONTAM93 User Manual", NISTIR 5385, NIST, Gaithersburg, MD.
[4] Peacock R. D., Jones W. W., Bulcowski R W., Forney C. L. (June 1991): "Technical Reference Guide for HAZARD I Fire Hazard Assessment Method", BFRL, NIST, Gaithersburg, MD.

9 FIRE PROTECTION IN MODERN COMPUTER NETWORKS CENTRES

M.T. AREITIO and J. AREITIO
Networks & Systems, Bilbao, Spain

Abstract

This paper analyzes new issues for fire protection in modern computer networks centres. Fire is the single most feared physical hazard that can strike a information processing facility. Fortunately major fires are rare occurrences due to the high level of cleanliness and the controlled environment which are to be found within the majority of installations. However, there are no completely non-combustible computers. Circuits boards, resistors, network interface cards, capacitors, transformers, wiring and other components can provide fuel for a fire. The risk of fire is increased when paper and plastic items are introduced into the computer networks room. It is essential that, should a fire start, it is detected and then extinguished rapidly in order that total disaster is avoided. This paper introduces a specific strategy in order to prevent fire in computer networks centres.
Keywords: Computer networks, detection, extinguishment system, fire protection, prevention, stategies.

1 Initial aspects

A computer networks centre can only function when all environmental controls and services are reliable and continuous. From a security standpoint, this paper shows how the environment is becoming increasingly important. For example, the situation could arise where an access control system, installed at considerable expense to prevent unauthorized access to vandals, burglars, etc was fully functioning but the entire computer networks centre was put out of action by a fire.

There are three keywords that come up in discussions of computer networks security: vulnerabilities, threats and countermeasures. A vulnerability is a point where a system is susceptible to attack. A threat is a possible danger to the system, the

Fire Engineering and Emergency Planning. Edited by R. Barham.
Published in 1996 by E & FN Spon. ISBN 0 419 20180 7.

danger might be an event (a fire, a flood,..), a person (a spy,...), a thing (a faulty piece of equipment) that might exploit a vulnerability of the system. Techniques for protecting your system are called countermeasures. Computer networks security is concerned with identifying vulnerabilities in systems and in protecting against threats to those systems.

Every computer networks system is vulnerable to attack. Security policies and products may reduce the likelihood that an attack will actually be able to penetrate your system's defenses, or they may require an intruder to invest so much time and so many resources that it's just not worth it, but there's no such thing as a completely secure system. Computer networks are very vulnerable to natural disaters and to environmental threats. Disasters such as fire can wreck your computer network and destroy your data. Natural and physical threats are the threats that imperil every physical plant and piece of equipment: fires, power failures, and other disasters. You can't always prevent such disaters, but you can find out quickly if one occurs (with fire alarms, temperature gauges, and surge protectors). You can minimize the chance that the damage will be severe (e.g., with certain types of sprinkler systems). You can institute policies that guard against hazards posing special dangers to computers (like smoking). You can also plan for disater (by backing up critical data off-site and by arranging for the use of a backup system (alternative computers network) that can be used if an emergency does occur).

Despite advances in computer security and communications security, physical security remains vitally important component of your total security plan. Physical security measures are tangible defenses that you can take to protect your facility, equipment, and information from theft, tampering, careless misuse, and natural disasters. In some ways, physical security is the easiest and the most rewarding type of security. It's very visible and reassuring. It's a tangible signal to employees and clients that you take security seriously. Building, computers room, computers, and media locks provide an important outer, physical perimeter of security. Within this perimeter, access controls and other types of security provide finer-grained protection of information.

Some important considerations:

1. Install smoke detectors near your equipment and check them periodically.
2. Keep fire extinguishers in and near your computer networks rooms, and be sure everyone knows they're there.
3. Enforce no-smoking policies, these are also important to controlling smoke, another hazard to computer networks.
4. Consider using specially formulated gases such as Argonite, which smothers fires and avoids the danger of water damage.

2 Prevention system

The facility should be constructed from fire-resistant materials, including floor-to-ceiling, fire-resistant barriers in order that the facility is completely separate from, for example, office areas. This is particularly important as such external areas normally receive a far lower level of fire protection than the computers room. A large proportion of recorded computer room fires started outside and burnt their way into the computers room. Items such as furniture, decorations, etc. should be fire-resistant. Strict limitations should be placed on the amount of combustible material which is unavoidably introduced such a manuals, books, stationery, cleaning fluids, aerosols, etc..

The computers room (i.e. server farm) must obviously be designated as a non-smoking area and all food and drink must be excluded. At the same time, acceptable rest and refreshment areas should be provided for staff. Regulations must be introduced for the removal of rubbish; the underfloor void must be cleaned at least quarterly. Despite the introduction of preventative measures, staff awareness and good housekeeping practices can achieve much in this area.

3 Detection system

In the event of a fire outbreak within the facility, the early detection of the fire is essential to prevent its spread and the resulting severe damage. A fire may be detected by one of the following methods:

1. Observation (artificial, human,..).
2. Heat Detectors.
3. Smoke Detectors.
4. Combustion Detectors.

The detection system must be linked to a central system which will sound an alarm and activate the extinguishing system. The reliance on human intervention to observe the outbreak of fire, possibly out of normal working hours, and then to raise the alarm should be ruled out as far as a computer installation in concerned. Automatic detectors must be installed. Their purposes could be:

1. To alert personnel to the fire in order that they could tackle it with portable extinguishers if safe to do so, prior to the operation of an automatic extinguishing system.
2. To summon the fire brigade.
3. To additionally initiate controls of special extinguishing systems such as air conditioning closedown, door closing, etc.

The usual method of detection is by the use of smoke detectors. The sensitivity of this type of detector is such that low levels of smoke particles are rapidly dispersed long before flames become visible or a significant temperature rise has occurred. Extinguishment at an early stage is able to take place. The system should be designed in such a way that more than one detector must confirm the presence of smoke prior to extinguihment release. In this way detection system fault can be avoided. Specialist advise should be sought for detector installation, but the following guidelines should be observed:

1. The efficiency of the detector could be considerably reduced if it is sited within the

airflow of an air conditioning unit. Smoke could be diffused to such a level that the fire would only be detected once it had grown to considerably proportions.

2. The detector should be capable of detecting different types of smoke. For example, smoke produced by burning plastic, often used in wiring insulation, may not be detected.
3. Smoke detectors should be installed in the periphery of the computers room in addition to within it. As previously discussed, a fire which starts outside the computers room can dangerously develop, so it is advisable to install a zone of detectors at least within the areas inmediately surrounding the facility. Such detectors should form part of the corporate fire protection system but there must be an indication of the special nature of a detected fire in these areas, prompting special action.
4. Smoke detectors should also be fitted in the underfloor void and within air conditioning ducts.

It should be noted that the activation of the detection system can have no benefit unless a response take place. When the system is planned, the detector zone should be kept as small as possible to minimize the damage a fire would cause. More important is the level of response to the alarm which may mean the different between a small fire being quickly extinguished and a major conflagration.

4 Extinguishment system

The agents used for fire suppression within the computing environment are:

1. Carbon Dioxide (CO_2).
2. Halon Compounds.
3. Water. It's not a good fire protector for computer systems rooms. In fact, more destruction has been done by sprinkler systems trying to stop fires in computers rooms than by fires themselves.
4. Argonite (50% Ar-50% N2). It's a clean agent. (No Observe Adverse Effect Level=40% Ozone Deplection Potential=Global Warning Potential=0).
5. Inergen (52%-N2, 40%-Ar, 8%-CO2). It's a clean agent.
6. Argon (100%-Ar). It's a clean agent.

During the past decade the use of CO_2 has generally been discontinued as it depends on the displacement of air to inhibit combustion and concentrations of up to 46% may be required. Used in occupied areas, the hazard of suffocating personnel would be present. However, small capacity, CO_2 extinguishers are in use although concern has been expressed regarding the thermal shock effect produced by the gas which could cause chips and circuit boards to crack. CO_2 has been replaced by Halon as the premise gas extinguishing agent because:

1. Halon interrupts the oxidation chain reaction, thus extinguishing the fire.
2. A concentration of only 6% is required, at which level it can be considered to be non-toxic.

3. Halon is non-conductive, non-corrosive and, when discharged, leaves no messy residue to clean up.
4. Halon can operate via an automatic release system, preferably equipped with a manual release override.

Halon compounds used in extinguishing systems are either Halon 1211 (Bromochlorodifluoromethane, or BCF) or Halon 1301 (Bromotrifluoromethane, or BTM). The differing physical properties of these two distate that 1301 system is suitable for fixed installation and 1211 more appropiate for portable extinguishers. A Halon 1301 system should be considered where there is:

1. The need to reduce equipment fire damage by way of early automatic fire extinguishment.
2. A critical need to facilitate a swift return to service of the equipment after the fire.
3. A critical need to protect data being processed.
4. The need to protect void spaces not suitable for water sprinkler protection.

It is argued that water sprinkler systems are more effective in extinguishing fire than the other available means. It is agree that the majority of Halon systems release the gas only once, or at most twice, and this may not be sufficient to put out a deep-seated fire. In addition, unless the system receives on-going expert attention, correct Halon concentrations are difficult to maintain. To ensure the effectiveness of a Halon system all ducting, doors and windows must be sealed, as, operating under pressure, the gas may dissipate rendering it ineffective and causing the fire to re-ignite. Despite all such precations a door or window left open could have the same effect, as could a fire which burnt its way into the computer networks room from outside. The Halon will escape through the entrance made by the fire. It is further argued that should the Halon system fail to supress the fire, more damage is inflicted by the indiscriminate use of fire hoses by the fire brigade.

Some organizations have banned the use of any other systems but water sprinklers. Sprinkler heads of the 'wet' type i.e. when water is held back by a heat activated valve at the head itself, have an extremely low failure rate and the introduction of 'dry' pipes, water retained in a remote tanks, have further contributed to their reliability. However, the concer of water coming into contact with electrical equipment may rule out their usage. Automatic equipment power-down should precede the activation of sprinklers. It is often the case that the surrounding areas, officies and storage, are protected by spriklers and the computer networks room itself is equipped with Halon.

A growing trend is the use of foam extinguishants. Previously widely used in aviation, marine and petrochemical industries, foam, particularly of the high expansion type, has now been used by a number of financial institutions. A use for a foam system could be for the protection of the underfloor void which could take place separately from the main extinguishing system, thus avoiding a total discharge situation.

Portable extinguishers must be made available both within and outside the computers facility. Those within the facility should be either Halon 1211 or CO2. It must be emphasized that dry chemical extinguishers should never be used in the computers room as the extinguishing agent contains corrosive substances which will attack electronic equipment. An inspection should take place, at least monthly, to ensure that the appliances are in their correct positions, have not been discharged or lost pressure. All staff should receive adequate training not only in the operation of fire extinguishers, but also which extinguisher to use.

5 Final considerations

A defense against fire is careful placement of a computing facility. A windowless location with fire-resistant access doors and nonflammable full-height walls can prevent a fire from spreading from adjacent areas to the computing room. With a fire- and smoke- resistant facility, personnel merely shut down the system and leave, perhaps carrying out the most important media.

Fire prevention is quite effective, especially since most computer goods are not especially flammable. Advance planning, reinforced with simulation drills, can help to make good use of the small amount of time available before evacuation is necesary.

6 Bibliography

1. Areitio, J. and Areitio, M.T. (1994) Aplicación multidimensional de la seguridad informática operacional. *Revista Española de Electrónica*, Barcelona, No. 481 pp. 43-48.
2. DataPro (1992) *Network Services*, McGraw-Hill, New York.
3. Ferry, T.S. (1988) *Modern Accident Investigation and Analysis*, (ed. T.S.Ferry), U.S.
4. National Fire Protection Association (1991) *Fire Protection Guide on Hazardous Materials*. NFPA, U.S.
5. National Fire Protection Association (1986) *Fire Protection Handbook*. NFPA, U.S.

10 DOMESTIC 'FIRST AID' FIREFIGHTING

C. REYNOLDS and K. BOSLEY
Fire Research and Development Group,
Fire and Emergency Planning, The Home Office, UK

Abstract

There were over 61,000 domestic fires reported to fire brigades in the UK in 1992. However, the British Crime Survey of 1988 indicates that approximately 9 out of 10 fires in the home go unreported. This suggests that many fires are being dealt with successfully by the householder, despite the fact that advice from the fire brigades and the Home Office is to leave the house and call the fire brigade.

The Fire Research and Development Group of the Home Office commissioned a study to assess the level and extent of first aid firefighting in the home, with the aim of providing information to review the current advice.

The study involved a detailed questionnaire completed by 1,000 members of the general public, another more specific questionnaire for 450 people who had experienced a fire in the home in the last 5 years, together with some practical tests of methods of tackling typical household fires.

This paper provides details of the findings of the study.

Fire Engineering and Emergency Planning. Edited by R. Barham.
Published in 1996 by E & FN Spon. ISBN 0 419 20180 7.

INTRODUCTION

In 1981 the number of injuries from fire in the home reported to the Home Office by fire brigades was 6,343. Since then, the number has increased steadily and the latest figure (1992) stands at 13,440.

Against this background, the 1988 British Crime Survey indicated that approximately 90% of fires in the home are not reported to fire brigades and therefore any injuries ensuing from those fires are not recorded in the annual fire statistics.

Standard advice from fire brigades and the Home Office is that in case of fire the public should evacuate buildings as quickly as possible and call the fire brigade to deal with the fire. The only exception to this is a chip pan fire for which the householder is given specific first aid firefighting advice.

It is clear that the annual statistics need supplementing to provide a more exact picture. Hence a project has been undertaken to determine the level and extent of first aid firefighting in the domestic environment and, where appropriate to help formulate best practice advice for dealing with fire in the home.

Two surveys were carried out for the Home Office by the opinion surveyors MORI. The first, the General Public Survey', involved approximately 1000 people chosen to be representative of the general population. The second, the `Experienced Fire Survey', involved approximately 450 people who had experienced a fire in the home in the last five years. Both surveys were conducted by field workers completing a questionnaire whilst talking to the respondents in their homes. These surveys were complemented by a limited set of practical trials undertaken at the Fire Experimental Unit.

GENERAL PUBLIC SURVEY

This survey involved approximately 1000 people and aimed to gauge their perceptions of the dangers of fire. The questionnaire consisted of over 70 questions on topics such as the perception of danger, fire safety measures, perceptions of ignition and training and information on firefighting. These are discussed in more detail below.

Perceived Risks

The results of this survey indicated that whilst domestic fires were relatively uncommon, the risk from fire was apparent to many people. Figure 1 shows that a fire in the home is high on the list of concern of those in the sample although its occurrence within the sample is relatively low.

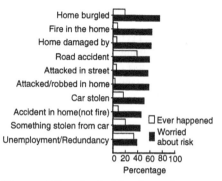

Figure 1 Perceived risks from fire (base 1053)

Fire Safety Equipment

As a consequence of this perceived risk, and also of recent advertising campaigns, many domestic dwellings now contain fire safety equipment.

Figure 2 shows that, according to the survey, by far the most common fire safety equipment in the home is a smoke alarm (70%), which is consistent with other recent surveys on smoke alarm ownership. Twenty three percent of households had planned their escape routes and ownership of fire extinguishers and fire blankets stood at 21% and 8%.

The majority of households now have some form of fire safety equipment, although knowledge about how to use it in the event of a fire is often incomplete. Thirty eight percent of those surveyed who owned a fire extinguisher, did not know what their fire extinguisher contained and many were unsure about which type of extinguisher was suitable for a particular type of fire. Figure 3 indicates the different colours of extinguishers owned by those surveyed and the extinguishing media the owners thought their extinguisher contained

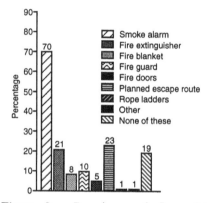

Figure 2:   Prevalence of fire safety measures (base 1,053)

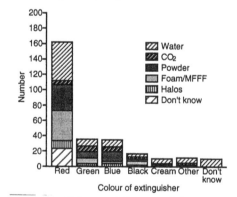

Figure 3: Colour and Presumed Contents of Household Extinguishers (base 135)

Training

Only 18% of those surveyed said that they had received training on how to deal with a fire, and 62% were unaware of a standard set of instructions, such as "Get out, call the fire brigade out and Stay Out". Figure 4 shows the basic, overall message that those who had been trained remembered from their training varied considerably.

Figure 4: Basic message of training (base 188)

EXPERIENCED FIRE SURVEY

The second survey involved approximately 450 people who had experienced a fire in the home in the last five years. They were asked to complete the same questionnaire as the general public survey, followed by an additional questionnaire which requested information on such things as, how fires started, whether anyone fought it, whether the fire brigade was called and whether the experience of fire had prompted a change in lifestyle or home.

How Fires Started

The results showed that one quarter of those questioned cited "Lack of concentration" as the initial cause of the fire, with a further 20% suggesting that carelessness was to blame. Many of those questioned sited diverse "other" reasons as the cause of the fire which did not fit into one of the main categories, shown in Figure 5.

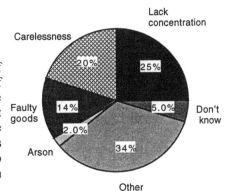

Figure 5: How, or why, fires started (base 449)

Causes of Fire

The majority (66%) of the fires in the survey involved cooking. Figure 6 displays the causes of the fires which had taken place in the homes of those questioned.

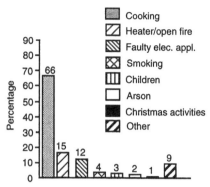

Figure 6: Causes of household fires in the last five years (base 449)

Timescales of Fires

The majority (69%) of those who did tackle the fire, did something straight away, without stopping to think about how best to respond to the situation. The majority of fires (53%) had been discovered within about a minute of starting and the average time to fight a fire was 2.5 minutes. (Figure 7).

Firefighting - Attitudes

Eighty percent of those who had experienced a fire did not call the fire brigade but tried to put it out themselves. Ten percent more men than women said they would tackle a fire.

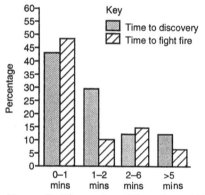

Figure 7: Timescales of fires (base 449)

The two most common initial actions were turning off the source of the heat (19%) and fighting the fire with a damp cloth (16%). However 8% said that they moved the item on fire, despite the fact that moving a burning object can be extremely dangerous. Fires involving electrical appliances caused the greatest concern when first discovered. They were also more likely to produce lots of smoke and to result in damage to property. Seventy three percent of the fires in this survey had flames less than one foot high at worst and 81% of those questioned described the fire as "small and easy to put out". This indicates that most people will try to fight fires less than a foot in height.

Firefighting - Techniques and Equipment

A range of items were used whilst tackling the fire. The most commonly used item was a damp cloth (15%), 9% said they had beaten out the fire, 6% said they had used a small container of water and 3% a bucket of water. Few mentioned using a fire blanket (1%) or a fire extinguisher (1%). Figure 8 provides details of the most common firefighting techniques and equipment.

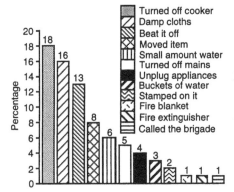

Figure 8: Techniques used to fight fire (base 331)

Most fires were extinguished using a damp cloth or water, and few people used a fire extinguisher or fire blanket. This indicates that using items readily to hand was a more instinctive reaction than using specific firefighting equipment.

Reporting of Fires

The fire brigade was called to only 19% of the fires experienced, but this seemed to be dependant upon the type of fire. Figure 9 shows that only 9% of cooking fires were reported, whilst the fire brigade were called to 51% of fires caused by heaters/open fires. This suggests that people are less likely to fight fire involving electrical equipment and open fires.

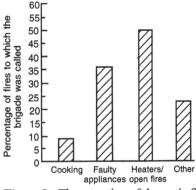

Figure 9: The reporting of domestic fires (base 449)

Consequences of Fire

Six out of ten people reported that since the fire they had become more aware of the danger of fire and had changed their home or lifestyle accordingly; for example, by unplugging electrical appliances that were not in use and buying smoke alarms (Figure 10). Many people had made minor "other" changes to their home or lifestyle which they felt also contributed to increased fire safety. In the course of fighting the fire, almost six in ten felt either calm or in control and the majority had put the fire out in an estimated two minutes. Only 3% said that their efforts had made the fire worse and only 5% of the fires had led to any kind of personal injury.

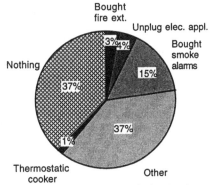

Figure 10: Changes made in the home after experiencing a fire (base 449)

Overall, three quarters (74%) said that the fire was quickly extinguished. Twenty eight percent of the fires caused damage to property estimated at greater than £50.

Publicity and Education
Over 70% of all questioned remembered seeing or hearing information about fires in the home. Most of these were from television, but others also mentioned newspapers, leaflets and fire brigade visits. Almost 90% still said that they would like more information on dealing with fires, the majority suggesting television as the most suitable media.

PRACTICAL FIRE TESTS
Practical fire tests were conducted to establish the most effective, practical method of extinguishing domestic fires. The fires were fought by staff who had not been trained in firefighting with a range of firefighting equipment likely to be found in the home.

Domestic Fire Scenarios
A series of practical fire tests were conducted involving typical domestic fire scenarios:

* electrical appliances such as irons and televisions,
* chip and grill pan fires,
* clothes and material on a clothes horse,
* fires in a mattress and sofa, and
* fires in a waste bin.

These fires were fought using a variety of techniques and equipment including:

* various amounts of water applied at different stages in the fire,
* smothering with a damp cloth,
* fire blankets, and
* dry powder fire extinguishers.

The fires were confined to a single source and the possible spread of the fire was not considered. Each type of fire was first allowed to burn freely in a room. The levels of smoke, air temperature and thermal radiation were assessed in the enclosed environment. This gave an indication of the environment that would be encountered by a householder discovering a fire.

Application of Water
For most categories of fire other than hot fat fires and electrical fires, applying water was found to be the quickest and cheapest option to extinguish the fire. In the tests, applying a small amount of water quickly controlled the fire even if it was not completely extinguished immediately. In a domestic scenario, this controlling action would allow the householder sufficient time for further, safer firefighting.

The tests highlighted that a small amount of water applied at an early stage in the fire controlled the potentially hazardous situation more effectively than applying a larger amount of water at a later stage. Once water had been applied to a fire, any unburned fire load was difficult to reignite and any burning areas were cooled.

Smothering with a Damp Cloth
For hot fat fires the Home Office advice for tackling chip pan fires is to use a damp tea towel. During the tests, this method proved to be the most effective method of tackling the fire in terms of safety, availability and ease of use. If the initial application was unsuccessful the first tea towel could be left in position, controlling the fire sufficiently to enable the application of another damp tea towel in relative safety.

Dry Powder Extinguishers
Multi purpose dry powder extinguishers successfully extinguished all fires except those involving hot fat. The dry powder extinguishers did however prove to be both messy and expensive and therefore were considered unlikely to be suitable for the domestic environment. Additionally, during the tests on hot fat fires, the extinguisher forced the hot fat out of its container, effectively adding to the hazard from the fire. The extinguishers' contents did not cool the fat sufficiently to prevent re-ignition.

Fire Blankets
The fire blankets used during the tests were found to have four main drawbacks for fighting domestic fires:

1. The material construction of the blankets meant that they were too rigid to form a seal around the fire.
2. The blankets were too big to apply easily to small fires that were not contained.
3. It was found to be very difficult to assess the state of the fire when a blanket had been placed over it.
4 The blankets did not cool the materials in and around the fire and therefore did not prevent re-ignition of the fire when the blanket was removed.

SUMMARY OF CONCLUSIONS
The findings of the study provide the following conclusions:

1. Concern about fire in the home is relatively high amongst the general public despite its low incidence.
2. The publicity campaigns for smoke alarms have been very successful.
3. The majority of fires were discovered within a minute of starting and the average time taken to tackle a fire was 2.5 minutes. This indicates that the both the discovery and fighting of fire is very quick, giving little time for thought.
4. Seventy three percent of the fires in this survey had flames less than one foot at worst and 81% of those questioned described the fire as "small

and easy to put out". This indicates that most people will try to fight fires less than a foot in height.

5.   Most fires were extinguished using a damp cloth or water, and few people used a fire extinguisher or fire blanket. This indicates that using items readily to hand was a more instinctive reaction than using specific firefighting equipment.

6.   Six out of ten people reported that since the fire they had become more aware of the danger of fire and had changed their home or lifestyle accordingly.

7.   Almost 90% of those surveyed said that they would like more information on dealing with fires, the majority suggesting television as the most suitable media.

8.   For most categories of fire other than hot fat fires and electrical fires, applying water was found to be the quickest and cheapest option. A small amount of water applied at an early stage in the fire controlled the potentially hazardous situation more effectively than applying a larger amount of water at a later stage.

9.   For hot fat fires, application of a damp cloth to test fires proved to be the most effective method of tackling the fire in terms of safety, availability and ease of use.

10.   Multi purpose dry powder extinguishers successfully fought all fires except those involving hot fat. They did, however, prove to be both messy and expensive and consequently were considered unlikely to be suitable for the domestic environment.

11.   The fire blankets used during the tests were found to have four main drawbacks for fighting domestic fires: they were too rigid and bulky, they obscured the progress of the fire and did not prevent re-ignition.

11 DYNAMIC MODELLING OF FIRES IN BUILDINGS

E. CAFARO, L. RANABOLDO and A. SALUZZI
Dipartimento di Energetica, Politecnico di Torino, Italy

Abstract

The transient fire growth in enclosed spaces is modelled using concepts of non linear dynamical systems theory and the theory of stochastic processes. The mathematical model, derived by a simplified thermodynamic approach to the problem, keeps into account two non linear effects on the burning rate of the fire: the radiation feedback from the hot layer to the fuel and the switch-over between fuel and ventilation control. The zone model used to derive the evolution of the temperature excess of the hot smoke layer is recast in the form of a *gradient type* dynamical system. A *swallowtail catastrophe function* is introduced to approximate the *fire potential function* and to define the boundaries of stable system behaviour. To perform a probabilistic analysis the temperature excess of the smoke layer is considered to be a stochastic variable the time evolution of which is modelled by a *Langevin equation*. The numerical solution of the corresponding *Fokker-Planck equation* allows to determine the probability density function of the stochastic process.

Introduction

Fires in enclosed spaces represent phenomena exhibiting a complex dynamical behaviour at variation of the ratio of air mass flow rate to volatilized fuel mass flow rate. Flashover and extinction jumps as well as hysteresis between the fuel volatilization rate and openings on the compartment have been experimentally observed and recast in mathematical models [1]. Deterministic and stochastic approaches have been used to model the fire growth process: the former tend to fall within the two categories of zone models and field models while the latter views fire spread as a percolation process [1, 2].

In a previous paper, by developing a thermodynamic approach introduced in [3], we derived a two layer type dimensionless model, keeping into account the radiation feedback effect and the transition effect from fuel control to ventilation control of the burning rate. Emphasis is placed on jumping phenomena and hysteresis [4]. The evolution of the temperature excess of the smoke layer has been modelled by a first order ordinary differential equation depending on a set of control parameters characterizing the heat and mass transfer processes in the compartment. The numerical integration of the model, performed by a fourth order Runge-Kutta method for time dependent cases and by a Gauss method for steady state conditions, shows the existence of two stable branches of solution and an unstable one depending on the values of control parameters.

In the present paper the derived model is recast in the form of a dynamical system by introducing a fire potential function recognized as a swallowtail catastrophe function [5,6]. To perform a stochastic analysis of the problem we perturb the system by a white noise term and transform the Langevin equation into a Ito-type stochastic differential equation. Finally we determine the probability density function for the temperature excess stochastic variable by numerical solution of the corresponding monodimensional time dependent Fokker-Planck equation [7,8].

Fire Engineering and Emergency Planning. Edited by R. Barham.
Published in 1996 by E & FN Spon. ISBN 0 419 20180 7.

Deterministic model

The dimensionless mathematical model is defined by the following first order ordinary differential equation [4] :

$$\frac{d\vartheta}{d\tau} = \frac{1+\vartheta}{V} \cdot \left[\mu(\vartheta) \cdot mf^* - (1+Nv) \cdot \vartheta\right] \tag{1}$$

where

$$\mu = 1 - \exp\left[\frac{Nv}{\lambda \cdot mf^*}\right] \tag{2}$$

$$mf^* = \alpha \cdot \left[\gamma + \varepsilon \cdot (1+\vartheta)^4\right] \tag{3}$$

The quantity θ represent the temperature excess of smoke layer, τ the dimensionless time, V the ratio of the smoke volume to the compartment volume, μ the dimensionless fuel-ventilation control function, Nv the control parameter defined as the product of the ventilation parameter and the inverse heat release parameter, λ the ratio of the combustion energy to a reference energy, mf* the dimensionless fuel mass flow rate, α the dimensionless area of the volatilized fuel, γ the radiation heat exchanged by flame and fuel, ε the global emissivity of the smoke layer.
The adopted in the model structure of dimensionless fuel-ventilation control function is recovered from correlation of some experimental results about the combustion efficiency of polymeric materials [9].
The model can be recast in the form of a dynamical system :

$$-\frac{dP}{d\vartheta} = \frac{d\vartheta}{d\tau} \tag{4}$$

where

$$P = -\int_0^\vartheta \frac{(1+\vartheta)}{V} \cdot \left[\left[1 - \exp(-\frac{Nv}{\lambda \cdot \alpha \cdot (\gamma + \varepsilon \cdot (1+\vartheta)^4)})\right] \cdot \alpha \cdot \left[\gamma + \varepsilon \cdot (1+\vartheta)^4\right] - (1+Nv) \cdot \vartheta\right] d\vartheta \tag{5}$$

The shapes of the *fire potential function P* for different values of control parameters are showed in figs. 1,2.

V=0.7   λ=0.1   α=0.12   ε=1   γ=10

Nv=4.5

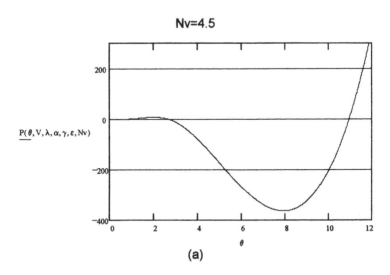

(a)

Nv=1.5

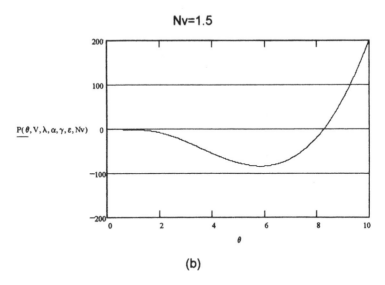

(b)

Nv= 50

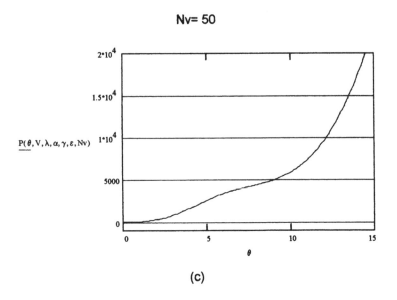

P(θ, V, λ, α, γ, ε, Nv)

θ

(c)

fig.1

V=0.5 λ=0.2 α=0.16 ε=1 γ=20

Nv=5.5

P(θ, V, λ, α, γ, ε, Nv)

θ

(a)

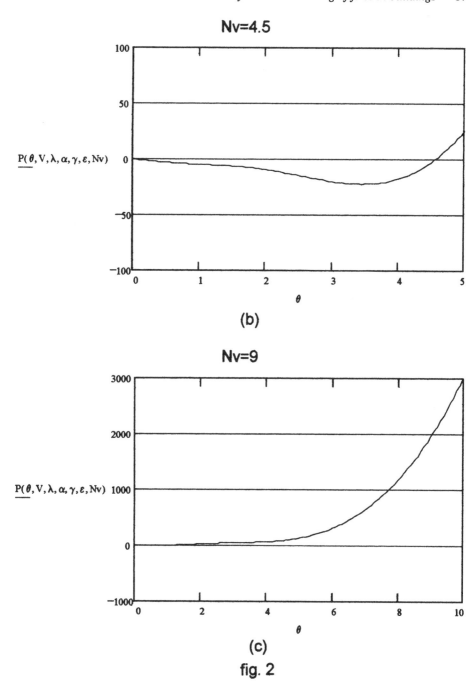

Nv=4.5

(b)

Nv=9

(c)

fig. 2

Comparing the figs.2 with the following fig.3 we notice that varying the Nv parameter the potential function P shows either three or one noteful points corresponding to the stationary solution branches of the temperature excess evolution equation.

The first point of minimum represent the extinction jump, the point of maximum the unstable solution branch, the second point of minimum the flashover jump. When the solution exhibits one noteful point only it represent the flashover jump.

$$V=0.5 \qquad \lambda=0.2 \qquad \alpha=0.16 \qquad \varepsilon=1 \qquad \gamma=20$$

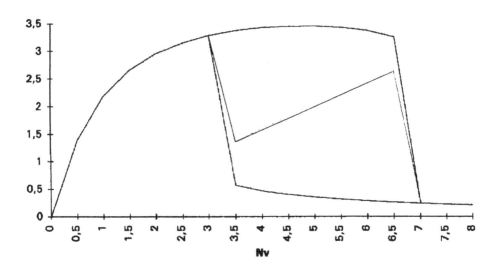

Stationary solutions of the deterministic model

fig.3

The fire potential function   can be approximated by means of the following *swallowtail catastrophe function* [5,6] :

$$P' = a \cdot \vartheta + b \cdot \vartheta^2 + c \cdot \vartheta^3 + d \cdot \vartheta^4 \qquad (6)$$

The coefficients of the polynomial have been determined by spline interpolation of the potential function.

Stochastic analysis

Let us consider the smoke layer temperature a stochastic variable the time evolution of which is governed by the *Langevin equation* [7] :

$$\frac{d\vartheta}{d\tau} = -\nabla P' + \xi(\tau) \qquad (7)$$

where

$$< \xi(\tau) > = 0$$

$$< \xi(\tau) \cdot \xi(\tau') > = D(\vartheta) \cdot \delta(\tau - \tau')$$

In the eq. (7) the quantity $\xi(\tau)$ is a white noise term with zero mean value and variable dependent variance function.
The Langevin equation can be rewritten as a Ito-Stratonovich stochastic differential equation [7] :

$$d\vartheta = -\nabla P' \cdot d\tau + \sqrt{D(\vartheta)} \cdot dw(\tau) \qquad (8)$$

where $w(\tau)$ is a standard Wiener process and D is the infinitesimal variance function which characterize the intensity of the noise.
The probability density function of a stochastic differential equation obeys a deterministic partial differential equation called Fokker-Planck equation [7,8] :

$$\frac{\partial}{\partial \tau} F(\vartheta, \tau) = \frac{\partial}{\partial \vartheta} \left[\nabla P' + \frac{1}{2} \cdot D(\vartheta) \cdot \frac{\partial}{\partial \vartheta} \right] F(\vartheta, \tau) \qquad (9)$$

As $\tau \to \infty$ the probability density function converges to a stationary density which is either a generalized function or a proper probability density function depending on the form of the potential function and the infinitesimal variance function.
The stationary solution of the Fokker-Planck equation when the infinitesimal variance function is constant reads :

$$F^*(\vartheta) = C \cdot \exp\left(-\frac{2P'(\vartheta)}{D} \right) \qquad (10)$$

where C is a normalization constant.
The eq. (10) implies:

$$\log(F^*) = \log(C) - 2\frac{P'(\vartheta)}{D} \qquad (11)$$

The eq. (11) defines an affine transformation of the potential function. Therefore, the whole apparatus of catastrophe theory, in particular, the classification of the degenerate singularities of the potential function, now applies without change to the logarithm of the stationary probability density function.

In the interior of its domain the stationary probability density function has differentiable relative maxima (modes) and minima (antimodes) which coincide with the relative minima and maxima of the potential function.

The modes and antimodes of the stationary probability density function are non trivially related to the relative minima and maxima of the potential function if the variance function is dependent from the stochastic variable. In this case, namely, the stationary probability density function is given by:

$$F*(\vartheta) = C \cdot \exp\left[2 \cdot \int^{\vartheta} \left(\nabla P'(s) - \frac{1}{2D(s)} \cdot \frac{\partial}{\partial s} D(s) \right) ds \right] \tag{12}$$

The eq. (12) can be rewritten as:

$$\frac{d}{d\vartheta}\left(\log(F^*) \right) = -2\nabla P' + \frac{1}{2} \cdot \frac{\partial D(\vartheta)}{\partial \vartheta} \tag{13}$$

Let's define the shape function of F^* to be

$$g(\vartheta) = \nabla P' + \frac{1}{2} \cdot \frac{\partial D(\vartheta)}{\partial \vartheta} \tag{14}$$

From the eqs.(13, 14) it can be seen that the modes and the antimodes of the stationary probability density occur at the zeroes of the shape function which do not necessarily coincide with the zeroes of the fire potential function.

The values of the control parameters selected in the numerical integration of the Fokker-Planck equation correspond to those used for determining the shape of the potential function showed in fig.1.a . Some results of the stochastic analysis are showed in figs. 4, 5, 6, 7.

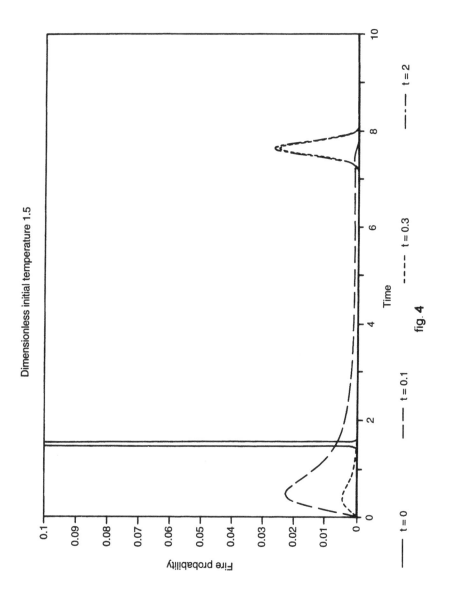

fig. 4

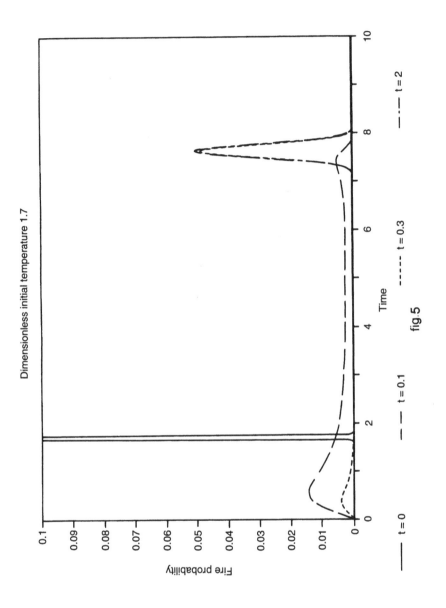

fig.5

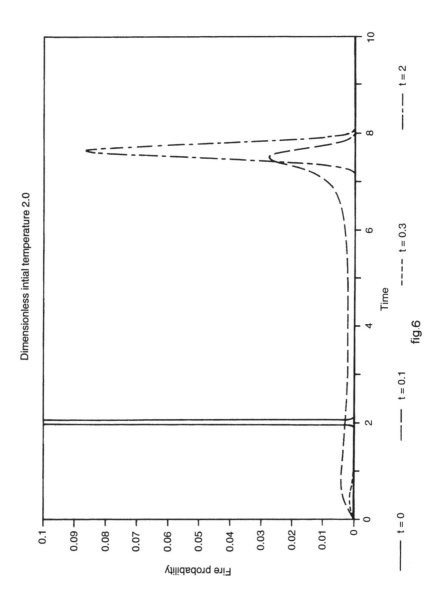

fig.6

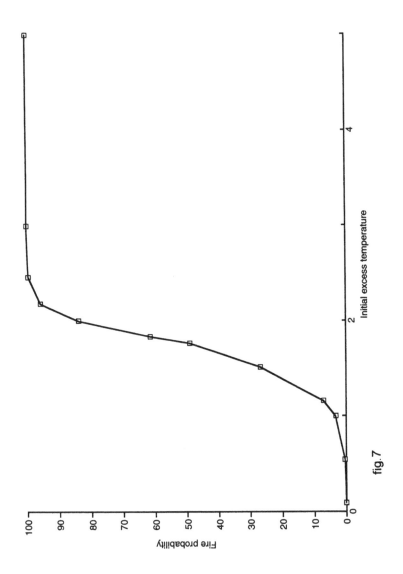

fig. 7

Conclusions

The analysis performed show how the catastrophe theory and the stochastic processes theory can be used to understand the main features of the complex dynamical behaviour of the fires in buildings. The reduction of fires growth process in compartments to a dynamical system of polynomial type allows to determine the exact scaling of the control parameters in physical simulation of the interesting process. The most relevant control parameters result to be the ratio of the air mass flow rate to the global heat exchange coefficient and the fuel mass flow rate.
The probabilistic analysis allows to affirm that stochastic systems with multiple stable equilibria may nevertheless exhibit unimodal stationary probability densities.

References

1) D.D.Drysdale, "An introduction to fire dynamics", Wiley, Chichester, 1985.

2) H. E. Mitler, "The Harvard fire model", Fire Safety Journal, 9, 7-16, 1985.

V. Bennardo, E. Cafaro, N. Inzaghi, "A zone model for the hot smoke layer evolution", Antincendio 12, 7-15, 1994.

N. C. Markatos, G. Cox, "Hydrodynamics and heat transfer in enclosures containing a fire source", Physico Chem. Hydrodynam. , 5, 57-66, 1984.

E. Cafaro, A. Saluzzi, "Computational Fluid Dynamics analysis for a fire detection and suppression in Mini Pressurized Logistic Module", Alenia-Spazio s.p.a, Grant, 1993-94.

T. Beer, "Percolation and fire spread", Combust. Sci. Tech., 72, 297-304,1990.

3) P.H. Thomas, M.L. Bullen, J.G. Quintiere, B.J. McCaffrey, "Flashover and instabilities in fire behaviour", Combust. Flame 38,159-171 ,1980.

4) V. Bennardo, E. Cafaro, N. Inzaghi, " Thermodynamic model for confined fires", Antincendio, 8, 7-13, 1994.

5) R. Thom, "Structural stability and morphogenesis", Benjamin, N.Y., 1975.

6) T. Poston, I.N. Stewart, " Catastrophe theory and its applications", Pittman, San Francisco, 1978.

7) T. T. Soong, " Random differential equation in science and engeneering", Academic, N.Y. , 1973.

8) E. Cafaro, C. Cima, A. Saluzzi, "Numerical analysis of non linear physical systems stochastically perturbed", 46° Congresso Nazionale ATI, V 189-200, 1991.

9) A. Tewarson, F.H. Jiang, T. Morikawa, "Ventilation controlled combustion of polimers", Comb. Flame, 95, 151-169, 1993.

12 AN HIERARCHICAL APPROACH TO FIRE RESISTANCE IMPROVEMENT OF COMPLEX TECHNICAL SYSTEMS

R.V. GOLDSTEIN
Institute for Problems in Mechanics,
Russian Academy of Sciences, Moscow, Russia

Abstract
An discrete-continual approach is developed for analysis of the problems on fireresistance estimation and improvement in hierarchical complex technical systems. The generalized characteristic of the catastrophic accidents scale, namely, the rank of the catastrophe is introduced. The conditions of the catastrophe transition from one rank to another are formulated. As an example of the approach application the model problem on improvement of fireresistance by means of protection distribution through the system structural levels is considered and the results of its analytical and numerical solution are given.
Keywords: Hierarchical complex technical system, fireresistance, protection distribution levels.

1 Introduction

Catastrophic accidents on complex technical systems, in particular accidents induced or accompanied by fire propagation, envelop, as a rule, many scale levels. These levels correlate with the scales of separate objects forming the complex technical system or their sets as well as with the scales inherent to appropriate physical processes.

Prediction of the catastrophic accidents and development of recommendations for improvement of resistance (e.g., fireresistance) of the complex technical systems to such catastrophes require description of conformities of their initiation and propagation in hierarchical systems taking into account the physical and/or chemical fields leading to the catastrophe and the properties of media being in contact with the complex technical system under consideration.

To analyze such problems we developed a multiscale approach which allows to model processes on various scales by discrete or continual manner [1]. The concept on the rank of the catastrophe relative to given physical field acting on the system is introduced. The rank is a generalized characteristic of the catastrophe scale. The conditions of the catastrophe transition from one rank to another are formulated.

As an example of the approach application we considered together with Dr. D.A.Onishcenko the problem on improvement of the fireresistance of a complex multiscale technical system by appropriate distribution

Fire Engineering and Emergency Planning. Edited by R. Barham.
Published in 1996 by E & FN Spon. ISBN 0 419 20180 7.

of protective means on various scales. It is shown that
the probability of the fire propagation in large scale
can be reduced strongly by an optimal distribution of
protective means on different scale levels of the
system. Some results of the numerical modeling of the
problem are obtained.

2 A discrete-continual approach

Assume that in the complex technical system one can
settle out elements and/or their groups having the
characteristic scales $L_1, \ldots, L_k$ ($L_1 < \ldots < L_k$). It is
possible but not necessary that elements of one or
several scales represent gemetrically identical or
similar objects.

   We will believe that the disturbance of the system
technical state is catastrophic if it is accompanied by
rise and propagation of dangerous field of action in
media surrounding the system and/or being in contact
with it. By the dangerous field we mean here the fields
dangerous from the ecological or mechanical point of
view (e.g., fire induced temperature, radiation and
pressure fields).

   Denote by $df_1, \ldots, df_p$ the dangerous fields caused
by a catastrophe.

   If the field df_k envelops the scale L_j one will say
that the catastrophe has the rank r_{jk} relative to the
field df_k.

   Let $r_{j1}, \ldots, r_{jp}$ be the catastrophe ranks of the
system under consideration relative to the fields $df_1,$
$\ldots, df_p$, respectively. Then the rank, r, of the
catastrophe of the system is by definition equal to
maximum of the ranks $r_{j1}, \ldots, r_{jp}$:

$$r = \max_{\substack{j \\ 1 \leq k \leq p}} r_{jk} \qquad\qquad (1)$$

   Variation of the catastrophe rank relative to the
field df_k, r_{jk}, is determined by kinetics and dynamics
of this field. Different fields df_m and df_k can be
conjugated and the appropriate ranks r_{jm} and r_{nk} inter
related. The relation between the ranks r_{jm} and r_{nk} can
be changed when the catastrophe rank increases.
Increase of the catastrophe rank, relative to one
field, (its scale) can be accompanied by decrease of
the rank relative to other dangerous field (e.g.,
localization of harmful contaminants issuing at
fracture caused by fire embrasing increasing scales).
   We will mainly consider later on catastrophs
connected with partual or total fracture of the complex

technical system caused by fire. The catastrophe rank r = i will be correlated with the scale of the fire embrased region L_i.

The catastrophe of i-th rank generation can be caused directly by the fire and fracture process of elements of the scale L_i or can be a result of formation of catastrophe zone of the scale L_i because of fire induced fracture of a certain set of elements of the scale L_{i-1}.

Description and modeling of the process of a definite rank catastrophe formation and extension as well as its rank increasing provide for availability of conditions which specify transition of an element of the system in the limit state, transition to the limit state of other elements of the same scale under the action of the perturbation field induced by the fire and/or fracture of an initial element as well as conditions of the next (larger) rank catastrophe formation.

The conditions of the first and second groups can be written similarly to nesessary and sufficient conditions of formation of an hierarchy of structures of fracture [2,3].

The peculiarity of the catastrophe extension process consists in its instability (dynamics). In this connection a specific role play media transmiting perturbations of different fields caused by fire and fracture of an element of the complex technical system. Namely these processes are ascartaining for formation of the second and third group conditions.

Indeed, a perturbation propagation depends upon both the situation in the neighborhood of the fructured element and the reaction of the system in larger scales or as a whole on the arised perturbation. The system reaction is essentially determined by its capacity to retain and maintain the regime of perturbation localization as well as to provide transition to regimes of propagation of distributed perturbations.

While it is convenient to consider in frames of a discrete approach conditions of catastrophic fire induced fracture of separate elements, the analysis of the system (or its subsystem) reaction as a whole on a local perturbations can be performed using a continual approach taking into account the system structure. By the way we obtain the possibility to separate the whole problem of the catastrophe formation and extension in multiscale system onto a series of inner and outer problems.

The analysis of the process on the scale L_i represents an inner problem relative to larger scales (L_{i+1}, etc.). This problem is solved by incorporating specific mechanisms of fire and fracture in the scale L_i and actions on the element under consideration from larger scales.

In turn, the problem on perturbation distribution in the complex technical system or in its subsystems of the scale L_{i+1} represents an outer problem relative to the scale L_i. In the outer problem one considers the process of propagation of a field induced by fire and/or fracture in an element of the scale L_i. Analysis of the outer problem can be performed in frames of a continual approach. The appropriate continuum can be modeled as an equivalent (generally nonlinear) composite medium (see [1]).

On can model perturbations caused by the processes in the scale L_i as specific body sources.

Note, that regimes with peaking at perturbation propogation can appear in such nonlinear medium with the body sources. Similar regimes have been described in [4] in the case of the heat field.

Unstable regimes, e.g. regimes with peaking, can lead to appearence of a correlation between processes in several elements of the scale L_i. This correlation, in turn, will promote increasing of the catastrophe rank.

The formulated approach allows to model the processes of catastrophe initiation and propagation in multiscale systems. It can be used also for consideration of the problems on the catastrophe protection and preventation leading to hierarchical principle of constuction of active and passive (or combined) protection.

3 A model of fire protection circuit in hierarchical complex technical systems

3.1 Main assumptions of the model
We will now consider a hierarchical complex technical system of the following type:
- on the rank o the system structure consists of elements of the same type;
- the system structure of rank (j + 1) is a set of blocks formed by pairwise join of j-rank blocks;
- the whole system coincides with the structure of rank n which can be represented as join of two blocks of (n − 1)-rank;

Note, that the described structure corresponds to a "fractal tree".

Assume that the intensity of an action depends on fire included structural level. Further, the system is provided with a fire protection. The protection resourses can be distributed through various structural levels of the system. The problem under consideration consists in searching for an optimal protection circuit. Here we will analyze the problem in assumption that all protection must be concentrated on a certain structural level and it is required to choice this level to achive maximum efficiency of the protection. More general case as well as detailed results of numerical calculations will be considered in separate publication with D.A.Onishcenko.

Let us formulate now the assumptions on the fire (and fracture) propagation in our hierarchical system.

The fire process can start from the structure of o-rank. Assume that a certain element of o-rank perished or failed due to fire. Then this element exerts an action of the given type on a conjugate element. Remind, that both these elements form one of the blocks of 1-rank. The conjugate element can also exert with certain probability which is determined by an appropriate distribution function and the action intensity. If this event arises one can say that a block of 1-rank is out of action.

Fracture of the block of 1-rank exerts an action on elements of a block conjugated to him. There exist two such elements for the block of 1-rank. Note, that the action caused by the failed block is constant up to fracture or perish of both of these elements forming the conjugate block. A refusal of these both elements is equivalent to a refusal of 1-rank block conjugate to failed one. Hence, two conjugate blocks of 1-rank are out of action. This means that a block of 2-rank is failed (or perished). The fire propagation by such a way leads to refusal of (n − 1)-rank block. The refusal of the whole system takes place if both (n − 1)-rank blocks fall out.

Let us assume now that a certain mechanism of reduction of an action on a given block exists on each level of the system hierarchy. Denote by r_j the value of the action reduction on j-rank block. Then the action on the given j-rank block, q_j, can be represented as follows

$$q_j = s_j - R_j \qquad (2)$$

where s_j is the action induced by the refusal of the conjugate j-rank block and R_j equals to the sum of the protection resources on all levels of the hierarchy from o-rank up to j-rank

$$R_j = \sum_{m=0}^{j} r_m \qquad (3)$$

3.2 A model problem on optimal distribution of fire protection resources

Assume that the protection volume is fixed and equals to V_o. For definiteness we will believe that the protection is achieved by using of special shell. The shell thickness determines the degree of the fire action reduction. Assume for simplicity that the shell is of cylindrical type when the elements of the system are placed in a plane square region and the shell is closed one when the elements are placed in a cubic region in 3D-space.

In a general case the protection resources can be distributed partly on various levels of the hierarchical structure of the system.

We will consider the simplest approach when the whole protection is concentrated on one level only. Then the problem on protection optimization consists in searching for number of the hierarchy level where the protection efficiency has maximum. We will assume that the probability of complete fracture of the system is a measure of the protection efficiency. Hence, the maximum protection efficiency corresponds to minimum of this probability.

First let us analyze the case when all system elements are placed in a plane.

Assume that the height of the protection wall equals to h independent on the structural level. Denote by t_j the thickness of the protection wall placed on j-th level.

Let us estimate the required extension of the protection wall. Suppose that it is equal to the total perimeter of all j-rank blocks. Then the protection volume V_j equals to

$$V_j = 4\, l_o \cdot r^{(n-j)/2} \cdot h \cdot t_j \tag{4}$$

where l_o is the characteristic size of o-rank element. Taking in mind that the protection volume is given one can equate $V_j = V_o$ in assumption that the total protection is concentrated on j-th level. Then from (4) we obtain

$$t_j = t_o\, 2^{j/2} \tag{5}$$

Note that in the case when the system elements are placed in a cubic region similarly

$$t_j = t_o\, 2^{j/3} \tag{6}$$

Assume that the protection degree is proportional to a certain power of the shell thickness. Then the sequence $\{r_j\}$ is increasing one such that

$$r_j = r_o \cdot a^j \quad , \quad a > 1 \tag{7}$$

The law of the action s_j increasing with the

structural level growth can be sufficiently arbitrary.
    We will now show on numerical examples that the
situations are possible when the protection
distribution on the o-th level is not optimal.
    Suppose that the strength distribution of o-th rank
elements is described by the Weibull one

$$F^{(o)}(x) = 1 - \exp(-x^2) \qquad (8)$$

We will analyze the two following variants:
1. the j-dependences of s_j and r_j are linear

$$s_j = s_o + j/3 \quad , \quad r_j = r_o + j/3 ; \qquad (9)$$

2. s_j is the quadratic function on j and r_j is the
linear one

$$s_j = s_o + [(j + 5)^2 - 25]/60$$

$$r_j = r_o + j/4 \qquad (10)$$

where $s_o = 0.6$, $r_o = 0.45$.
    The hierarchy level n equals to 10 in the both
cases.
    The probabilities of total fracture of the system
were calculated according to the formulae (11)-(13)
given in subsection 3.3. The results of calculation are
presented below

Table

Protection		Variant	
		1	2
with no protection		$2.21 \cdot 10^{-1}$	$3.02 \cdot 10^{-1}$
with protection on the level	0	$1.32 \cdot 10^{-3}$	$1.27 \cdot 10^{-3}$
	1	$4.09 \cdot 10^{-4}$	$1.98 \cdot 10^{-4}$
	2	$8.60 \cdot 10^{-4}$	$1.49 \cdot 10^{-4}$
	3	$4.24 \cdot 10^{-3}$	$9.08 \cdot 10^{-4}$
	4	$1.50 \cdot 10^{-2}$	$1.18 \cdot 10^{-2}$
	5	$3.35 \cdot 10^{-2}$	$7.86 \cdot 10^{-2}$

It is seen from the Table that in Variant 1 the optimal protection is achieved if the protection is concentrated at o-rank elements while in Variant 2-at 2-rank blocks. Moreover, in the last Variant the probability of total fracture of the system is almost one order of magnitude smaller with protection at 2-rank blocks than that for the protection at o-rank elements.

3.3 Probability of refusal of the total system

In this subsection we will give a formula for calculation of the probability of refusal of the total system. This formula follows directly from some general results obtained in [5]. Omiting details we will only formulate the final result.

Denote by $D^{(j)}$ the probability of refusal of j-rank block having initially a refused element. In particular, $D^{(o)} = 1$ since the initial fire damage of o-rank element is accepted as a certain event. The value $D^{(n)}$ is equal to the desired probability of refusal of the total system.

Let $p^{(j)}(x)$ be the integral distribution function of block "strength" under the local action of the intensity x. This function depends on the distribution function of o-rank element strength, $F^{(o)}(x)$, the scale variation of the action, s_j, and on the distribution of the protection resources through the structural levels, r_j.

Under the above assumptions one can write the following formula for the value $D^{(n)}$ of the total system refusal

$$D^{(n)} = \prod_{k=1}^{n-1} p^{(k)}(q_k) \qquad (11)$$

where

$$q_k = s_k - \sum_{m=o}^{k} r_m \quad , \quad k = 0,\ldots, (n-1) \qquad (12)$$

and

$$p^{(j)}(q) = 2\, p^{(j-1)}(q + s_{j-1}) - [p^{(j-1)}(q)]^2$$

$$j = 1, \ldots, (n-1) \qquad (13)$$

4 References

1.   Goldstein   R.V.   (1993)   *About   an structural-continual approach in mechanics of catastrophic fracture of complex technical systems.* Reports of the Russian Academy of Sciences, 330, 45-47 (in Russian).
2.   Goldstein R.V., Osipenko N.M. (1978) *Fracture and structure formation.* Reports of the USSR Academy of Sciences, 249, 829-832 (in Russian).
3.   Goldstein R.V., Osipenko N.M. (1978) *Structures of fracture (Conditions of formation. Echelons of cracks).* Preprint N 110. Institute for Problems in Mechanics, USSR Academy of Sciences (in Russian).
4.   Danilov V.G. (1991) *Asymptotic finite solutions of degenerate quasilinear parabolic equations with small diffusion.* Mathematical notes, 50, 77-88 (in Russian).
5.   Onishcenko D.A. (1992) *Refusalresistant building structures, in "Reliability, lifeteme and safety of technical systems",* SPb, 1994, pp. 102-107 (in Russian).

13 A SOFTWARE PACKAGE FOR DETERMINISTIC AND STOCHASTIC MODELLING OF FIRES IN BUILDINGS WITH A CAD-BASED GRAPHICAL USER INTERFACE

A. LANCIA, L. BORDIGNON and M. SINI
TRI srl, Scanzorosciate (BG), Italy

G. GALLINA
CNR – ICITE, Sesto Ulteriano (MI), Italy

ABSTRACT : Fire risk assessment in buildings can be performed by running a representative set of simulations including modelling for fire dynamics, smoke transport and occupants movement. A software package is described allowing the user to define the modelling data while working in a friendly CAD environment. The analyst can choose to perform a certain simulation in a deterministic way or to indicate statistical distributions for input parameters and then generating and running a given number of simulations (stochastic mode). A graphical and numeric statistical post-processor is also added to help in the evaluation of stochastic modelling results. Fire and smoke transport modelling is accomplished by the CFAST (W:Jones et al.) multi-compartment zone model while evacuation is simulated by a rule-based proprietary code. A sample application of the package is given for a 13 compartments building hosting an applied research laboratory. Typical expected applications of the developed system such as cost/benefit analysis, performance oriented design and support for fire engineering courses are discussed.

1. Introduction to the *TRISTAR* project

The TRISTAR project was started in 1989 with the aim of bridging the gap between fire modelling research and its application to practical problems in fire safety engineering for large buildings where such an approach can be useful and potentially cost effective.
The project was executed by the R&D group of the company Tecsa SpA within the framework of "Progetto Finalizzato Edilizia" which is a large R&D actions (total budget about 50 million ECU) promoted by the Italian National Council of Research (CNR) through their ICITE Institute (the body in charge of building technology). The project was completed in 1994 by TRI srl which is a company formed in 1993 by detaching from Tecsa SpA the R&D and the technology departments. A close co-operation with the researchers of ICITE was maintained during the work and a joint effort is still in progress with the aim of refining the software and applying it on a series of test cases.

The following basic needs were identified from the beginning of our work:
* allowing the software to be applied by fire safety engineers in a user friendly way
* taking into account a representative set of fire scenarios and the variability of their development
* incorporating a fire & smoke transport modelling software with a sufficient accuracy but within the execution time constraints related to the need to run the models thousand of times with a reasonable time and computing budget
* incorporating a suitable model for the prediction of the occupants movement
* producing results in a form useful to fire safety engineers

The problem of user friendliness is a general one for engineering packages. In the case of multi-compartment fire models the main discouraging aspect for the users is the long and boring procedure which is necessary to prepare the data set which is describing the building

Fire Engineering and Emergency Planning. Edited by R. Barham.
Published in 1996 by E & FN Spon. ISBN 0 419 20180 7.

and the input variables needed for the computation. The decision was therefore taken in the project to use a CAD environment as the main MMI (man-machine interface) and developing a series of procedures to input the required data while interacting with the graphical building layout.

A major issue in the project was the need to take into account for each fire scenario the variability of its development. Regardless the complexity and the accuracy of the fire model being used, a single deterministic computation is just producing a set of output data that cannot be taken as the meaningful answer to the problem of forecasting the consequence of a certain fire initiation. The main reasons for this are the following:
- the description of the building and its contents is necessarily incomplete
- a significant uncertainty often exists in the actual input data
- the combustible items in the building are often subject to changes in their position within compartments
- any of the doors and windows are never always open or close at different times
- models are more or less precise but they are never reproducing exactly the actual phenomena
- the behaviour of peoples movement is not deterministic
- the variability of people behaviour affects the chances to escape safely but also the fire scenario dynamics (e.g. they open or close the doors)
- the response of fire detection and fire suppression systems cannot be modelled sufficiently well and such systems are characterised by a rather good but limited reliability

Therefore a meaningful evaluation of the outcome of an initial fire scenario requires to carry out a large number of calculations where the scenario development is different by reflecting the input data variability and the branching processes related to the decisions of occupants and to the consequences of such decisions on fire and smoke transport dynamics.

A basic choice had to be made initially on the type of model to be used for fire dynamics and smoke transport. The use of CFD models (field models) was excluded for the impossibility to run such codes automatically and for a large number of times within reasonable computing resources. The only major limitation in choosing zone models was that they are applicable only to sets of small or medium size compartments and therefore they cannot deal with large atria, underground stations, industrial spaces with single large volumes, etc. This is not however a very important problem since most of the buildings we are interested in are made of a number of small compartments (offices, hospitals, schools, etc.).

Concerning evacuation modelling we chose to exclude any deterministic model based on the optimisation of evacuation or on the straightforward application of queue theory on predetermined pathways. The reason for this is that such models are useful only for preparing emergency evacuation plans or in those cases where the building is accessible only to certain persons and all of them have been trained to follow predetermined escape ways (e.g. the case of military ships, certain research centres, etc.). For our purpose we wanted an evacuation model that could simulate the actual behaviour of the occupants taking into account the results of the recent studies on the behaviour of people in fires (e.g. J.Sime et al.).

Running multiple fire simulations is necessary for our purpose but the huge set of output data is impossible to be evaluated directly regardless the forms in which such results are produced. Therefore we decided to design a module to merge the results of the individual simulations and to produce a series of tables and graphs in order to show the overall results and to evaluate the statistical distributions, the correlation and the sensitivity of the outcome to certain input parameters.

Fig. 1 shows a simplified diagram of the first conception of the TRISTAR system. A module for deterministic modelling was included to offer the possibility to use the data input by CAD for running without large efforts some modelling programs such as EVACNET and CFAST.

The first year of the CNR TRISTAR project was devoted to the reviewing of available modelling programs and to collect additional information for designing the software. The system design was carried out in the second year and the software was developed in its first version during the third year.

This section contains a summary of the functional specification of the TRISTAR software package version as it was implemented in 1994.

Fig. 1 shows the structure of the system indicating the main software modules and the principal sets of data files.

2. General information on the system

TRISTAR provides the user with the following main functions:
- *entering the description of the building model* by defining a "virtual building description" (VBD) within an application using a standard architectural CAD
- *executing the fire & smoke dynamics models*
- *performing the evacuation simulation for a defined VBD*
- *displaying in graphical form* the results of modelling
- generating series of varied VBD and executing the related modelling in order to perform *sensitivity analysis and randomised studies*
- *analysing the results of the execution of multiple VBD modelling*

The drawing of the building can be prepared or imported in the CAD system and can be used as a graphical background for defining the VBD.

Fire and smoke dynamics modelling is performed by the program CFAST by W.Jones at al. (US-NIST). The VBD structure is defined by nodes and arcs. VBD nodes and arcs directly correspond to the nodes (compartments and ventilation nodes) and arcs (links between nodes) used by CFAST. A description of CFAST can be found in the documents produced by NIST and available through the US NTIS.

Evacuation modelling is performed by a proprietary TRI software hereby indicated as EMS (evacuation modelling system). A subset of VBD nodes and arcs corresponds to the occupants movement network used by EMS. The EMS model is based on the assumption that occupants at a node make a decision on what they should do and then they execute the selected move until they reach another node where they will make next decision. Decision making for the occupants is based on a set of rules that are used to set statistical likelihood for the possible alternative decisions (i.e. moving to a certain node, waiting, etc.). The actual move is determined by a random choice weighted on the calculated likelihood. The rules can be chosen by the user and they reflect the current knowledge on occupants behaviour. The data used for computing the rules are the relevant perception of the occupant such as the presence of an escape way sign, the view of some other occupants on one end of a corridor, the presence of smoke, an audible alarm, etc. The actual movement of the occupants from one node to the other is based on the usual algorithms used in deterministic modelling of people movement in buildings. This model is still being refined at TRI.

The user defines the VBD by creating, deleting and editing nodes and arcs which are corresponding to certain graphic patterns within CAD and to which a list of editable attributes is associated. The nodes and arcs attributes comprise all the information related to modelling by CFAST and EMS plus certain data indicating if and how the individual information must be varied to generate the stochastic data sets (when the user is asked for data which can be randomised to generate a set of input data files, he is given the possibility to define - in addition to a base value - a statistical distribution for the input data).

The user navigates in the system without accessing the operating system command level by interacting with a supervisor system software.

The system has been developed on an IBM RS6000 workstation using as a main developing shell the Microstation CAD system by Intergraph. The modules not developed within Microstation are written in C language.

In the first program version the models were running on a PC connected by a TCP-IP ethernet link while the current version is fully implemented on the workstation following the kind decision of NIST to make the CFAST source files available.

3. Applications of the *TRISTAR* system

TRISTAR can be applied to different problems in fire safety engineering. We mention here the principal ones with a few short comments.

3.1 cost vs. benefit analysis
A common problem in fire safety engineering is deciding which is the most cost effective way to reduce risk in a certain building. This can be accomplished by selecting a set of reference fire scenarios (they might be the "worst cases") and then running a stochastic simulation for each of the possible changes introduced to improve the safety level (e.g. one more escape way, fire detectors, smoke venting, stairwell pressurisation, etc.).
A engineering decision can be made by comparing the expected improvements and taking into account the cost of the alternate solutions. In some case the cost will just be limited by a maximum budget and the system will help to select the best solution or combination of solutions within such economic constraint.

3.2 evaluation of the fire safety impact of changes in building structure or in its use
In this case the reference fire scenarios are identified and the stochastic simulations are run with and without taking into account a planned modification to the building or its contents and use. The risk histograms produced from the simulation will help in deciding if authorising the change or conditioning the permission to the implementation of better prevention and/or protection means.

3.3 performance oriented design
The idea of performance oriented design consists in setting a safety goal and then leaving the building designers and the fire safety engineers free to adopt the solutions they wish, providing that the safety goal is reached. In the most common opinion this will not however cause the relaxation of a series of basic rules such as the maximum distance between fire exits, the availability of fire extinguishers etc. Performance oriented design looks to be the only way to allow architects and engineers to create new buildings without rule based constraints only but within the respect of safety. Cost benefit analysis concepts are frequently used within this kind of application.

3.4 risk analysis

A system like TRISTAR can be used for the evaluation of risk (to human lives or to property). In this case the basic approach consists in:

- selecting or generating a representative set of fire scenarios
- running stochastic simulation for each scenario
- adding the risk from each simulation after multiplying it for the expected frequency of the relevant scenario

An absolute risk evaluation can be useful in a number of cases such as the negotiation of insurance costs and conditions.

3.5 training

TRISTAR could be used in training courses for fire safety engineers, fire brigades officers, civil engineers, etc. The software can be a tool to be used by a teacher in the classroom with the help of a computer screen projector and / or accessed by the trainees for carrying out exercises aiming to explore through modelling a series of notions such as how the building design features affect fire dynamics and smoke movement.

3.6 research in fire safety engineering

The TRISTAR system is designed in such a way to allow the replacement of the fire model with a limited effort. Thus the researchers can use the system as a test bench for their models allowing the user friendly input of building data, carrying out sensitivity analysis and comparing the predictions by different models.

4. An example application of the *TRISTAR* system

In order to give an example of the use of TRISTAR we report in this section a relatively simple exercise demonstrating the kind of application mentioned in 3.2.

This example is explained in great detail in Volume II of the final TRISTAR report to CNR and here we will just give a minimum of information to illustrate the application example.

The building used for the exercise is a two storey rectangular premise used by a department of a research institute for the development of special electro-optics instrumentation.

The building layout is represented in the annexed printouts of computer screens documenting the data input process. Two rows of offices and laboratories are stacked along one of the longer sides of the building while the other part is occupied by a large hall conceived to carry our experiments and miscellaneous work without constraints, with a flexible use of space and with the possibility to move large items in and out through the large doors at the two extremities of the hall. Structural beams are dividing the building and particularly the ceiling of the test hall into three sections. The hall was therefore simulated by three large rooms with open intercommunicating tall doors.

A total of 13 persons are normally present in the building (their work, functions and habits are described in the CNR report)

The example calculation we give here is related to the problem of deciding if authorising or not the regular use of flammable substances in the test hall. This problem arises because former simulations demonstrated that a fire in the test hall can make the evacuation of the upper rooms impossible in a few second unless costly changes are made such as the installation of smoke vents on the ceiling or the closure of the balcony and the stairwell e.g. with a series of air tight glass panels.

The fire scenario used for this example arises at an apparatus for generating atmospheric samples containing organic pollutants. A two litres glass bottle containing a mixture of alkyl-nitryles dissolved in n-hexane falls and breaks. The vapours are ignited and a spill fire rapidly develops. The fire area is however limited by confinement to about 1 square meter. Hexane combustion is completed within 40 sec but fire continues due to the combustion of plastic parts of the equipment including polyurethans, polyethylene and polypropylene. The generation of CO and HCN was considered and their dispersion modelled by CFAST. In this fire the heat release is quite high for a short time followed by a variable but mild combustion. Simulations were carried out for a fire time of 1000 sec.

The most critical element in the stochastic simulation are the status of doors and the distribution of occupants. The following values and statistical distribution were used in the exercise.

input variable	type of distribution	parameters of the distribution
door status 5-9	binary	P(0.00)= 0.5 ; P(1.00)= 0.5
door status 6-9	binary	P(0.00)= 0.9 ; P(1.00)= 0.1
door status 10-7	binary	P(0.00)= 0.9 ; P(1.00)= 0.1
door status 11-8	binary	P(0.00)= 0.9 ; P(1.00)= 0.1
door status 12-8	binary	P(0.00)= 0.7 ; P(1.00)= 0.3
door status 13-9	binary	P(0.00)= 0.9 ; P(1.00)= 0.1
n. occupants cmpt. 1	4 values int.	P(0)= 0.1 ; P(1)= 0.8; P(2)= 0.1 ; P(3)= 0.0
n. occupants cmpt. 5	4 values int.	P(0)= 0.2; P(1)= 0.3 P(2)= 0.4; P(3)= 0.1
n. occupants cmpt. 6	4 values int.	P(0)= 0.1; P(1)= 0.8; P(2)= 0.1; P(3)= 0.0
n. occupants cmpt. 10	4 values int.	P(0)= 0.1; P(1)= 0.3; P(2)= 0.3; P(3)= 0.3
n. occupants cmpt. 11	4 values int.	P(0)= 0.1; P(1)= 0.3 : P(2)= 0.6 ; P(3)= 0.0
n. occupants cmpt. 12	4 values int.	P(0)= 0.1 ; P(1)= 0.3 ; P(2)= 0.6 ; P(3)= 0.0
n. occupants cmpt. 13	4 values int.	P(0)= 0.1 ; P(1)= 0.1 ; P(2)= 0.2 ; P(3)= 0.6

The exercise is addressing the evaluation of the usefulness of an automatic fire detection system that could be installed in order to allow the use of flammable substances in the hall. Therefore the stochastic simulation is repeated for the following three data sets.

data set	alarm tripping time
A	15 sec i.e. the detection system is correctly functioning
B	120 sec due to the undetected failure of smoke detector in cmpt. 8
C	2000 sec - the system doesn't work or it is not installed

The exercise was carried out before the models were ported to the RS6000 workstation and therefore the number of modelling run was limited to 30 due to the time required to run the modelling on a 66 MHz 80486 based PC (each CFAST calculation took between 12 and 150 minutes). Even though smooth converged distributions are not reached, the limited number of runs is sufficient to demonstrate the technique.

The three tables in Fig. 9, 10 and 11 collect the door status and the number of occupants of 7 compartments for each of the data input files in each set of simulations. The results are reported in by specifying the number of successfully evacuated occupants for each run and by giving the statistical frequency distribution of total evacuees from each room for the set of simulations.

Fig. 12 compares the (total) occupants frequency histogram for the three data sets in order to verify that there were no large differences between the three sets.

Fig. 13 summarises the results of the three stochastic simulations in the form of frequency distributions of the number of non-evacuees (potential victims) in the individual simulations.

- In the A case only exceptional victims are reported (corresponding to the case in which some of the occupants do not start moving following the alarm)
- Case B demonstrates that 2 minutes of detection response time is sufficient to cause a relevant number of victims due to the fast descent of the smoke layer on the balcony communicating with the test hall
- Case C shows a potential very severe loss of lives.

The analysis of the details of the simulations for data set C shows that the only occupants of the upper floor that escape safely are the ones with the door open since smoke acts as an alarm for them.

5. Current status of the *TRISTAR* project and planned developments

TRISTAR is at the moment a research tool being used for pilot studies on the use of modelling technique in fire safety engineering. No commercialisation was attempted yet but a first series of releases to external pilot users could start soon following the completion of some refinements being done on the EMS model (subject to the availability of funding).

The main planned developments for the future include the following main enhancements:
- full integration of CFAST and EMS (now EMS is run following CFAST while we are working on the insertion of EMS as a procedure within CFAST so that the occupants actions will have the possibility to change the CFAST parameters - e.g. door opening - dynamically)
- enhancement of the EMS model by adding a more complex set of rules and validating / calibrating the parameters on the basis of evacuation case histories and experiments
- implementation of "important sampling" techniques as an alternative to Montecarlo randomisation in order to reduce the overall run time
- implementation of functions activating alarms, smoke management systems and fire suppression systems within the modelling software
- implementation of functions to generate and manage a set of scenarios in order to carry out a complete risk analysis process without a manual housekeeping of scenarios

6. Acknowledgements

The authors wishes to thank CNR for funding the TRISTAR project and the modelling team and the directors of US NIST for the making the CFAST source code available to the project team.

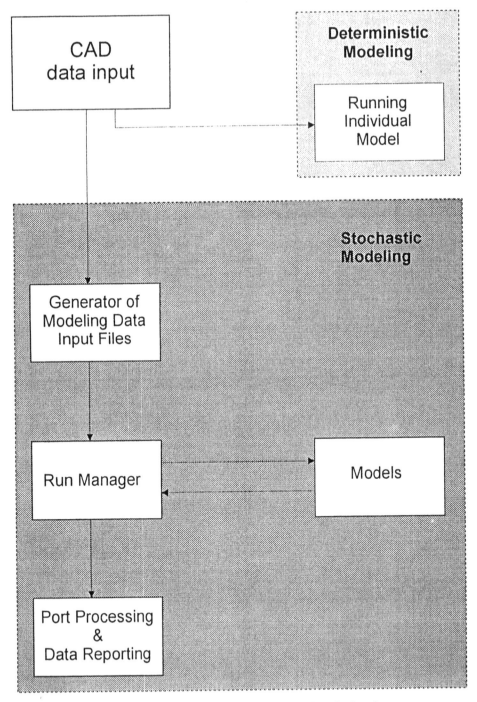

1. Basic scheme of the TRISTAR original design

Structure of TRISTAR Package

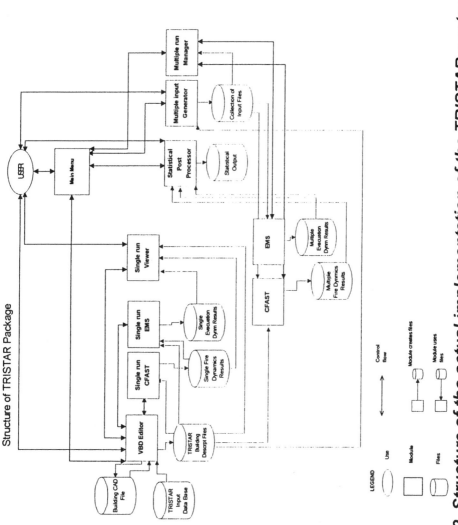

LEGEND

Use

Module

Files

Control flow

Module creates files

Module uses files

2. Structure of the actual implementation of the TRISTAR system

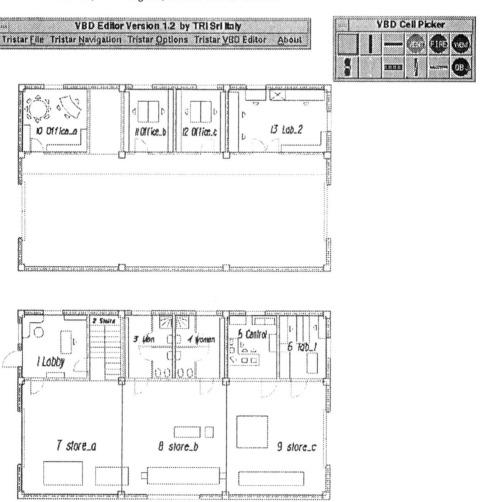

3. Printout of a TRISTAR screen while inputing data.

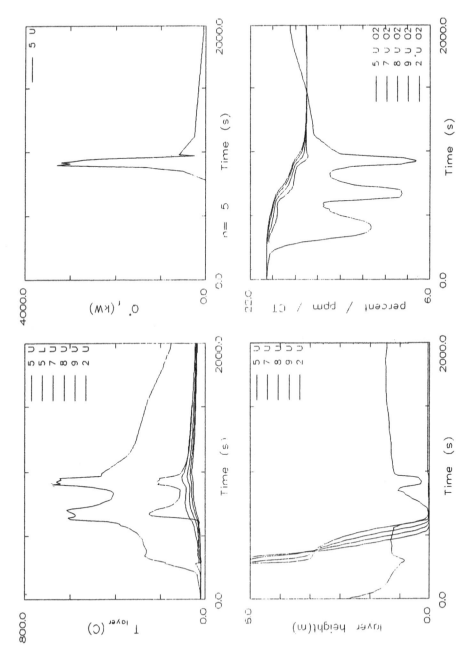

4. *Example of module display of variables vs. time for a CFAST simulation*

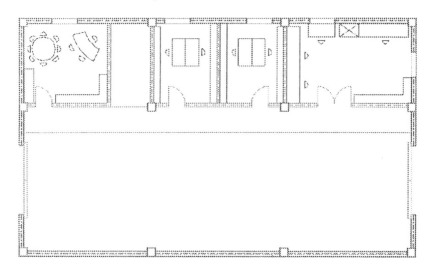

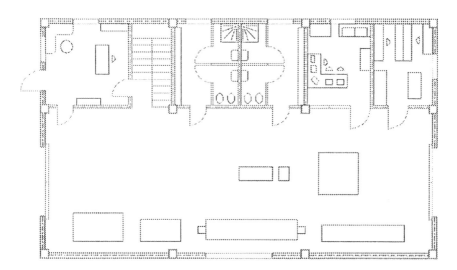

5. CAD drawing of test building imported into the system (before data input begins)

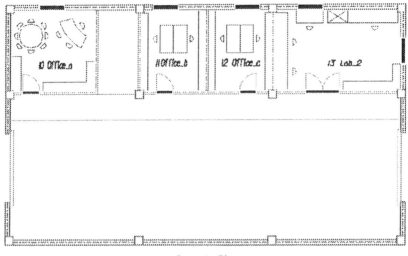

Seconda Piano

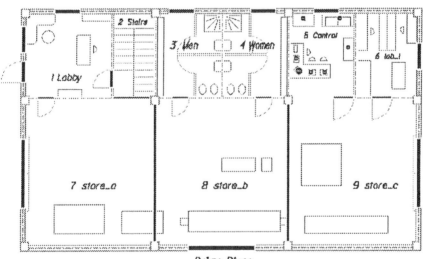

Primo Piano

6. CAD drawing of test building imported into the system (following the completion of data input begins)

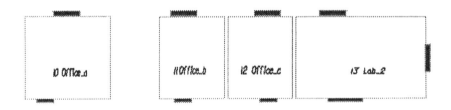

Seconda Piano

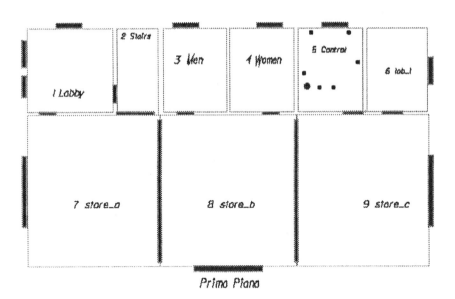

Primo Piano

7. TRISTAR CAD drawing of test building representation without the background view of the original drawing (VBD visualisation)

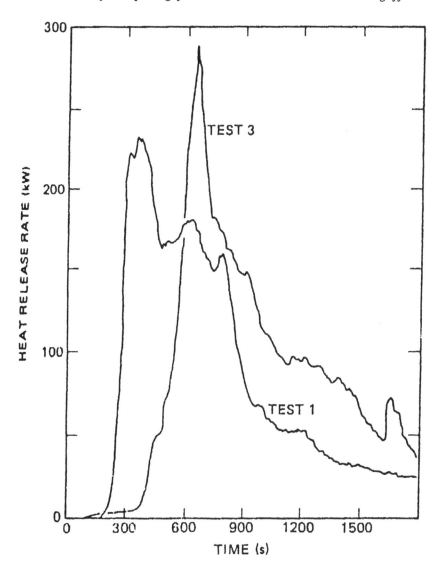

TEST1: usato per il computer 2
TEST3: usato per il computer 1

8. Examples of experimental HRR (heat release rate curves) used to simulate the heat relead of two different computers in some of the fire modeling test exercises

Data set A - Table of randomized input and evacuation success

Run #	Compt. 1			Compt. 5			Compt. 6			Compt. 10			Compt. 11			Compt. 12			Compt. 13			totals		
	door	occ	evac	door	occ	evac	door	occ	evac	door	occ	evac	door	occ	evac	door	occ	evac	door	occ	evac	occ	evac	trap
1	1	1	1	1	1	1	0	1	1	0	1	1	0	0	0	0	1	1	0	3	3	8	8	0
2	1	1	1	0	2	2	0	2	2	0	2	2	0	1	1	0	2	2	0	2	2	12	12	0
3	1	1	1	0	2	2	0	1	1	0	0	0	0	2	2	0	1	1	1	3	3	10	10	0
4	1	1	1	1	0	0	1	1	1	0	3	3	0	2	2	1	2	2	0	0	0	9	9	0
5	1	0	0	1	2	2	0	1	1	0	2	2	0	1	1	0	1	1	0	3	3	10	10	0
6	1	1	1	0	1	1	0	1	1	1	1	1	0	2	2	0	0	0	0	3	2	9	8	1
7	1	1	1	1	2	2	0	0	0	0	1	1	0	1	1	1	2	2	0	2	2	9	9	0
8	1	1	1	0	1	1	1	1	1	0	3	3	0	2	2	0	1	1	0	1	1	10	10	0
9	1	1	1	1	1	1	0	1	1	0	2	2	1	1	1	0	2	2	1	3	3	11	11	0
10	1	2	2	0	0	0	0	1	1	0	2	2	0	2	2	0	2	2	0	3	3	12	12	0
11	1	1	1	1	3	3	0	1	1	0	0	0	0	2	2	0	1	1	0	3	3	11	11	0
12	1	1	1	1	2	2	0	1	1	1	1	1	0	0	0	0	2	2	0	3	3	10	10	0
13	1	1	1	0	1	1	0	1	1	0	3	3	0	2	2	1	2	2	0	3	3	13	13	0
14	1	0	0	0	3	3	0	0	0	0	2	2	0	2	2	1	2	2	0	0	0	9	9	0
15	1	1	1	1	2	2	0	1	1	0	1	1	0	1	1	0	0	0	0	3	3	9	9	0
16	1	1	1	0	0	0	0	1	1	0	2	2	0	1	1	0	2	2	0	3	3	10	10	0
17	1	2	2	0	2	2	0	1	1	0	2	2	0	2	2	0	1	1	0	2	2	12	11	1
18	1	1	1	1	2	2	0	0	0	0	0	0	0	2	2	0	1	1	0	2	2	8	8	0
19	1	2	2	1	1	1	1	1	1	0	3	3	0	0	0	0	2	2	0	0	0	9	9	0
20	1	1	1	1	3	3	0	1	1	0	3	3	1	2	2	1	2	2	0	1	1	13	13	0
21	1	1	1	0	1	1	0	1	1	0	1	1	0	2	2	0	2	2	0	3	3	11	11	0
22	1	1	1	0	2	2	0	1	1	0	2	2	0	1	1	0	2	2	0	3	3	12	12	0
23	1	1	1	1	2	2	0	1	1	0	3	3	0	2	2	0	2	2	0	3	3	14	14	0
24	1	1	1	0	1	1	0	1	1	0	3	3	1	1	1	1	1	1	0	2	2	10	10	0
25	1	1	1	1	0	0	1	1	1	0	1	1	0	2	2	0	2	2	0	3	3	10	10	0
26	1	1	1	1	1	1	0	1	1	0	2	2	0	2	2	1	2	2	0	3	3	12	12	0
27	1	0	0	0	2	2	0	1	1	0	1	1	0	0	0	0	2	2	0	3	3	9	9	0
28	1	1	1	1	2	2	0	1	1	0	1	1	0	1	1	0	1	1	0	3	3	10	10	0
29	1	2	2	0	3	3	0	1	1	0	3	3	0	2	2	1	2	2	0	2	2	15	15	0
30	1	1	1	1	1	1	0	2	2	0	2	2	0	1	1	0	2	2	0	3	3	12	12	0
F(0)	0.00	0.10	0.10	0.47	0.13	0.13	0.87	0.10	0.10	0.93	0.10	0.10	0.90	0.13	0.13	0.73	0.07	0.10	0.93	0.10	0.10	0.00	0.00	0.93
F(1)	1.00	0.77	0.77	0.53	0.33	0.33	0.13	0.83	0.83	0.07	0.30	0.30	0.10	0.33	0.33	0.27	0.30	0.27	0.07	0.07	0.07	0.00	0.00	0.07
F(2)	0.00	0.13	0.13	0.00	0.40	0.40	0.00	0.07	0.07	0.00	0.33	0.33	0.00	0.53	0.53	0.00	0.63	0.63	0.00	0.20	0.23	0.00	0.00	0.00
F(3)	0.00	0.00	0.00	0.00	0.13	0.13	0.00	0.00	0.00	0.00	0.27	0.27	0.00	0.00	0.00	0.00	0.00	0.00	0.00	0.63	0.60	0.00	0.00	0.00
F(4)	0.00	0.00	0.00	0.00	0.00	0.00	0.00	0.00	0.00	0.00	0.00	0.00	0.00	0.00	0.00	0.00	0.00	0.00	0.00	0.00	0.00	0.00	0.00	0.00
F(5)	0.00	0.00	0.00	0.00	0.00	0.00	0.00	0.00	0.00	0.00	0.00	0.00	0.00	0.00	0.00	0.00	0.00	0.00	0.00	0.00	0.00	0.00	0.00	0.00
F(6)	0.00	0.00	0.00	0.00	0.00	0.00	0.00	0.00	0.00	0.00	0.00	0.00	0.00	0.00	0.00	0.00	0.00	0.00	0.00	0.00	0.00	0.00	0.00	0.00
F(7)	0.00	0.00	0.00	0.00	0.00	0.00	0.00	0.00	0.00	0.00	0.00	0.00	0.00	0.00	0.00	0.00	0.00	0.00	0.00	0.00	0.00	0.00	0.00	0.00
F(8)	0.00	0.00	0.00	0.00	0.00	0.00	0.00	0.00	0.00	0.00	0.00	0.00	0.00	0.00	0.00	0.00	0.00	0.00	0.00	0.00	0.00	0.07	0.10	0.00
F(9)	0.00	0.00	0.00	0.00	0.00	0.00	0.00	0.00	0.00	0.00	0.00	0.00	0.00	0.00	0.00	0.00	0.00	0.00	0.00	0.00	0.00	0.23	0.20	0.00
F(10)	0.00	0.00	0.00	0.00	0.00	0.00	0.00	0.00	0.00	0.00	0.00	0.00	0.00	0.00	0.00	0.00	0.00	0.00	0.00	0.00	0.00	0.27	0.27	0.00
F(11)	0.00	0.00	0.00	0.00	0.00	0.00	0.00	0.00	0.00	0.00	0.00	0.00	0.00	0.00	0.00	0.00	0.00	0.00	0.00	0.00	0.00	0.10	0.13	0.00
F(12)	0.00	0.00	0.00	0.00	0.00	0.00	0.00	0.00	0.00	0.00	0.00	0.00	0.00	0.00	0.00	0.00	0.00	0.00	0.00	0.00	0.00	0.20	0.17	0.00
F(13)	0.00	0.00	0.00	0.00	0.00	0.00	0.00	0.00	0.00	0.00	0.00	0.00	0.00	0.00	0.00	0.00	0.00	0.00	0.00	0.00	0.00	0.07	0.07	0.00
F(14)	0.00	0.00	0.00	0.00	0.00	0.00	0.00	0.00	0.00	0.00	0.00	0.00	0.00	0.00	0.00	0.00	0.00	0.00	0.00	0.00	0.00	0.03	0.03	0.00
F(15)	0.00	0.00	0.00	0.00	0.00	0.00	0.00	0.00	0.00	0.00	0.00	0.00	0.00	0.00	0.00	0.00	0.00	0.00	0.00	0.00	0.00	0.03	0.03	0.00
TOT F	1.00	1.00	1.00	1.00	1.00	1.00	1.00	1.00	1.00	1.00	1.00	1.00	1.00	1.00	1.00	1.00	1.00	1.00	1.00	1.00	1.00	1.00	1.00	1.00
AVG	1.00	1.03	1.03	0.53	1.53	1.53	0.13	0.97	0.97	0.07	1.77	1.77	0.10	1.40	1.40	0.27	1.57	1.53	0.07	2.37	2.33	10.63	10.57	0.07
STDEV	0.00	0.49	0.49	0.51	0.90	0.90	0.35	0.41	0.41	0.25	0.97	0.97	0.31	0.72	0.72	0.45	0.63	0.68	0.25	1.00	0.99	1.75	1.77	0.25
MIN	1.00	0.00	0.00	0.00	0.00	0.00	0.00	0.00	0.00	0.00	0.00	0.00	0.00	0.00	0.00	0.00	0.00	0.00	0.00	0.00	0.00	8.00	8.00	0.00
MAX	1.00	2.00	2.00	1.00	3.00	3.00	1.00	2.00	2.00	1.00	3.00	3.00	1.00	2.00	2.00	1.00	2.00	2.00	1.00	3.00	3.00	15.00	15.00	1.00

9. Data set A -Table of randomized input and evacuation success

	Compt. 1			Compt. 5			Compt. 6			Compt. 10			Compt. 11			Compt. 12			Compt. 13			totals		
Run #	door	occ	evac	door	occ	evac	door	occ	evac	door	occ	evac	door	occ	evac	door	occ	evac	door	occ	evac	occ	evac	trap
1	1	1	1	0	2	2	0	1	1	0	3	2	0	2	0	0	2	2	0	3	1,	14	9	5
2	1	1	1	0	1	1	0	1	1	0	1	1	0	2	2	0	1	0	0	0	0	7	6	1
3	1	1	1	1	2	2	0	1	1	0	2	1	0	2	1	0	1	0	0	3	1	12	7	5
4	1	1	1	1	2	2	0	1	1	0	3	1	0	1	1	0	2	2	0	2	2	12	10	2
5	1	1	1	1	2	2	0	0	0	1	2	0	0	0	0	1	2	1	0	3	0	10	4	6
6	1	1	1	0	2	2	0	1	1	0	1	1	1	2	2	1	2	0	0	1	2	10	9	1
7	1	2	2	1	1	1	1	1	1	0	3	3	0	2	2	1	1	1	0	2	1	12	11	1
8	1	1	1	0	2	2	0	1	1	0	2	1	0	1	1	0	2	1	0	3	1	12	8	4
9	1	1	1	1	1	1	0	1	1	0	1	0	0	2	2	1	2	2	0	2	0	10	7	3
10	1	1	1	0	3	3	0	2	2	0	0	0	0	1	1	0	1	0	0	3	2	11	9	2
11	1	1	1	1	0	0	0	1	1	1	0	0	0	2	0	0	2	2	0	3	1	9	5	4
12	1	2	2	0	3	3	0	1	1	0	3	2	0	2	0	0	2	0	0	0	0	13	8	5
13	1	0	0	1	0	0	0	1	1	0	2	2	0	1	1	0	2	1	0	2	2	8	7	1
14	1	1	1	0	1	1	0	1	1	0	1	0	0	2	2	1	0	0	0	3	2	9	7	2
15	1	1	1	1	2	2	0	1	1	0	1	1	1	2	1	0	1	0	1	3	0	11	6	5
16	1	0	0	1	2	2	0	1	1	0	2	1	1	0	0	0	2	0	0	3	3	10	7	3
17	1	1	1	0	3	3	0	0	0	0	3	3	0	2	2	1	1	0	0	1	0	11	9	2
18	1	1	1	1	1	1	1	1	1	1	3	2	0	2	2	0	0	0	0	2	1	10	8	2
19	1	1	1	0	1	1	0	1	1	0	1	1	0	1	1	1	2	2	0	3	0	10	7	3
20	1	2	2	0	2	2	0	2	2	0	0	0	0	2	0	0	2	1	0	3	1	13	8	5
21	1	1	1	1	2	2	1	1	1	0	2	1	1	2	2	0	2	2	0	2	1	12	10	2
22	1	1	1	0	0	0	0	1	1	0	1	1	0	2	1	1	1	0	0	3	1	9	5	4
23	1	1	1	1	2	2	0	1	1	0	3	0	0	0	0	0	2	1	0	3	2	12	7	5
24	1	2	2	1	3	3	0	2	2	0	1	1	0	2	0	1	1	1	1	1	1	12	10	2
25	1	1	1	1	1	1	0	1	1	0	2	1	0	1	0	0	2	0	0	3	1	11	5	6
26	1	1	1	0	2	2	1	2	2	0	0	0	0	1	1	0	0	0	0	3	0	9	6	3
27	1	0	0	1	1	1	0	1	1	0	3	3	0	2	1	0	2	2	0	3	1	12	9	3
28	1	1	1	0	0	0	0	1	1	0	1	1	0	2	2	1	1	0	0	1	1	7	6	1
29	1	1	1	1	1	1	0	1	1	0	2	0	0	1	1	0	2	2	0	2	0	10	6	4
30	1	1	1	0	2	2	0	1	1	0	1	1	0	2	1	0	1	1	0	3	2	11	9	2
F(0)	0.00	0.10	0.10	0.47	0.13	0.13	0.87	0.07	0.07	0.90	0.13	0.30	0.87	0.10	0.30	0.67	0.10	0.47	0.93	0.07	0.30	0.00	0.00	0.00
F(1)	1.00	0.77	0.77	0.53	0.30	0.30	0.13	0.80	0.80	0.10	0.33	0.47	0.13	0.27	0.40	0.33	0.33	0.27	0.07	0.13	0.43	0.00	0.00	0.17
F(2)	0.00	0.13	0.13	0.00	0.43	0.43	0.00	0.13	0.13	0.00	0.27	0.13	0.00	0.63	0.30	0.00	0.57	0.27	0.00	0.23	0.23	0.00	0.00	0.27
F(3)	0.00	0.00	0.00	0.00	0.13	0.13	0.00	0.00	0.00	0.00	0.27	0.10	0.00	0.00	0.00	0.00	0.00	0.00	0.00	0.57	0.03	0.00	0.00	0.17
F(4)	0.00	0.00	0.00	0.00	0.00	0.00	0.00	0.00	0.00	0.00	0.00	0.00	0.00	0.00	0.00	0.00	0.00	0.00	0.00	0.00	0.00	0.00	0.03	0.13
F(5)	0.00	0.00	0.00	0.00	0.00	0.00	0.00	0.00	0.00	0.00	0.00	0.00	0.00	0.00	0.00	0.00	0.00	0.00	0.00	0.00	0.00	0.00	0.10	0.20
F(6)	0.00	0.00	0.00	0.00	0.00	0.00	0.00	0.00	0.00	0.00	0.00	0.00	0.00	0.00	0.00	0.00	0.00	0.00	0.00	0.00	0.00	0.00	0.17	0.07
F(7)	0.00	0.00	0.00	0.00	0.00	0.00	0.00	0.00	0.00	0.00	0.00	0.00	0.00	0.00	0.00	0.00	0.00	0.00	0.00	0.00	0.00	0.07	0.23	0.00
F(8)	0.00	0.00	0.00	0.00	0.00	0.00	0.00	0.00	0.00	0.00	0.00	0.00	0.00	0.00	0.00	0.00	0.00	0.00	0.00	0.00	0.00	0.03	0.13	0.00
F(9)	0.00	0.00	0.00	0.00	0.00	0.00	0.00	0.00	0.00	0.00	0.00	0.00	0.00	0.00	0.00	0.00	0.00	0.00	0.00	0.00	0.00	0.13	0.20	0.00
F(10)	0.00	0.00	0.00	0.00	0.00	0.00	0.00	0.00	0.00	0.00	0.00	0.00	0.00	0.00	0.00	0.00	0.00	0.00	0.00	0.00	0.00	0.23	0.10	0.00
F(11)	0.00	0.00	0.00	0.00	0.00	0.00	0.00	0.00	0.00	0.00	0.00	0.00	0.00	0.00	0.00	0.00	0.00	0.00	0.00	0.00	0.00	0.17	0.03	0.00
F(12)	0.00	0.00	0.00	0.00	0.00	0.00	0.00	0.00	0.00	0.00	0.00	0.00	0.00	0.00	0.00	0.00	0.00	0.00	0.00	0.00	0.00	0.27	0.00	0.00
F(13)	0.00	0.00	0.00	0.00	0.00	0.00	0.00	0.00	0.00	0.00	0.00	0.00	0.00	0.00	0.00	0.00	0.00	0.00	0.00	0.00	0.00	0.07	0.00	0.00
F(14)	0.00	0.00	0.00	0.00	0.00	0.00	0.00	0.00	0.00	0.00	0.00	0.00	0.00	0.00	0.00	0.00	0.00	0.00	0.00	0.00	0.00	0.03	0.00	0.00
F(15)	0.00	0.00	0.00	0.00	0.00	0.00	0.00	0.00	0.00	0.00	0.00	0.00	0.00	0.00	0.00	0.00	0.00	0.00	0.00	0.00	0.00	0.00	0.00	0.00
TOT F	1.00	1.00	1.00	1.00	1.00	1.00	1.00	1.00	1.00	1.00	1.00	1.00	1.00	1.00	1.00	1.00	1.00	1.00	1.00	1.00	1.00	1.00	1.00	1.00
AVG	1.00	1.03	1.03	0.53	1.57	1.57	0.13	1.07	1.07	0.10	1.67	1.03	0.13	1.53	1.00	0.33	1.47	0.80	0.07	2.30	1.00	10.63	7.50	3.13
STDEV	0.00	0.49	0.49	0.51	0.90	0.90	0.35	0.45	0.45	0.31	1.03	0.93	0.35	0.68	0.79	0.48	0.68	0.85	0.25	0.95	0.83	1.71	1.76	1.59
MIN	1.00	0.00	0.00	0.00	0.00	0.00	0.00	0.00	0.00	0.00	0.00	0.00	0.00	0.00	0.00	0.00	0.00	0.00	0.00	0.00	0.00	7.00	4.00	1.00
MAX	1.00	2.00	2.00	1.00	3.00	3.00	1.00	2.00	2.00	1.00	3.00	3.00	1.00	2.00	2.00	1.00	2.00	2.00	1.00	3.00	3.00	14.00	11.00	6.00

Data set B - Table of randomized input and evacuation success

10. Data set B-Table of randomized input and evacuation success

Data set C - Table of randomized input and evacuation success																								
	Compt. 1			Compt. 5			Compt. 6			Compt. 10			Compt. 11			Compt. 12			Compt. 13			totals		
Run #	door	occ	evac	door	occ	evac	door	occ	evac	door	occ	evac	door	occ	evac	door	occ	evac	door	occ	evac	occ	evac	trap
1	1	1	1	0	0	0	0	1	1	0	2	1	0	2	0	0	2	0	0	2	0	10	3	7
2	1	1	1	1	2	2	0	1	0	0	1	0	1	2	1	0	2	0	0	3	0	12	4	8
3	1	1	1	0	1	1	0	0	0	1	2	1	0	2	0	0	2	0	0	3	1	11	4	7
4	1	1	1	0	2	2	0	1	1	0	3	0	0	2	1	1	1	1	0	1	0	11	6	5
5	1	1	1	1	2	2	0	1	0	0	1	0	0	2	0	1	0	0	0	3	2	10	5	5
6	1	1	1	1	0	0	0	1	1	0	2	1	0	1	1	0	2	1	0	3	0	10	5	5
7	1	1	1	1	1	1	0	1	1	0	2	0	0	2	1	0	1	1	1	2	1	10	6	4
8	1	1	1	0	2	2	0	0	0	0	3	0	0	2	0	0	2	1	0	2	1	12	5	7
9	1	1	1	1	2	2	0	1	1	0	3	2	1	1	0	0	1	0	0	3	0	12	6	6
10	1	1	1	1	1	1	0	1	0	1	2	2	0	2	0	1	2	1	0	1	1	10	6	4
11	1	0	0	1	2	2	0	2	0	0	2	0	1	0	0	0	0	0	0	3	0	9	2	7
12	1	1	1	1	0	0	0	1	1	0	3	0	0	2	1	0	2	0	0	3	0	12	3	9
13	1	1	1	0	2	2	0	1	1	0	0	0	0	2	2	0	2	2	0	0	0	8	8	0
14	1	2	2	0	1	1	0	1	1	0	2	1	0	1	0	1	1	1	0	2	1	10	7	3
15	1	1	1	1	3	3	0	0	0	0	3	0	0	2	0	0	1	0	0	3	2	13	6	7
16	1	1	1	1	1	1	0	1	1	0	1	1	0	1	0	0	0	0	0	3	0	8	4	4
17	1	1	1	0	1	1	1	1	1	0	2	1	0	2	0	0	2	0	0	0	0	9	4	5
18	1	0	0	1	2	2	0	2	2	1	1	1	0	1	0	0	2	1	1	2	2	10	9	1
19	1	1	1	1	0	0	1	1	0	0	1	0	0	1	1	1	2	0	0	1	0	7	2	5
20	1	2	2	0	1	1	0	1	1	0	2	0	0	0	0	0	1	1	0	3	1	10	6	4
21	1	1	1	1	2	2	0	0	0	0	3	0	0	2	1	1	2	1	0	3	0	13	5	8
22	1	1	1	0	2	2	0	1	1	0	2	1	0	1	1	0	2	1	0	3	0	12	7	5
23	1	1	1	1	3	3	0	1	0	1	3	3	1	2	1	0	0	0	0	2	1	12	9	3
24	1	1	1	0	2	2	0	1	0	0	1	1	0	2	0	1	2	0	0	3	2	12	6	6
25	1	0	0	0	1	1	0	1	1	0	2	0	0	2	2	0	1	0	0	3	0	10	4	6
26	1	1	1	1	2	2	1	1	1	0	0	0	0	1	0	0	2	1	0	0	0	7	5	2
27	1	1	1	0	0	0	0	2	1	0	3	0	0	0	0	0	2	0	0	2	0	10	2	8
28	1	1	1	1	2	2	0	1	0	0	1	1	0	1	1	0	2	1	0	3	2	11	8	3
29	1	1	1	1	2	2	0	1	1	0	3	1	0	2	2	0	1	0	0	2	0	12	7	5
30	1	1	1	0	2	2	0	1	1	0	1	1	0	1	1	0	2	0	0	3	1	11	6	5
F(0)	0.00	0.10	0.10	0.43	0.17	0.17	0.90	0.13	0.40	0.87	0.07	0.50	0.87	0.10	0.53	0.77	0.13	0.57	0.93	0.10	0.57	0.00	0.00	0.03
F(1)	1.00	0.83	0.83	0.57	0.27	0.27	0.10	0.77	0.57	0.13	0.27	0.40	0.13	0.33	0.37	0.23	0.27	0.40	0.07	0.10	0.27	0.00	0.00	0.03
F(2)	0.00	0.07	0.07	0.00	0.50	0.50	0.00	0.10	0.03	0.00	0.37	0.07	0.00	0.57	0.10	0.00	0.60	0.03	0.00	0.27	0.17	0.00	0.10	0.03
F(3)	0.00	0.00	0.00	0.00	0.07	0.07	0.00	0.00	0.00	0.00	0.30	0.03	0.00	0.00	0.00	0.00	0.00	0.00	0.00	0.53	0.00	0.00	0.07	0.10
F(4)	0.00	0.00	0.00	0.00	0.00	0.00	0.00	0.00	0.00	0.00	0.00	0.00	0.00	0.00	0.00	0.00	0.00	0.00	0.00	0.00	0.00	0.00	0.17	0.13
F(5)	0.00	0.00	0.00	0.00	0.00	0.00	0.00	0.00	0.00	0.00	0.00	0.00	0.00	0.00	0.00	0.00	0.00	0.00	0.00	0.00	0.00	0.00	0.17	0.27
F(6)	0.00	0.00	0.00	0.00	0.00	0.00	0.00	0.00	0.00	0.00	0.00	0.00	0.00	0.00	0.00	0.00	0.00	0.00	0.00	0.00	0.00	0.00	0.27	0.10
F(7)	0.00	0.00	0.00	0.00	0.00	0.00	0.00	0.00	0.00	0.00	0.00	0.00	0.00	0.00	0.00	0.00	0.00	0.00	0.00	0.00	0.00	0.07	0.10	0.17
F(8)	0.00	0.00	0.00	0.00	0.00	0.00	0.00	0.00	0.00	0.00	0.00	0.00	0.00	0.00	0.00	0.00	0.00	0.00	0.00	0.00	0.00	0.07	0.07	0.10
F(9)	0.00	0.00	0.00	0.00	0.00	0.00	0.00	0.00	0.00	0.00	0.00	0.00	0.00	0.00	0.00	0.00	0.00	0.00	0.00	0.00	0.00	0.07	0.07	0.03
F(10)	0.00	0.00	0.00	0.00	0.00	0.00	0.00	0.00	0.00	0.00	0.00	0.00	0.00	0.00	0.00	0.00	0.00	0.00	0.00	0.00	0.00	0.33	0.00	0.00
F(11)	0.00	0.00	0.00	0.00	0.00	0.00	0.00	0.00	0.00	0.00	0.00	0.00	0.00	0.00	0.00	0.00	0.00	0.00	0.00	0.00	0.00	0.13	0.00	0.00
F(12)	0.00	0.00	0.00	0.00	0.00	0.00	0.00	0.00	0.00	0.00	0.00	0.00	0.00	0.00	0.00	0.00	0.00	0.00	0.00	0.00	0.00	0.27	0.00	0.00
F(13)	0.00	0.00	0.00	0.00	0.00	0.00	0.00	0.00	0.00	0.00	0.00	0.00	0.00	0.00	0.00	0.00	0.00	0.00	0.00	0.00	0.00	0.07	0.00	0.00
F(14)	0.00	0.00	0.00	0.00	0.00	0.00	0.00	0.00	0.00	0.00	0.00	0.00	0.00	0.00	0.00	0.00	0.00	0.00	0.00	0.00	0.00	0.00	0.00	0.00
F(15)	0.00	0.00	0.00	0.00	0.00	0.00	0.00	0.00	0.00	0.00	0.00	0.00	0.00	0.00	0.00	0.00	0.00	0.00	0.00	0.00	0.00	0.00	0.00	0.00
TOT F	1.00	1.00	1.00	1.00	1.00	1.00	1.00	1.00	1.00	1.00	1.00	1.00	1.00	1.00	1.00	1.00	1.00	1.00	1.00	1.00	1.00	1.00	1.00	1.00
AVG	1.00	0.97	0.97	0.57	1.47	1.47	0.10	0.97	0.63	0.13	1.90	0.63	0.13	1.47	0.57	0.23	1.47	0.47	0.07	2.23	0.60	10.47	5.33	5.13
STDEV	0.00	0.41	0.41	0.50	0.86	0.86	0.31	0.49	0.56	0.35	0.92	0.76	0.35	0.68	0.68	0.43	0.73	0.57	0.25	1.01	0.77	1.61	1.92	2.13
MIN	1.00	0.00	0.00	0.00	0.00	0.00	0.00	0.00	0.00	0.00	0.00	0.00	0.00	0.00	0.00	0.00	0.00	0.00	0.00	0.00	0.00	7.00	2.00	0.00
MAX	1.00	2.00	2.00	1.00	3.00	3.00	1.00	2.00	2.00	1.00	3.00	3.00	1.00	2.00	2.00	1.00	2.00	2.00	1.00	3.00	2.00	13.00	9.00	9.00

11. Data set C-Table of randomized input and evacuation success

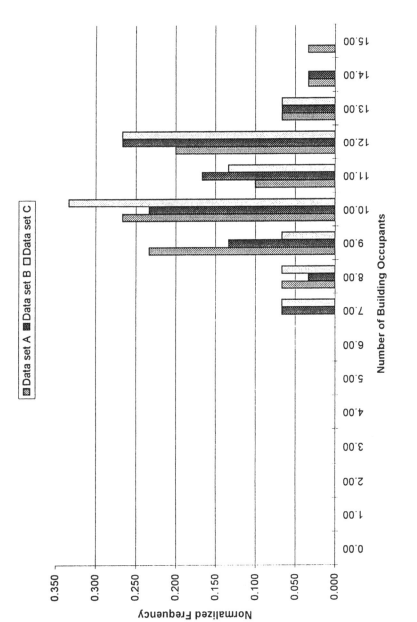

12. *Frequency hystogram comparing the number of occupants in stochastic run A, B, and C*

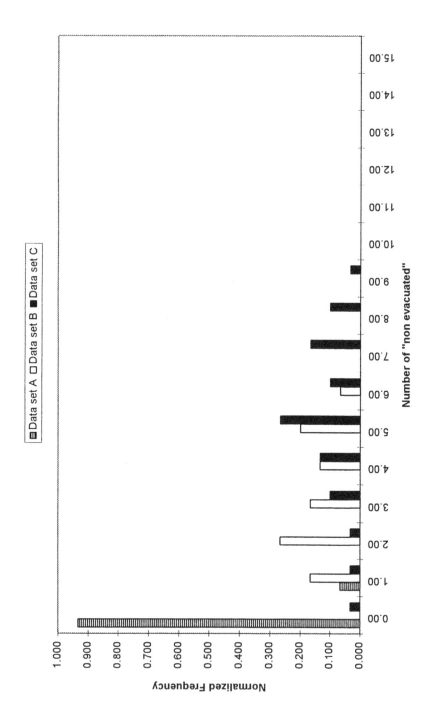

13. Frequency hystogram of "non evacuated occupants" in stochastic run A, B, and C

```
VERSN    2CNR PFE TRISTAR - NEWCO LABS TEST CASE - RUN A001
TIMES   1800     0     5    10     0
TAMB    293.  101300.     0.
EAMB    288.  101300.     0.
HI/F    0.00    0.00    0.00    0.00    0.00    0.00    0.00    0.00    0.00    3.00    3.00    3.00    3.00
WIDTH   5.00    2.40    3.80    3.80    3.80    3.80    7.70    8.00    7.70    5.00    3.80    3.80    7.60
DEPTH   5.00    5.20    5.00    5.00    5.00    5.00    8.20    8.20    8.20    5.00    5.00    5.00    5.00
HEIGH   2.70    6.00    2.70    2.70    2.70    2.70    6.00    6.00    6.00    3.00    3.00    3.00    3.00
HVENT   1   2  1   1.000    2.000    0.000
HVENT   1   7  1   1.000    2.000    0.000
HVENT   1  14  1   1.200    2.000    0.000    0.000
HVENT   2   7  1   2.400    6.000    0.000
HVENT   3   8  1   1.000    2.000    0.000
HVENT   4   8  1   1.000    2.000    0.000
HVENT   5   9  3   1.200    2.000    0.000
HVENT   5  14  1   1.500    2.400    1.000   -1.000
HVENT   6   9  1   1.200    2.000    0.000
HVENT   7   8  1   7.900    5.000    0.000
HVENT   7  10  1   1.000    5.000    3.000
HVENT   7  14  1   4.000    4.000    0.000    0.000
HVENT   8   9  1   7.900    5.000    0.000
HVENT   8  11  1   1.000    5.000    3.000
HVENT   8  12  1   1.000    5.000    3.000
HVENT   8  14  2   2.000    2.400    1.000    1.000
HVENT   9  13  2   1.000    5.000    3.000
HVENT   9  14  1   4.000    4.000    0.000    0.000
CVENT   1   2  1   0.00    0.00    0.00    0.00    0.00    0.00    0.00
CVENT   1   7  1   0.00    0.00    0.00    0.00    0.00    0.00    0.00
CVENT   1  14  1   0.00    0.00    0.00    0.00    0.00    0.00    0.00
CVENT   2   7  1   1.00    1.00    1.00    1.00    1.00    1.00    1.00
CVENT   3   8  1   0.00    0.00    0.00    0.00    0.00    0.00    0.00
CVENT   4   8  1   0.00    0.00    0.00    0.00    0.00    0.00    0.00
CVENT   5   9  3   0.00    0.00    0.00    0.00    0.00    0.00    0.00
CVENT   5  14  1   0.00    0.00    0.00    0.00    0.00    0.00    0.00
CVENT   6   9  1   0.00    0.00    0.00    0.00    0.00    0.00    0.00
CVENT   7   8  1   1.00    1.00    1.00    1.00    1.00    1.00    1.00
CVENT   7  10  1   0.00    0.00    0.00    0.00    0.00    0.00    0.00
CVENT   7  14  1   0.00    0.00    0.00    0.00    0.00    0.00    0.00
CVENT   8   9  1   1.00    1.00    1.00    1.00    1.00    1.00    1.00
CVENT   8  11  1   0.00    0.00    0.00    0.00    0.00    0.00    0.00
CVENT   8  12  1   0.00    0.00    0.00    0.00    0.00    0.00    0.00
CVENT   8  14  2   0.00    0.00    0.00    0.00    0.00    0.00    0.00
CVENT   9  13  2   0.00    0.00    0.00    0.00    0.00    0.00    0.00
CVENT   9  14  1   0.00    0.00    0.00    0.00    0.00    0.00    0.00
VVENT  14   7  0.01    2
VVENT  14   8  0.01    2
VVENT  14   9  0.01    2
CEILI GYPSUM     CONCRETE  CONCRETE   CONCRETE   CONCRETE   CONCRETE   CONCRETE   CONCRETE   CONCRETE   GYPSUM
GYPSUM     GYPSUM     GYPSUM
WALLS GYPSUM     GYPROCK   BRICK      BRICK      GYPSUM     GYPSUM     GYPROCK    GYPROCK    GYPROCK    GYPSUM
GYPSUM     GYPSUM     GYPSUM
FLOOR BRICK      CONCRETE  BRICK      BRICK      BRICK      BRICK      CONCRETE   CONCRETE   CONCRETE   HARDWOOD
SOFTWOOD   SOFTWOOD   BRICK
CHEMI    16.   40.   10.0    15900000.   300. 388.    0.
LFBO    5
LFBT    2
FPOS    3.50    0.38    1.00
FTIME    375.    520.    670.    740.   1100.   1400.
FMASS   0.0000 0.0005 0.0053 0.0193 0.0120 0.0060 0.0053
FHIGH    1.00    1.00    1.00    1.00    1.00    1.00    1.00
FAREA    0.00    0.00    0.00    0.00    0.00    0.00    0.00
FQDOT   0.00    7.95E+03 8.43E+04 3.07E+05 1.91E+05 9.54E+04 8.43E+04
CJET OFF
CO      0.020 0.020 0.020 0.020 0.020 0.020 0.020
OD      0.010 0.010 0.010 0.010 0.010 0.010 0.010
HCN     0.000 0.000 0.000 0.000 0.000 0.000 0.000
HCL     0.000 0.000 0.000 0.000 0.000 0.000 0.000
STPMAX  5.00
DUMPR HISTORIA.HI
DEVICE 1
WINDOW 0     0.     0. 1280. 1024. 1100.
GRAPH   1  100.    60.    0.  600.  475.   10.  3 time smoke_height_[m]
GRAPH   2  100.   550.    0.  600.  940.   10.  3 time temperature_[C]
GRAPH   3  750.    60.    0. 1250.  475.   10.  3 time fire_HRR_[kW]
```

14. Example of a CFAST input file generated automatically by the TRISTAR VBD editor

7. APPENDIX : Structure of the *TRISTAR* system

The user interacts with five functional modules:
- Main Menu
- VBD Editor
- Single Run Viewer
- Multiple Input Generator
- Statistical Post-Processor

The following part of this section gives a concise account of the commands and functions available in the TRISTAR system.

7.1 Main Menu
This is the top level navigation node where the user can choose between:
- Accessing VBD Editor
- Accessing Single Run Viewer
- Accessing Multiple Input Generator
- Activating the Multiple Run Manager
- Accessing Statistical Post-Processor
- Exiting from the system

7.2 VBD Editor

7.2.1 TRISTAR File
- Open : Opens a previously saved VBD file.
- New : Creates a new VBD file.
- Save : Saves the current VBD file.
- Save as : Saves the current VBD file with a user given name.
- Statistics : Displays a window containing the main information on the present VBD.
- Summary : Displays a scrollable window containing the complete information on the present VBD.
- View Virtual Building : Displays a 3-D view of the virtual building structure representing nodes and arcs with solid symbols (cubelets, spheres and pipes).

7.2.2 TRISTAR Navigation
- Exit: Exit from the VBD editor environment and returns control to main menu.
- Run CFAST : Executes the CFAST program on current VBD and then allows to check the output main file.
- Run EMS : Executes the EMS program on current VBD and using the result from the execution of CFAST. Then allows to check the output main file.
- View Results : Transfers control to an application displaying the modelling results in tabular and in graphics form.

7.2.3 TRISTAR Options
- Edit Options : Gives the user the means for setting a series of parameters related to how the VBD is represented (colours, etc.)
- Select Database Files : The user is prompted to select which database files to use for building the VBD (e.g. thermal properties, objects, etc.)
- Edit Defaults : Gives the user to set the default values used in the data entry windows (e.g. the height of compartments, etc.)

7.2.4 TRISTAR Virtual Building Editor

The Virtual Building Editor Functions are the ones used to construct and update a VBD and to assign all the data required for modelling.

- New : This command is used to ask in a menu based way to insert a new element in the VBD. The immediate effect is displaying a list of the possible types of elements to be inserted. The user selects the type of element to be inserted and this causes the popping up of the relevant dialog box. After filling the dialog box and accepting the contents (pressing OK) the element is inserted in the data structure and displayed at the upper left corner of the drawing window. The user is then allowed to relocate and resize the element on the screen.
- Show VDB cell picker : This command causes the appearance of a small window containing the pictorial symbols of all the possible elements that can be added to the VBD. This window can be closed using its standard button on the left of the title bar. This tool provides an alternate and generally more handy way of doing the "new" operation. The user clicks on an element symbol and gets the symbol "attached" to the mouse pointer in order to position it. Positioning is done by clicking the 1st mouse button at the proper place in the drawing window. Following the placement of the image, the relevant window is popped up asking for the elements parameters. The object sizing can be done following the clearance of the dialog box input.
- Select by mouse : This command gives the possibility to select an element by the mouse in order to edit its properties or to delete it. The user is provided with the mouse pointer function of object selection and following the clicking over an element displays the relevant dialog box allowing the user to change the field values or to delete the element.
- Select by List : The command is equivalent to the former one and differs in the fact that the user is prompted by a list of types of elements (including the ones for which a graphical representation does not exist). The user selects the type of element and this causes the list of such element to be displayed allowing the selection of the individual element to be edited or deleted through the dialog box.
- Show Links : This command has the effect of generating coloured lines which shows the linkages between the object on the screen. Different colours are used for the compartment arcs and for the ventilation connections. The function is used to verify that the connections are correct.
- Hide Links : This command clear the connection lines drawn by the previous command.

7.2.5 List of the VBD elements

- This item contains the list of the elements of the VBD which are managed by the above VBD editing functions.
- Compartments Nodes : Compartment Nodes are the actual rooms, corridors, landings, etc.
- TRISTAR has in addition to CFAST the data on the position of the room in the building, a room name and data on occupancy.
- Compartments Arcs : Compartment Arcs are corresponding to the openings between compartments and are relevant for both fire & smoke modelling and evacuation simulation. They correspond to the vents defined in CFAST by the keywords CVENT, HVENT and VVENT. More data are present in TRISTAR than in CFAST for the vents in order to represent the physical position of the vent in the compartment and to give information related to the dynamic of evacuation through the arc.
- Ventilation Nodes : Ventilation Nodes are used to represent the mechanical ventilation system and are just connection points characterised by a height. They are defined in CFAST by the keyword INELV. A physical representation is used in TRISTAR for ventilation nodes.
- Ventilation Openings : Ventilation Openings are the openings in a compartment to which a Ventilation Node is attached. They are the objects defined by the keyword MVOPN in

CFAST. More data are present in TRISTAR than in CFAST for the Ventilation Openings in order to represent their physical position in the compartment.

- Ventilation Ducts : Ventilation Ducts are the ducts connecting Ventilation Nodes. They correspond to the MVDCT keyword defined in CFAST. More data are present in TRISTAR than in CFAST for the ventilation ducts in order to represent their physical position and pathway.
- Ventilation Fans : Ventilation Fans are the ones defined by the keyword MVFAN in CFAST and they are attached to Ventilation Nodes. More data are present in TRISTAR than in CFAST for the Ventilation Points in order to represent their physical position in the building.
- Fire Node : A certain single symbol is used for the user to position the fire node and a dialog box is provided to edit the data required by CFAST through the keywords CHEMI, CJET, FAREA, FHIGH, FMASS, FPOS, FQDOT, FTIME, LFBO, LFBT, HCN, HCL, CT, HCR, O2, OD and CO.
- Objects : CFAST "objects" can be positioned at certain points. Their name is specified and their burning behaviour is fetched from a database. It is the equivalent of the CFAST keyword OBJECT.
- Environment : A dialog box is available in order to edit the environmental data which are currently required by CFAST by the keywords EAMB, TAMB, WIND.
- Modelling Parameters : The user has the possibility to enter and edit the following information: Restart file information; Run Version Identification; Time intervals for simulation, print, history, display; Printing control information; Output files names

7.3 Single Run Viewer

This is a second application running in the CAD environment. It is called by the VBD editor and works on the output files generated by running CFAST and the EMS program executed from the VBD editor.

The run viewer resembles a tape recorder that allows to take a dynamic look at the modelled fire history, to take snapshot stills and to browse back and forth in time.

Further functions are provided to define the building view and the data to be displayed.

7.3.1 Set Modelling Movie View

This menu section is devoted to setting how the building view will appear and which data will be displayed, how and where.

compartments colours coding

The user is first asked to select the parameter being used for the colour coding of the virtual building compartments. Then the user is asked to define the range limits corresponding to each colour code and the colour to be used for each of such ranges.

data display tags

the user is prompted with the following sub-menu:

- general setting
- new
- edit
- delete

If **general setting** is selected a dialog box is displayed giving the chance to the user to set automatically the display of data tags for all the compartments.

If **new** is selected the user selects the relevant compartment by the mouse on the drawing or selecting it in a list displayed in a dialog box. Then the user interacts with the Data Display Dialog Box to select the data to be displayed and its format (pls. refer to Dialog Box section in this document). At last the user uses the mouse to position the data display tag on the drawing.

If **edit** the user selects the relevant compartment by the mouse on the drawing or selecting it in a list displayed in a dialog box. Then he is asked to select the Data Display Tag between the ones defined for that compartment in a list displayed in a dialog box. At last the user interacts with the Data Display Dialog Box (pls. refer to Dialog Box section in this document) and is finally allowed to move the Data Display Tag to a new position.

If **delete** is selected the user selects the relevant compartment by the mouse on the drawing or selecting it in a list displayed in a dialog box. Then he is asked to select the Data Display Tag to be deleted in a list displayed in a dialog box between the ones defined for that compartment.

7.3.2 data vs. time graphs

the user is prompted with the following sub-menu:

- graph auto arrange
- new
- edit
- delete

If **graph auto arrange** is selected a dialog box is displayed giving the chance to the user to arrange a certain number of graphs according to standard patterns (e.g. a single row on the display bottom, etc.). The user can define for each graph which data to show.

If **new** is selected the user selects the data to be displayed in a list inside a dialog box and specify the graphs details (scale etc.). Then the user interacts uses the mouse to position the Graph Box on the drawing.

If **edit** the user selects the relevant graph box by the mouse or within a list in a dialog box. The user is allowed to modify the display parameter of to reposition it by the mouse.

If **delete** is selected the user selects by the mouse on the drawing or selecting it in a list displayed in a dialog box which is the Graph Box to be deleted.

legend

The user is asked to define the format of the legend to be displayed in a title box and then position it by the mouse on the display.

titling

The user is asked to define the text to be displayed in a title box and then position it by the mouse on the display.

7.3.3 Modelling Movie Recorder Control

This menu offers a choice of control commands

- run : starts the movie view from the beginning or after a pause
- pause : pauses the movie view
- step forward : advances to next one time interval in pause mode
- step backward : goes back to previous time interval in pause mode
- fast forward : advances at maximum run speed
- fast backward : goes backward at maximum run speed
- rewind : resets the movie to its beginning entering in pause mode
- go to time : goes to pause mode at a time given by the user
- movie speed : sets the actual display speed of the simulation movie
- sequence start and stop times : defined the time interval of interest in the complete movie file
- dump view file : creates a graphic image file of the present view
- print / plot : prints or plots a graphic image of the present view

7.3.4 Selected Data Output

view result files : enters a file viewer to examine the tabular output of the simulation programs

view data vs. time graph : enables the user to display a sizeable window where he can display any single or multiple line graph of any variables as a function of time

7.3.5 TRISTAR Navigation

exit : returns control to the main TRISTAR menu

VBD editor : returns to the VBD editing application

7.3.6 Options

save settings file : saves all the set parameters for movie viewing into a file

load settings file : loads all the set parameters previously saved for movie viewing into a file

print / plot selection : selects the output physical device and allows to set the relevant parameters

graphic display defaults : prompts the user with a dialog box allowing the definition of a number of default choices for the movie images (colours, etc.)

7.4 Multiple Input Generator

This program is used to generate series of VBD files where certain parameters are varied so that the user can perform sensitivity studies and randomised calculations.

In practice the program just asks to the user how may files to generate and they common name. The information for the base case and for the variation of each field is contained in the VBD file prepared in the CAD environment.

Additionally the user specifies which data to keep for the modelling results in order to avoid the generation of a monstrous output data set.

7.5 Statistical Post-Processor

This program is used to examine the data produced by the multiple run manager.

The user has a number of tabular and graphics utilities including the following ones:

- statistics on a certain output parameter including histograms
- correlation analysis and correlation (scatter) plots for any two output parameters
- correlation analysis and correlation (scatter) plots for one input and one output parameter
- multivariate analysis of input and output data sets
- generation of summary tables

14 THE PROBABILITY OF PROGRESSIVE FIRE PROPAGATION IN COMPLEX SYSTEMS

D.A. ONISHCENKO
VNIPIMORNEFTEGAS, Moscow, Russia

Abstract
This paper presents a model describing progressive fire propagation in a discrete multicomponent system with combustible elements. Each element of the system can catch fire with some probability which depends on fire intensity and element's fire resistance, where the last is supposed to be random.

The propagation of fire over the system is characterized by means of a stepwise procedure. The ignition of the next element or elements results in load increase for elements that are not yet in flames. If the intensity of fire load exceeds the corresponding value of fire resistance for some elements, these ones are considered to be ignited at this procedure step. With a view of obtaining conservative results, the conjecture of instantaneous ignition is adopted. If at any step none of the elements catches fire, this means the fire localization; otherwise, the progressive fire covers all the system.

The principal characteristic of fire emergency for the system is the probability of progressive fire propagation. It is shown that an analytical recurrence relation for the required probability can be found. Previously, similar approaches were used in the progressive collapse analysis of redundant engineering structures. As an example, the results of calculations for model systems are presented. Some directions for future research are also proposed.
Keywords: analytical estimation, complex system, fire path, probabilistic approach, progressive fire propagation.

Fire Engineering and Emergency Planning. Edited by R. Barham.
Published in 1996 by E & FN Spon. ISBN 0 419 20180 7.

1 Introduction

The process of fire propagation in a complex system composed of many elements (objects) is often characterized by successive ignition of the constituents. For systems in which the origin of fire is a negative phenomenon, it is desirable to have a localization property, e.g. fire stop following the ignitions of some elements. When analyzing complex systems, the probabilistic approach is used as adequate one. In this case the fire resistance of a system is characterized by the probability of fire localization, while its fire emergency is characterized by a complementary probability, with respect to 1, of the beginning of global fire covering all the system.

Note that the probabilistic nature of fire propagation process can be caused by various reasons, including random character of loads, scattering of element fire resistance, the presence of randomness in an ignition criterion and so on. Besides, if the system consists of similar elements, the probabilistic setting of the problem is one of the approaches used for proper description of multivariant progressive failure process in complex systems [6].

In many cases, the process of fire propagation and, more general, the process of progressive failure has a hierarchical feature. It is caused by natural and artificial hierarchy of system's structure as well as internal properties of the process [7]. As a result, some intricate and interesting problems arise. An example of the study of such a problem may be found in [5], where one model system is analysing. The present paper proposes a probabilistic model which describes a fire propagation process in complex system with combustible elements under the condition that all elements are of the same hierarchy level. The parameter of element fire resistance is supposed to be random. A recursion relation for the calculation of the desired probability is analytically found. Some results of model system analysis are also presented.

2 The principal features of the model

Consider a system being a collection of n elements arbitrary enumerated. Let us assume that the elements of the system are combustible and that after their ignition they influence on non burning elements by means of fire loading. We assume that the loads have prescribed values which may differ for different elements. When some elements are in flames, the corresponding loads are summarized.

Let us suppose further that every element has the property to withstand fire loading. The corresponding quantitative measure is characterized by fire resistance that we will designate r. Under some load, the ignition of non burning element can only occur if the load intensity q exceeds the fire resistance r. So, the ignition criterion takes the form of

$$r < q . \qquad (1)$$

Now describe the process of fire propagation over the system. Let us assume that the system, in its initial state, is subjected to any external influence which may cause the ignition of the elements with some definite probability. Note that in the capacity of such influence we can also consider the action from the element that accidentally caught fire.

The process of fire propagation will be analyzed with the help of step by step techniques. At the first step we select all those elements whose parameters of fire resistance are smaller then the intensity of the load on them. According to the criterion (1), such elements must catch fire. At the next step we determine loads on non burning elements. Their values equal the sum of the initial load and the loads caused by inflamed elements. It is clear that the new values of the loads are, at any case, no less then those at the preceding step. We find again those non burning elements that are overload in accordance with (1), and then they are regarded to be in fire. And so on.

The stepwise procedure described may come to its close in two ways. In the first variant, the consecutive ignition of all elements will occur after a number of steps. In another one, the step will be found where the criterion (1) will not be satisfied for any non burning element, i.e. the process of fire propagation will stop and the fire will be localized. Such states we will call the stationary ones.

Let us make an important remark. For the sake of simplicity, we will consider the case when the ignition criterion takes into account the intensity of loading only, but not its duration. In this paper the subject of inquiry is the systems with instantaneous ignition of elements, that is the limiting case. The analysis of such systems has usually resulted in conservative estimations when one investigate the question of fire resistance of complex systems.

For the convenience of further presentation, refer a sequence of the elements ignitions as a fire path. Then introduce the notion of a critical state of the system. First of all, the ignition of all elements of a system is, obviously, a critical state. Besides, there may exist such states that, by virtue of some reasons, can be considered as identical to the full system collapse. The critical states can be characterized by the fact of the ignition of the most important elements or by the event in which the ignition of preassigned, generally speaking, great enough number of elements take place. Below, we suppose the collection of critical states to be defined in advance.

We say now that the fire resistance of a system is ensured under given external load and for given values of elements fire resistance, if the relevant fire path will not lead to any critical state. Otherwise, the fire resistance is considered to be not ensured.

The above presented model is not of any interest, when all parameters of the system and the loading are determinate quantity. On the other hand, if at least some of the parameters are random, then the problem of the system fire resistance estimation complicates significantly.

We will not treat the question of random loading and will regard the loads as determinate parameters. As to fire resistance parameters, let us suppose that they are random and are characterized by given distribution functions.

In this case, a fire path is a random entity. Indeed, at any step of the procedure describing the process of fire propagation every non-burning element may catch fire - with some probability. Hence, the consequence of elements ignitions is random. It means that the fire path can reach any critical states with some probability only. Within this approach, system fire resistance may be quantitatively characterized by the probability of the event that the fire path will come to a stationary state.

Note that the described statement of a question is similar to a considerable extent to that widely used in the study of engineering structures reliability [1,3]. The common feature of these two problems is the presence of 'redundancy factor': ignition (failure) of one or several elements does not immediately lead to system collapse, since the system continues to operate in a damaged state. The outlined probabilistic model of fire propagation is similar, in its common features, to the models that were earlier applied by the author to the analysis of carrying capacity and reliability of complex technical systems [8,9].

It may be shown that, within the scope of the model presented, the required probability can be found with the help of analytical methods in a form of a recursion relation (See Appendix A). This relation defining the quantity complementary with respect to 1, i.e. the probability of global fire emergence, may be written as

$$q^{(n)} = L(q^{(n-1)}, q^{(n-2)}, ..., q^{(1)}) , \qquad (2)$$

where L is a linear function, and $q^{(n-j)}$ is the probability of global fire emergence for a subsystem in the initial state of which j elements are already in flames. The coefficients in (2) depend only on the probability of elements ignition when the system is in its initial undamaged state. The formula (2) is recurrent, and the probabilities $q^{(n-j)}$ for all j are determined by appropriate initial states.

It should be emphasized that despite a relatively simplicity of the deciding relation (2), the execution of the relevant calculations will entail great difficulties, a part of which are, however, typical, when one programmers recurrent formulae. First of all, as n increases, the number of treatment of the formula (2) is enlarged abruptly. For example, the number of different system states considered as initial is of order 2^n, that raises the required computer time up to the inaccessible level when a straightforward algorithm is used even on n = 20-30. The other obstacle is associated with the dangerous of loss of significant figures. This is caused by the next reason. The terms in the right side of (2) have alternating signs and large absolute values. At the same time their sum, being a probability, should be of order 1 and below. Because register length and, hence, number precision in computers are bounded (for instance, the latter equals 20 for Turbo Pascal), some significant digits may be lost even on moderate values of n .

The elaboration of appropriate stable and effective algorithms for n large enough demands an additional investigation. Nevertheless, some recommendations can be stated just now. Firstly, the cases are often met when if the ignition of several elements occurs, the load on the other elements abruptly increases. The process of fire propagation becomes of avalanche type. So, one may assume that if the fire covers more then some definite number of elements (for example, 10 or more from total 50), then all the probabilities $q^{(m)}$ with m < 40 may be taken at once equal to 1 instead of their direct computation. The necessary enumeration of various system states will be principally reduced.

Secondly, the qualitative analyses of possible fire paths in a given system may lead to a conclusion that some of them can't be realized at all, or may be realized with very little probability in comparison with others. In this case there are good reasons to put the corresponding quantities $q^{(m)}$ equal to 0, since their full accounting would give a little correction only. Note that such states may be numerous. The similar approach has been earlier applied repeatedly in analyses of structural reliability [4,10].

3   The results of a model system analysis

As an illustration, we will present some results concerning the systems composed of uniform elements. The uniformity we will treat here as an identity of all determinate parameters of the elements. As to random characteristic of fire resistance, we assume that for all elements it obeys to the same probability distribution.

Determine some parameters of the system analyzed. Let the system contains n elements. We define loads on the elements in the following way: 1) the initial load has the intensity s_0; 2) when an arbitrary element catches fire, the load on each of the remaining (non-burning) elements increases by the value s_1. Thus, if m elements are in flame, the remaining n - m elements are influenced by fire load with the intensity $s_0 + ms_1$. Let F(x) with $0 \le x < \infty$, be a distribution function of elements fire resistance which is a random parameter.

In such a situation the common relation (2) may be rewritten in a considerably more short form:

$$q^{(k)}(x_k) = \sum_{i=1}^{k} (-1)^{i-1} C_k^i [F(x_k)]^i q^{(i-1)}(x_{i-1}), \quad k = 1, ..., n;$$

$$q^{(0)} = 1,$$

(3)

where $q^{(k)}(x_k)$ is the probability of global fire emergence in a system consisting of k elements under initial load x_k; C_k^i are the binomial coefficients equal to $k!/[i!(k-i)!]$, and x_k equals $s_0 + (n-k)s_1$.

Note an interesting analogy with the model describing the classical problem on the strength of fiber bundles resolved by H.Daniels [2]. It was found [11] that the relation, derived previously by Daniels, may be written in another way, similar to (3). In Daniels model the quantities $q^{(k)}$ equal the probability of rupture of a bundle consisting of n fibers; the function F(x) describes the strength distribution for individual fiber, and x_k are defined slightly differently:

$$x_k = (n/k)s_0,$$

where s_0 is a specific load (per fiber) on a bundle. The total load is equal then to ns_0. The main qualitative difference between the two models is that as fire (failure) propagates, the load on remaining elements increases linearly in our model, while in Daniels model - in accordance with the hyperbolic law. Below, some numerical results in comparison for both models are presented.

Let $s_1 = s_0$ in the fire model; this corresponds to the case when the initial external influence is caused by some burning element. In this case, the initial loads in both models are the same. As a distribution of random parameter of fire resistance (strength) of elements we take the Weibull one:

$$F(x) = 1 - \exp[-(x/x^*)^m],\tag{4}$$

where x^* and m are some constants. Let for definiteness $x^* = 1$, $m = 2$. In Table 1 and corresponding Fig.1 some relevant results of the calculations are given.

Table 1. The probabilities of system collapse

Initial load	Fire model			Model of a bundle		
	n = 1	n = 10	n = 40	n = 1	n = 10	n = 40
0.1	$9.95 \cdot 10^{-3}$	$1.26 \cdot 10^{-1}$	$2.22 \cdot 10^{-1}$	$9.95 \cdot 10^{-3}$	$<10^{-4}$	$<10^{-4}$
0.2	$3.92 \cdot 10^{-2}$	$1.67 \cdot 10^{-1}$	$7.85 \cdot 10^{-1}$	$3.92 \cdot 10^{-2}$	$<10^{-4}$	$<10^{-4}$
0.3	$8.61 \cdot 10^{-2}$	$5.56 \cdot 10^{-1}$	$9.36 \cdot 10^{-1}$	$8.61 \cdot 10^{-2}$	$9.20 \cdot 10^{-3}$	$1.60 \cdot 10^{-4}$
0.4	$1.48 \cdot 10^{-1}$	$7.93 \cdot 10^{-1}$	$9.85 \cdot 10^{-1}$	$1.48 \cdot 10^{-1}$	$1.21 \cdot 10^{-1}$	$9.79 \cdot 10^{-2}$
0.5	$2.21 \cdot 10^{-1}$	$9.18 \cdot 10^{-1}$	$9.90 \cdot 10^{-1}$	$2.21 \cdot 10^{-1}$	$4.31 \cdot 10^{-1}$	$7.34 \cdot 10^{-1}$
0.6	$3.02 \cdot 10^{-1}$	$9.73 \cdot 10^{-1}$	1.00	$3.02 \cdot 10^{-1}$	$7.56 \cdot 10^{-1}$	$9.89 \cdot 10^{-1}$

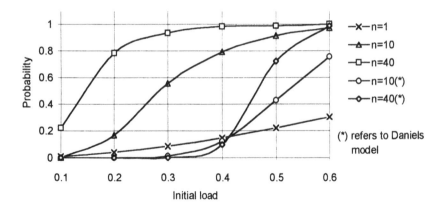

Fig. 1 The probabilities of system collapse

It should be noted, firstly, that under various loads the probability of system collapse for fire model are essentially greater then for Daniels model, in particular, in the range of small probabilities. Secondly, as to the fire model, one can see that as the system size n increases, the distribution of the system fire resistance gradually narrows (the distribution function becomes steeper) and shifts to the left. This means that over the examined range of the parameters s_0 and n, the raising of the probability of global fire emergence is observed when n is growing. Is this tendency holds or not up to the smallest values s_0, will be established subsequently.

In connection with this note that in Daniels (bundle) model there exists the asymptotic threshold value s^* of specific initial load. When n is large enough, the probability of bundle rupture is close to 0 for $s_0 < s^*$ and to 1 for $s_0 > s^*$.

4 Conclusion

In this paper, the model is presented which describes the process of fire propagation in a system, consisting of discrete combustible elements. Fire resistance of elements is specified by means of probabilistic distribution, so the system fire resistance is characterized by the probability of fire localization. An analytical calculation formula is found for the probability of global fire propagation. As an example, some results of calculations for a model system consisting of similar elements are presented. A comparison was made with the well-known Daniels model of fiber bundle, when the fibers have random strength.

In the future the following questions are planned for treatment:
1. elaborate a computer code realizing the calculations on the formula (2) in a general case;
2. investigate the process of fire propagation in complex systems with regard for fire load reducing as a distance is increasing;
3. consider a case of fireproofing setting;
4. carry on the analysis of possible stable regimes of fire propagation in uniform systems with a great number of elements;
5. develop a fire propagation model with consideration for influence of fire duration on the elements of the system.

References

1. Bennett, R.M. and Ang, A.H.-S. (1986) Formulations of structural system reliability. **Journal of Engineering Mechanics**, 112(11), 1135-1151.
2. Daniels, H.E. (1945) The statistical theory of the strength of bundle of threads. I. **Proceedings of Royal Society of London**, 183(A), 405-435.
3. Ditlevsen, O. and Bjerager, P. (1986) Methods of structural systems reliability. **Structural Safety**, 3(3-4), 195-229.
4. Feng, Y. (1988) Enumerating significant failure modes of a structural system by using criterion methods. **Computations and Structures**, 30(5), 1153-1157.
5. Goldstein, R.V. (1995) An hierarchical approach to fireresistence improvement of complex technical systems, in **Proceedings of European Symposium on Research and Applications in Fire Engineering and Emergency Planning.**
6. Liebowits, H. (Ed) (1968) **Mathematical Fundamentals**, vol. 5 of **Fracture**. Academic Press, New York and London.
7. Marshall, V.C. (1987) **Major Chemical Hazards**. Ellis Harwood Ltd. Publisher, Chichester.
8. Onishcenko, D.A. (1994) Analytical estimations for the failure probability of a redundant complex technical system in quasistatic approximation. **Problems of Safety**, in press (in Russia).
9. Onishcenko, D.A. (1992) Fault-tolerant building structures - analysis and research methods, in **Reliability, Survival and Safety of Technical Systems** (Proceedings of scientific seminar, 3-4 March, 1992, St. Petersburg), St. Petersburg, 1992, pp. 102-107 (in Russia).
10. Rashedi, R. and Moses, F. (1988) Identification of failure modes in system reliability. **Journal of Structural Engineering** (USA), 114(2), 292-313.
11. Vizir, P.L. (1981) An estimation of reliability of parallel structure with regard for load redistribution. **Structural Mechanics and Analysis of Structures**, (1), 15-18 (in Russia).

Appendix A. The defining formula

The model of fire propagation presented here is quite similar, from the mathematical point of view, to that describing the developing process of complex technical system failure, the investigation of which was carried out by the author in [8,9]. If one takes advantage of the approach there stated, the following relation defining the probability of global fire emergence may be obtained for the case when fire load on the elements that are not yet in flames increases:

$$q(I_n) = \sum_{i_1=1}^{n} a_{i_1} q(I_n \setminus \{i_1\}) - \sum_{i_1,i_2=1}^{n} a_{i_1} a_{i_2} q(I_n \setminus \{i_1,i_2\}) + \ldots + (-1)^{n+1} a_1 \cdots a_n. \quad \text{(A1)}$$

Here:

- the symbol $\sum_{i_1,\ldots,i_m=1}^{n}$ means that in the sum all the collections of m mutually exclusive integer from the set I_n must be taken;

- $a_j = F_j(s_{0j})$, $j=1,\ldots,n$, where s_{0j} is the load on the element j in the initial system state, and $F_j(x)$ is a distribution function of fire resistance for this element;

- $q(I_n \setminus \{i_1,\ldots,i_m\})$ is the probability of global fire emergence, under appropriate load, in the system consisting of the elements with the numbers from the set $I_n \setminus \{i_1,\ldots,i_m\}$;

- formula (A1) is recursion. To calculate the quantities $q(I_m)$ with $I_m = \{i_1,\ldots,i_m\}$, where $m < n$, it is necessary to substitute the set I_m in place of I_n if I_m does not relate a critical state, and otherwise to put $q(I_m) = 1$.

STRATEGIC EMERGENCY PLANNING

15 EMERGENCY PLANNING

T.J. SHEILDS
Fire SERT Centre, University of Ulster, Jordanstown, UK

Abstract

Planning for a disaster or emergency is complex since they are largely unscheduled. In this context, issues are raised and questions posed regarding the current status of disaster planning. Since current planning relies heavily on training exercise designed to cope with known types of disaster, it is suggested that they may be too stereotyped and as such not sufficient or acceptable.

It is also proposed that since the age of quality control has arrived, disaster emergency planning be made more transparent and rendered capable of being assessed.
Keywords: disasters, crisis management, emergency planning, training, quality assessment.

1 Introduction

Emergency planning for the purposes of this symposium can be considered as a process by which civil authorities prepare for a range of possible disasters and will include:

- identification and analysis of the potential hazards and if possible the mitigation or elimination of their consequences,
- analysis of the resources available to cope with any potential disaster, and
- post disaster response and recovery planning.

The primary purpose of emergency planning is not the acquisition of additional resources but rather the anticipation of foreseeable problems and development of achievable possible solutions within existing constraints. Basically the process includes:

- analysis of hazards and resources,
- identification of necessary post disaster tasks,
- allocation of these tasks to organisations and individuals,

Fire Engineering and Emergency Planning. Edited by R. Barham.
Published in 1996 by E & FN Spon. ISBN 0 419 20180 7.

• co-ordination of the planning of all responding groups to ensure a cohesive and effective response.

In this connection, disasters may be categorised as:

• natural,
• technological,
• ecological.

The problems which arise with respect to emergency planning are all the more acute because the cause, timing, location and extent of a future disaster is by definition unknown. Further and thankfully, the vast majority of emergency plans are never subjected to the heat of battle in a real life emergency. It follows that the mere existence of an emergency plan in itself, may not be sufficient to cope with a real emergency event. In one report of an emergency, it was concluded that '**local governments continue to be surprised when standard procedures in their lengthy plans prove irrelevant in the real disaster**'. Here it must be understood that emergency planning operates at several levels, eg, when national security is threatened.

2 Frequency of Disasters

Through the 1980's and early 1990's in the United Kingdom it would appear that the occurrence of serious incidents has significantly increased. The collapse of syndicates at Lloyds of London would support this hypothesis. Given such a trend, four questions arise:

• are disasters predictable?
• what are the underlying causes?
• were the emergency plans adequate, effective and sufficient?
• can disasters be rated in an acceptable universal scale?

3 Disaster Conceptualization

In the actual preparation of emergency plans how do the planners identify, organise and hence conceptualise an otherwise collection of rare diverse events? Unless the planners have actual experience of a disaster it is probable that the disaster scenarios envisaged will be conceptualised in terms familiar to the planner, ie, accidents. This perhaps explains why after "successful" trials of emergency procedures the planners are inevitably pleased and optimistic. Are planners prepared to actually countenance failure, and acknowledge that events can and will occur, with which they may not be able to cope? Acknowledging that the impact of a potential disaster may have a local or wide zone of influence, it may be necessary to consider that the nucleus of the emergency services may have been located at the epicentre of the emergency. Since the majority of disasters are unscheduled and capable of causing various levels of chaos, it is left to the imagination as to the frenetic nature of the activities contingent upon the occurrence of such an event. Thus stereotyped emergency planning may not be sufficient or acceptable in today's modern world.

4 Disaster Management Syndrome

There are few who are not aware of the theatrical expression "it will be alright on the night". This notion suggests that rehearsal after rehearsal with practically everything going awry, the cast and the entire company will somehow on opening night contrive to give a faultless performance.

On the other side of this particular coin is, that a disaster might have been averted, or its effects minimised if the crisis manager had not believed that his emergency plan was entirely flawless. Is it a matter of "those who believe do not need to understand"?

Closely associated with the disaster management syndrome is an increasing dependence on technologies which in effect sideline the crisis manager for periods of time during which crucial decisions may have to be taken. How many such managers will in the future be confident enough to hit the manual override button and take control? With the increasing use of computers in emergency response planning, how many back up systems are required and how confident are we that the software has been completely debugged?

5 Issues of Concern

5.1 Information systems
There is much ongoing work with regard to the development of different types of information systems. Questions arise such as:

- do the eventual end users of information systems understand the objectives of the software?
- do the end users of information systems understand the algorithms contained in the software?
- does the software force decisions and are the decisions always right? Who decides?
- who is sufficiently competent to assimilate, interpret, cumulated information which on occasions may be extremely sparse?
- in the context of crisis management is good news always welcome?

5.2 Training
Is it possible in training to overcome optimistic mind sets?, ie:
- is training actually realistic - can it be?
- are worst case scenarios used?
- why assume competency in front line management whose training may have little or no bearing on the reality?

- is there is a need to develop agreed structures for the assessment of training methods?

5.3   Crisis Management

It is clear from the literature that the division of responsibilities at national, regional and local level can affect the nature and effectiveness of the immediate response in an emergency. There are many issues here to be addressed:

- which Government department or departments are responsible and for precisely what?
- how is co-ordination effected?
- is the management system hierarchical?
- do people operate on a need to know basis?
- what if a link in the system is missing?
- at national/regional/local levels, who is actually in control?
- is control vested in one individual? There is evidence to suggest that crisis management teams are more effective.
- how are tasks allocated; roles and responsibilities determined?
- how is incoming information collated, assessed and disseminated in digestible "information packages"?.
- how are the "information packages" prioritized? Too much information sometimes leads to confusion.
- there is clear evidence that in many situations there is an appalling failure to share information - in one case in Melbourne, Australia, there was a better exchange of information between some people on the ground in Melbourne and experts in Norway than with colleagues in Melbourne.
- is it sufficiently understood that "normal management systems are too inflexible and too slow to cope with a disaster"?
- there is evidence to suggest that leadership and management styles change during a disaster - are such learning curves (if that's what they are) acceptable/affordable?

5.4   Media Management

Most authorities tend to agree that crisis management should include a strategy of dealing with the media and communications. In any crisis situation, factual information is preferable to rumour and it is clear in the literature, that the supply of misinformation to the media has often back-fired with sometimes devastating affect on those involved. However, it appears that there has been little attempt to identify aspects of crisis management where the media can positively contribute. Further, the overall impact of such contributions in mitigating the potential outcomes of a disaster have not been quantified.

5.5   Emergency Planning and Crisis Management

The literature is almost silent on the contribution of the planning for, and the actual management of, an emergency to the final outcomes. This silence suggests that good planning and management are assumed. In many cases were it so, this was soon to be an assumption too many. Notwithstanding, there is no universally agreed method of assessment for emergency planning and crisis management. In an age which has seen the advent of total quality control, such a basic requirement should not be left unsatisfied.

6 Concluding Remarks

The purpose of this paper is to raise obvious issues of concern and to hopefully generate useful discussion. The issues raised due to transnational complexities are necessarily general and as such serve to illustrate underlying national and regional complexities. For example, is it well known that if the weather conditions had been different, the Lockerbie air disaster might well have happened somewhere in Ireland. One final issue to be addressed is simply that many disasters, for example, the Challenge space shuttle, are predictable but commercial pressures determine that the risk is worth taking. In many of our normal activities, who actually decides to take the risk of a disaster occurring?

16 EXPO '92: A REVIEW OF ITS EMERGENCY PLANNING INTEGRATED DESIGN

R. BARHAM
Department of Built Environment,
University of Central Lancashire, UK
R. FERNÁNDEZ-BECERRA
Jefe del Gabinete Técnico de Protección contra Incendios,
Gerencia Municipal de Urbanismo, Sevilla, Spain

ABSTRACT

This paper reviews the emergency planning provision of the EXPO'92 site in Seville. The site was intensively developed ab initio with a completely new infrastructure and a comprehensive set of new and unique buildings. Playing host to 110 nationalities and with some 85 individual building constructions, the developing authority took special steps to ensure that, in the event of a civil emergency, a natural disaster or a major outbreak of fire, its estimated 500,000 plus daily visitors would have the benefit of a well planned response. The constructed provision included new river crossings, evacuation areas, segregated traffic, segregated water supplies and unique fire regulations and materials specifications.

Introduction

In 1976, the idea of holding a universal exposition in 1992, to celebrate the discovery of America by Columbus, was put to the Paris Bureau International des Expositions (Bureau of International Expositions). The Bureau decided, originally, to split the universal exposition onto two sites and took as the subject the Age of Discovery - from the discovery of America to the latest technological discoveries. Chicago was allocated the exhibition for technological discovery and Seville that for historical and artistic discoveries. However, by 1985 local politics in the American town resulted in its withdrawal from the proposed twin venue EXPO and Seville was asked to take on the whole exposition. By 1983, Seville had almost completed its design for the smaller more compact cultural EXPO and so, when Chicago dropped out in 1985, it was necessary to redesign and expand the original proposals with very little time remaining. The change also placed responsibility for the whole universal exposition onto the local development team in Seville which suddenly found it necessary to revise its design provision of buildings from 300,000m² to 650,000m² to accommodate the increase in participating nations from 60 to 111.

Fire Engineering and Emergency Planning. Edited by R. Barham.
Published in 1996 by E & FN Spon. ISBN 0 419 20180 7.

Background to the Site

The first priority for Seville was to find an appropriate site. For site selection, there are always at least two alternatives: i) a location where the infrastructure is already in place and all that is necessary is to remodel and/or recondition it, or ii) to go somewhere where there is nothing and start from scratch. Where there is infrastructure, there is usually life. To expropriate land for a development is socially very difficult, so Seville chose the second option.

Seville exists because of its river, the Guadalquivir. Like in London or Paris, the river is what makes the city, but in contrast to Paris and London where the river is an urban route, Seville has its back to the river - it is no longer an economic axis; the port of Seville moved down to the south of the city many years ago and large ships no longer come upriver.

Floods in 1973 had re-opened a previously silted up dead meander and left a large triangular "island" to the south of the city. This so-called island, of some 3,000,000m², was a public area (no man's land) and was the property of the city. It was unused and since it was subject to occasional flooding, there was nothing there. No-one had built; there were no trees, no water; no electricity; no natural gas network and no sewerage. The only structure on the site was the Monastery of Santa Maria de las Cuevas, a monastery of the Carthusian or charterhouse monks, built about 1460. As a result, Seville decided to create the universal exposition there - on the "island" of La Cartuja on which, co-incidentally, was the monastery where Columbus had prepared for his trips.

To create an EXPO on such a site, is a challenge to any professional. In 1987, the team met for the first time. Thirty-five strong, it consisted of exposition experts, architects, designers, landscape architects and, of course, fire safety experts. This executive team relied, for detailed advice, development and implementation of the emergency planning and fire protection proposals, on the Directorate del Programma del Construccion y Protección contra Incendios.

Emergency Planning

Coping with civil emergencies and disasters, security against fire (including fire safety for a building or a set of buildings) starts at the design stage. So, for example, if we design a city so that the primary concept of its traffic circulation routes is to provide ease of access for fire fighting, we would gain a fifty percent increase in fire safety. This would also make the response to other emergency situations much easier. In just such a way, the island chosen for the development of the EXPO'92 gave the possibility of designing the road layouts ab initio with emergency planning in mind.

The first decision was that the island of La Cartuja should be reconnected to the main area of the City of Seville by a new causeway. This causeway would split the island into two areas - the main expositions area and the area for support and expansion. That was one of the primary design constraints identified at the very beginning and imposed because the EXPO would introduce an extremely large number of people to an area where the seismic prediction of earthquake was up to 7 on the Richter Scale. In the event of an earthquake at any time during the six month's of EXPO, it would have been very difficult to evacuate the island because the only evacuation routes would have been over a small number of bridges and there would have been a collective mix situation - vehicular and pedestrian traffic. The bridges would have been bottlenecks. Any mass evacuation of the EXPO site would have needed to be of massive proportions and would have created havoc on the bridges. An alternative had to be found where people could be "evacuated" from the area developed with buildings without having to cross the water or pass through constricted egress points.

One of the primary access points to the island necessitated the construction of a new bridge. It was decided, therefore, to bisect the island site with an access road leading from this bridge, thereby segregating the public access area, which would be developed with the exposition's buildings, from the open and undeveloped area into which people could be evacuated. (*See Figure 1*) The original proposal from the highway engineers was to construct this road at ground level but, on the grounds of safety, it was constructed as an elevated highway so that people could pass under it and flee into the evacuation area through underpasses without having to fight their way through evacuating vehicular traffic. With the provision of a sufficiently large number of underpass connections people could easily be evacuated in that direction. That was the first and primary emergency planning consideration. The undeveloped evacuation area was also planted with saplings so that the resultant "forest" could provide a green "lung" for the locality.

The second major decision by the emergency planning team was to provide a helipad or heliport. This was considered to be essential for special evacuations. Whilst not mandatory, it was considered essential because the only medical facility planned for the site was a first aid unit (albeit a very large first aid unit). There was to be no purpose- built hospital and, therefore, it was decided that a helicopter pad was needed for evacuations to hospital. To clarify this point: in Spain, the provision of such a facility is not mandatory in either national or local law. Spain has a national ordinance for fire protection and another one for civil defence. The civil defence regulations do not require any specific provision, such as a helipad, for any particular type of development. However, it does provide a philosophy for the security of new developments and the protection of existing facilities. That being so, each developer is left to apply that philosophy (indeed, is required to apply the philosophy) to its own proposals. As a result, the design team for EXPO'92 concluded that the inclusion of medi-vac facilities was necessary and included the helipad area in the preliminary layout.

Road layout

These two primary decisions set the basic constraints or limits to the layout design within which the planners and the architects would have to work. Following a design competition, the master plan for the exposition was finalised in 1987 and this set out the formal and functional areas and produced the finally accepted road layout.

The plan was very simple: the internal site roadways, and the service roads or service lanes, mesh together in a so-called comb. The site is divided, internally, by a large canal (which has a sluice or lock down by the river) leading from an ornamental lake and a major roadway, called the Way of Discoveries, the main access which approximately divided the Spanish pavilions from the international pavilions. These divisions resulted in the creation of two zones: A and B, (the monastery itself was considered as a separate area, C). Around the lake was a crescent of pavilions occupied by the self-governing regions of Spain together with a couple of the theme pavilions and parallel to the river and central canal were the Pavilion of the Future, the Pavilion of Major Discoveries, and the Seafaring or Navigation Pavilion.

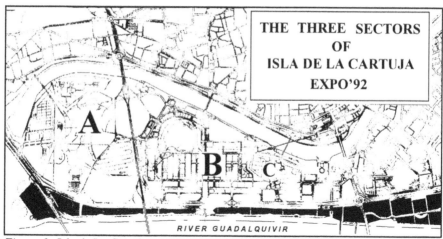

Figure 1 Isla de La Cartuja and the EXPO'92 site

In the international area (*see Figure 2*), it can be seen that the road layout is based on a double comb, two combs each fitting into the other, but each tooth of each comb having, to all intents and purposes, a dead end. One comb forms the basis of the pedestrian areas - traffic free avenues. The other comb forms the access for service - service lanes for vehicles and service access for pipes, etc. which then run into the backs of the pavilions. In this way, there is no interference between the two systems - pedestrian visitors and service vehicles, delivery trucks and emergency vehicles (ambulances, fire tenders, etc.) - the service warehouses, the fire station and the security offices, etc. were at the rear of the site.

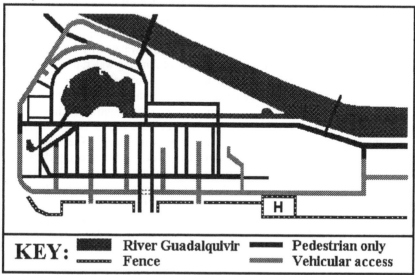

Figure 2   The segregated road layout in the international area

Fire-fighting Water Supply

In making a comprehensive development of a virgin site, it is possible to let the safety consultant step in at the design stage. In this way it is not only possible to control the design but is also possible to control budget costs. That is what was done in the run up to EXPO'92. A comprehensive emergency provision costs less money than if each building has its own safety standard because there are many things that can be done overall instead of building by building and thus economies of scale can be achieved.

As has been explained, the first major consideration was traffic. The second was the fire-fighting water. All cities have a fire fighting network but normally they share it with their drinking water supply and this latter is provided subject to the local water pressure available. For EXPO'92 it was decided that the network should be fire exclusive. So what was provided was a segregated network, drinking/fire, - in this way it was possible to give it sufficient pressure so that the fire hydrant could be independent and fires could be fought from the hydrant itself without having to through a pump vehicle or via an independent building-located emergency water supply storage. It was decided that, for a scheme like EXPO, it would be preferable to have 7 kilo of pressure - a minimum of 7 kilo. The unsegregated water supply in the city was not 7 kilo; it was, at most, about 4½ or 5 kilo. With a segregated supply of fire-fighting water, supplied via a ring main, it is possible to maintain pressure even if one wants to set one area or close it down to do repairs or maintenance (testing, etc.). Good design was necessary because of the decision to use hydrant-supplied water for fire-fighting. It was necessary, therefore, to locate fire hydrants so that all pavilions could be more than adequately covered by their output.

A further consideration, in the planning stage, was the quality of the water uptake for fire-fighting supply. It was decided, in the end that it should be the same quality as the local drinking water supply. The design team started off thinking about using the river water but analysis showed that it contained too many colloids. As a result, the sprinklers would be susceptible to becoming clogged up with impurities, mainly lime. If the quality required was the same as that of drinking water, why not use drinking water? The city of Seville had an adequate supply of drinking water; the only problem was that it was not at the required pressure and, because of leaks, etc., maintenance of constant pressure could not be guaranteed. This problem was resolved by the use of pumps to reinforce the pressure - to bring it up to the required 7 kilos pressure. Because a constant pressure was required, a fail-safe plan was needed. Contingency panning, therefore demanded the installation of back-up pumps. The final back-up installation consisted of two diesel oil sets and two electric sets and two jockey pumps - so that automatically on a reduction of the pressure in the ring main they would go into action. If the diesel sets were unable to deliver the required pressure, the electric pumps would cut in and, if there was co-incidentally, an electricity failure, then there was still a third set of pumps to provide the ring main with the required pressure of 7 kilos. The pumps, and the fire-fighting ring main, were provided with a secondary water supply from an 11 cubic metre reservoir fed permanently from the external water supply.

As a special safety precaution, in case there was an external interference with the water supply (e.g. by terrorist activity or major accident, etc.) and the supply of mains water was cut off, a second safety measure was a connection to the river water - so there would never be a lack of water, even though in this last eventuality it would not be clean. And in the case of all the pumps failing on the mains supply, which would be virtually impossible, the river supply could also be used, in which case the pressure available through the small riverside pumping station would be at 5 kilos. Consideration was given to providing filtration facilities for the river water but it was decided that the water coming in from the west side comes in relatively clean and having to use that facility was a very remote possibility. Because the normal feed for the fire-fighting ring main was to be clean water (from a reservoir) of drinking quality, the only filters that were installed on the water supplies on the site were put in for a different safety reason. They were installed to filter the cooling water for air conditioners. The fire-fighting ring main, itself, actually consists of a number of separately identifiable rings. They vary in diameter from 360 to 340 to 300cm. The main advantage of using a ring main is that it was then not necessary to provide any independent emergency water storage facilities in any of the pavilions. In this way, the cost was much lower than having to build reservoirs in the buildings. No pavilion had to do any pressure pumping because they had water at the required pressure already, thus saving a considerable amount on the initial construction expense.

At the planning stage, a worst case scenario was assumed. This was considered to be a fire-fight requiring the use of three hydrants running at full capacity and, in addition, two buildings with all their sprinkler installations also fully operational, i.e. three hydrants in use attending to the fire in a building, with its sprinklers also going, and the sprinklers fully in use in one of the adjoining buildings. Normally, according to the design, one hydrant would have had enough capacity to deal with a single building fire.

Conclusion

The emergency precautions built into the EXPO development scheme were not called into use during the six months that the exposition was open to the public; EXPO'92 passed without major incident. The one major fire that did occur, and which totally destroyed one of the pavilions, occurred during the construction phase. It was not, at that time, subject to the fire prevention controls of the EXPO Authority. Control and responsibility of the pavilion construction site still remained with the builder, as on all constuction sites in the majority of European countries. The problem, for the fire-fight, was that the service installations were not completed; the occupational health and safety law in Spain places responsibility on building companies to provide adequate fire prevention fire supression facilites on the construction site up to the time of building hand-over. In the circumstances the builder's provision proved inadequate

This incident apart, it can be seen that the contingency planning for EXPO was very detailed. The reason for this was a realisation by the EXPO Authority and its development team that, because they were dealing with an International Universal Exposition, any incident on the site, no matter how minor, would be international news. In the normal course of events, if there was to be a fire in Seville with a resultant fatality, it would be reported in the local press and, possiblly, it would also recieve a small mention in a Madrid newspaper. But in reation to Spain's staging of the EXPO'92, if there was to any type of incident, especially one resulting in a fatality, the Authoriy was concious of the fact that there would be some 18,000 international newspapermen on site, bored and looking for a good news story. It was, therefore, prudent to ensure total safety both from a technical point of view and also from the social and political point.

17 TOWARDS THE QUANTIFICATION OF EMERGENCY EGRESS CAPABILITIES FOR DISABLED PEOPLE

K.E. DUNLOP, T.J. SHIELDS and G.W.H. SILCOCK
Fire SERT Centre, School of the Built Environment,
University of Ulster, UK

Abstract

The numbers of disabled people using buildings to which the public have access has increased significantly in recent years. However, the provision of accessible means of escape from fire for disabled people has not been adequately addressed and in general it is hoped that the traditional codified approaches to the provision of means of escape are sufficient. It is not unreasonable to assume that with ongoing societal and cultural changes in attitudes to 'disability' more disabled people will gain access to and use buildings at will. Hence, the need for the provision of adequate and accessible means of escape will become more acute.

In order to address this emerging situation it is essential to have accurate knowledge of the numbers of disabled people using different types of buildings, the nature of their disabilities and their corresponding capabilities with regard to effecting their escape in the event of an emergency. This paper describes a programme of work designed to obtain this essential information.

Keywords: disability, egress, capability, occupancy profiles, categorisation, locomotion, dexterity, seeing, hearing

1 Introduction

Over the past decades there has been growing concern for the life safety potential in a fire emergency of people with disabilities. Studies have identified [1,2,3 4]:

- the lack of useful data with respect to the numbers and capabilities of disabled persons and the manner in which they interact with each other and their environment,
- that building codes, in the absence of such information are not adequately addressing the problem,

Fire Engineering and Emergency Planning. Edited by R. Barham.
Published in 1996 by E & FN Spon. ISBN 0 419 20180 7.

- that communication systems are not compatible with the needs of end users,
- that fire emergency plans do not adequately address the needs of end users, and
- the lack of education and training of staff/management with respect to the needs of disabled persons.

The contextual backcloth for these concerns lies in the increasing provision of access to buildings over the past two decades for people with disabilities.

This paper describes a programme of research designed to produce information with regard to the numbers of disabled people using particular types of buildings, the nature of their disabilities and the corresponding egress capability profiles.

2 Prevalence of Disability

In 1988, the Office of Population Censuses and Surveys (OPCS) published a series of reports [5] giving information on the prevalence of disability among adults and children in Great Britain. According to the definitions and measurements used in the surveys, it was estimated that over 6 million, ie, 14.2% of adults in the Great Britain have a disability. Similar surveys, commissioned by various government departments were conducted in Northern Ireland during 1989 and 1990 by the Policy Planning and Research Unit, (PPRU) [6,7]. From these studies the overall rate of disability amongst adults in Northern Ireland was estimated to be 174 per thousand.

3 Experimental Programme

The focus of this research [8] was to quantify the physical capabilities of people which affect their ability to evacuate a fire threatened space to a place of safety. The research was not concerned with establishing pre-movement times for different evacuation scenarios. From consideration of the necessary evacuation activities and different disabilities defined in [5,6,7], five areas of disability were considered to have an impact on the capability of persons to escape a fire threatened building. Four of these disabilities, namely, locomotion, dexterity/strength, seeing and hearing, formed the contextual basis for the corresponding experimental design.

3.1 Sample Design
Day care centres were identified as the best potential providers of subjects to participate in the experimental programme; five day centres were chosen on the basis that they provided the largest number of potential participants, with the widest possible range of disabilities, severity of disability and age. Within each day centre, participants were selected randomly. In order to extrapolate from this sample population to the disabled population of Northern Ireland as a whole (or indeed the UK), each participant was surveyed prior to commencement of the experiments to determine characteristic data such as age, sex, nature and severity of disability, and locomotion aid normally used.

3.2   Experiment 1 - Locomotion

This experiment was designed to test individual capability with respect to movement on both horizontal and inclined surfaces, ie stairs and ramps.

3.2.1 Horizontal Movement

Each participant was asked to move in a prompt manner along a horizontal route within the day centre. The general design parameters for the horizontal route were that it should:

- be approximately 50 m long,
- contain at least one 90° or 180° turn, and
- contain at least one door.

Individuals not capable of walking 50 m were only invited to walk a distance along the route which was commensurate with their ability, and participants who normally required assistance to walk on level ground were assisted throughout the experiment.

The following measurements were made:

- time required to traverse each direct horizontal section of the route,
- time required to pass through the door (measured from a distance of 1 m either side),
- time required to turn through the 90° angle (measured 1 m either side angle), and
- frequency and duration of rest periods.

In addition, the following observations were made:

- effective width of participant (and assistor),
- portion of route utilised, and
- extent of use of handrails.

3.2.2 Inclined Movement - Stairs

Each participant was asked to ascend and descend a stairway in the day care centre. After ascending the stairs, participants were allowed to rest before beginning their descent. The desired parameters for the stairs were that it should incorporate a landing and comprise, where possible, not less than 12 steps.

The following measurements were made:

- times required to ascend stairs,
- times required to descend stairs,
- times required to negotiate landing,
- time, position and duration of rest periods.

In addition, the following observations were made:

- the effective width of participant (and assistor),
- the portion of the stairs utilised,
- the extent of use of handrail, and
- the continuity of movement.

Once again, participants who normally required assistance to ascend/descend stairs were assisted throughout the experiment.

3.3.3 Inclined Movement - Ramps

Each participant was asked to traverse a ramp in the day care centre; participants who normally required assistance to traverse ramps, were assisted throughout.

The outputs obtained from this experiment included the:

- time taken to traverse the ramp of known length, downwards and upwards,
- time, position and duration of rest periods,
- effective width of participant (and assistor).

In addition the following observations were made:

- the portion of the ramps utilised, and
- the extent of use of handrail.

3.3   Experiment 2 - Dexterity/Strength

In order to assess individual capability with respect to dexterity as pertaining to evacuation, it was necessary to design a suite of experiments to measure capabilities with regard to:

- turning door knobs,
- operating lever type door handles,
- unlocking a door,
- applying sufficient force to operate a manual fire alarm system, and
- opening doors with varying opening resistances.

In this experiment the following measurements were made:

- maximum torque which can be applied by individuals using a knob type handle,
- maximum force which can be applied by individuals using a lever type handle,
- maximum force which can be applied using a finger/thumb,
- maximum force which can be applied using a hammer,
- time required to lift a key out of break glass box and open lock.

In order to measure the capability of people with respect to negotiating doors subjected to a range of closing forces, a fully instrumented, demountable door assembly was constructed.

Each participant was asked to open and move through the door for each door setting and mode of operation, ie, pushing and pulling. The door closing forces were varied randomly such that the participants were not aware of the relative degree of difficulty of consecutive tasks.

The outputs obtained from this experiment included the:

- opening angle of the door at the point when each individual was deemed to have passed through the door, ie when the door was able to close freely,
- total time to open and move through the door , and
- characteristics opening patterns for each individual.

3.3   Experiment 3 - Hearing
Each participant was invited to listen to a tape recording of alarm bells over the range 55 dBA to 80 dBA with incremental intervals of 5 dBA. The experiments were conducted in quiet rooms in order to eliminate background noise.

The output obtained from this experiment was the lower threshold of hearing of each participant with respect to the alarm bell.

3.4   Experiment 4 - Seeing
Each participant was invited to locate and read three types of exit signs ie, ordinary (complying with BS 5499 Part 1), illuminated and LED. The outputs obtained from this experiment included the:

- maximum distance at which the participants could <u>locate</u> each exit sign (ie, the distance at which they are aware of the position of the sign) and
- maximum distance at which participants could read the exit sign.

4 Analysis of Results

Initially the results of the experimental programme were analyzed to produce descriptive statistics by presence or absence of a particular, relevant disability. For example, with respect to the locomotion activities, descriptive statistics were produced for speed on horizontal, ramps and stairs by presence/absence of a locomotion disability, and also by level of assistance given. For each area of measurement, statistical analysis of the differences between the capabilities of disabled and non-disabled subjects was conducted.

Given that the initial analysis revealed considerable variation among the samples of disabled persons with respect to each activity, it was considered necessary to identify and characterise distinct sub-groups within each of the disability areas. In this respect, further analysis of the results produced:

- descriptive statistics for speed on horizontal, ramps and stairs by locomotion aid used and level of assistance given,
- descriptive statistics for times to negotiate 90° and 180° bends by locomotion aid used and level of assistance given,
- probability of successful negotiation of a door subjected to a range of closing forces by persons with locomotion/dexterity disabilities,
- given successful negotiation of the door, descriptive statistics for the time required to negotiate a passage through the door by locomotion aid and presence/absence of a dexterity disability,
- descriptive statistics for the distances at which subjects could read and locate exit signs by seeing severity, and
- descriptive statistics for the dBA level at which subjects could hear an alarm bell by hearing severity.

Statistical analysis of the differences between respective sub-groups in each disability area indicated that:

- the speed of movement of persons with locomotion disabilities can conveniently be described in terms of the mobility aid which they may or may not use; in this respect five distinct locomotion sub-categories, representing different capabilities and space requirements can be identified;
- the distances at which persons with seeing disabilities could read and locate exit signs was related to the severity of their seeing disability (as defined in [5,6,7]; in this respect two distinct seeing sub-categories can be defined;
- the dBA level at which persons with hearing disabilities can hear an alarm bell is related to the severity of their hearing disability (as defined in [5,6,7]; in this respect two distinct hearing sub-categories can be defined.

5 Building Occupancy Profiles

As mentioned earlier a comprehensive survey of disability in Northern Ireland was conducted by the Policy Planning and Research Unit, N. Ireland in 1990. Although the main aim of the Northern Ireland Disability Survey was to estimate the overall prevalence of disability in Northern Ireland, useful information was also collected with respect to the degree of mobility of disabled persons, the specific disabilities that they possess, the severity of those disabilities, the aids (locomotion, seeing, hearing) that they use and the extent of their involvement in leisure and social activities. Recognising the potential of this data to provide information with respect to the buildings that disabled persons might be expected to occupy, the authors obtained and conducted detailed analysis of the data to provide information with respect to the numbers of disabled persons who are mobile in the community and the numbers of persons in each disability category (as defined through the experimental programme) likely to be found in:

- private and communal dwellings,
- cinema/theatres,
- hotels/boarding houses,
- leisure centres/sports clubs,
- educational establishments, and
- places of employment.

In some instances, it has been possible to predict actual percentages of disabled persons in particular building populations. The data thus provided represents occupancy profiles for particular building types which could be used as input to fire risk assessment modelling.

6 Concluding Remarks

The results obtained from the experimental programme indicate that within each disability area, distinct sub-categories, each representing different capabilities with respect to emergency egress activities, can be defined. The consolidated data derived for each sub-category is considered sufficiently detailed for use in fire engineering risk assessment models.

In this programme of research it has also been possible to predict the percentage of disabled persons in each disability category likely to be out and about in the community and present in any building to which the public have access. It some instances it has also been possible to predict the percentages of disabled people by disability category likely to be found in a particular building occupancy at any time, eg theatres. It is considered an added advantage that the types of buildings for which data has been elicited easily maps onto the building purpose groups used in building regulations.

For the purposes of risk assessment, it is now possible to predict with some confidence:

- the likelihood of the presence of disabled people in buildings by disability category, and
- the capability of each category of disabled to perform the various emergency egress activities.

Regulators, designers and those charged with the management of buildings in use, now have sufficient knowledge regarding the likelihood of the presence of disabled people in buildings, to make, what they consider to be necessary and sufficient provisions to ensure equitable life safety options for all building populations.

7 References

1. Marchant E W (Ed) (1975) *Proceedings of Seminar: Fire Safety for the Handicapped,* University of Edinburgh.
2. Levin B M (Ed) (1980) *Fire Safety and Life Safety for the Handicapped: Conference and Preparatory Workshop Reports,* NBSIR 80-1965, National Bureau of Standards, Department of Commerce, Washington DC.
3. Shields T J (Ed) (1993) *Proceedings Engineering Fire Safety for People with Mixed Abilities, CIB W14 International Symposium and Workshops, Volume 1,* University of Ulster.
4. Shields T J (1993) *Fire and Disabled Persons in Buildings,* BR 231, Building Research Establishment Report, 1993.
5. OPCS Surveys of Disability in Great Britain (1988) *Report 1: The Prevalence of Disability Among Adults,* HMSO, London.
6. McCoy D, Smith M (1992) *The Prevalence of Disability Among Adults, Report 1, of the PPRU Surveys of Disability,* Policy Planning and Research Unit, Statistics and Social Division.
7. McCoy D, Smith M (1993) *The Prevalence of Disability Among Children, Report 2, of the PPRU Surveys of Disability,* Policy Planning and Research Unit, Statistics and Social Division.
8. Private Communication to Fire Research Station, Building Research Establishment, 1995.

18 ASSESSMENT AND SIMULATION OF CROWD EVACUATION ISSUES

N. KETCHELL, D.M. WEBBER, S.K. COLE,
P.J. STEPHENS and N. HIORNS
AEA Technology Consultancy Services, Risley, Cheshire, UK

Abstract

Ensuring the safety of people in the built environment requires that the issues of fire initiation, fire and smoke spread, and evacuation are effectively addressed. With increasingly complex structures, and in the light of recent accidents, the ability to simulate likely events is therefore becoming progressively more important. The paper discusses the methods of assessing the evacuation, from applications of the appropriate standards to detailed simulations using the AEA EGRESS code developed by the authors. The issues of dealing with large buildings and large crowds, and taking account of the fire and smoke spread and human response to an emergency will be discussed, along with an overview of the modelling and examples of simulations.
Keywords: Evacuation, simulation, crowd, modelling

1 Introduction

The need to evacuate people from structures, particularly during fires, continues to be an essential requirement for safety.

For office buildings in the UK, for example, evacuation standards are set in the British Standard [1], and Building Regulations. These standards are based largely on the ability to evacuate people to a protected area (eg outside, or a protected stairwell) within 2½ minutes. This leads to constraints on building populations dependent on the available stairwells. This and other standards provide an excellent baseline against which to assess evacuation provisions.

With increasingly complex building designs however, it becomes important to consider the evacuation provisions in relation to the possible fire spread within the building. This is the area where simulation can make a significant contribution.

The authors have developed the AEA EGRESS evacuation modelling computer

Fire Engineering and Emergency Planning. Edited by R. Barham.
Published in 1996 by E & FN Spon. ISBN 0 419 20180 7.

code for this purpose.

Evacuation modelling can enhance safety assessments in a number of ways including:

- design changes can be assessed before any detailed design or construction is carried out,
- the effect of different numbers of people and scenarios can be studied in a simple manner,
- there is no interruption to existing systems or services,
- modelling can speed up the assessment, resulting in significant time and cost reductions.

AEA EGRESS has not been developed as a fire or smoke simulation code. However the results of such analyses can be input as scenarios which may affect the evacuation process.

2   Modelling crowd movement

In generating any evacuation model, it is important to understand the available experimental data which can be used for calibration and validation. Figure 1 gives an overview of the flow rates which can be calculated from the work of three well known authors in the field. This collection of data is by no means exhaustive, however Figure 1 serves to show the possible variations in flow rates which may be produced if different models are used.

The data in Figure 1 is taken for essentially uniform density crowds. In reality, and in detailed simulations, crowd densities are rarely spatially or temporally uniform, which can also call the use of such data into question. Another assumption, frequently made, is that the maximum flow rate can always be achieved.

To overcome these shortcoming a more sophisticated modelling approach has been adopted in AEA EGRESS. AEA EGRESS models "people" as individuals located on a hexagonal grid (Figure 2) which covers the floors of the building structure. The technique is based on the use of cellular automata. At each "timestep" the "people" move from cell to cell based on the throw of a weighed die. The usefulness of the technique is determined by the ability of the model to represent the experimental evidence. In the case of AEA EGRESS the weights required for the die can be calibrated against information on speed, or flow, as a function of density, so that the experimental data can be adequately represented where it is valid.

In all grided codes there is some asymmetry between on-axis and off-axis directions. Directions 1, 2 and 3 in Figure 2 are on-axis. Direction 4 is off-axis; movement in this secondary direction is achieved by the shortest permissible, zig-zag (dotted line), path with a length which is a factor of $2/\sqrt{3}$ longer than the direct distance. The time taken to move in this direction is therefore approximately 13% longer than in directions 1, 2 and 3. This uncertainty is small, or comparable, with the other uncertainties involved. If a rectangular grid had been used then this error would have been significantly larger.

Based on this simple underlying technique the model is able to cope with thousands of people, distributed over areas of over a square kilometre, running on an IBM compatible PC running Windows.

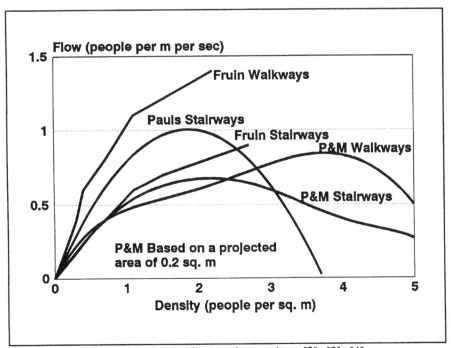

Fig. 1.   Illustration of flow rates from various authors [2], [3], [4]

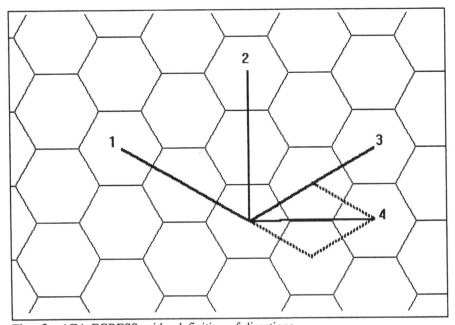

Fig. 2.   AEA EGRESS grid - definition of directions

The use of the automaton system makes the model very flexible, as all decisions can be made at the person level. This allows group behaviours to be easily input and the building plan to be changed with time (due to an incident) and affect the people.

3 Verification and validation

Verification and validation of any simulation model is essential to have confidence in its predictions; verification ensures that the code operates in accordance with the underlying mathematical models, validation concerns itself with the accuracy of the model when compared with real life situations.

Verifications of unimpeded movement speeds and standard deviations, and flows through openings have conducted [5]. For the first case a discrepancy of 4% was observed, which is consistent with the approximations in the implementation of the algorithm. For flows down corridors good agreement was similarly observed with the optimum experimental flow rates. However for flows through doorways, it is possible to achieve model flows of 60% higher than the optimal uniform density flows. This is caused by the rapid change of density across a doorway, and the ability in the model to achieve an unrealistically optimal approach of people to the doorway; in practical simulations such conditions are rarely achieved. This highlights one of the problems in the experimental data noted above, in that it relates to uniform density; if the density is rapidly varying it is difficult to accurately define an appropriate uniform density. Experimental data does show an increased flow through doorways [4], and whilst this is not to the same extent the AEA EGRESS calculations are reasonable when problems with the definitions, and realistic situations are considered.

For validation we have chosen to use a range of available data, then simply draw up the configurations and use all the default parameters (in some cases these may not be the most appropriate) rather that fine tune the model on any specific example [5]. Validation simulations have been conducted for an aircraft, a double-decker bus, two theatres [5], and other cases. The general agreement between code and measured evacuation times was of order $\pm$ 20%. An exception being for an over-wing aircraft evacuation which was affected by the use of a difficult exit, which was not modelled; in line with the philosophy above. Given the complexity of the evacuations, and some of the data in Figure 1, this degree of agreement is very encouraging.

4 Example Simulation

An overview of a simulated evacuation from a small hotel is presented below, as an example. The hotel in question has only a single staircase, in the centre, which would be in breach of a number of design codes. The hotel has four floors, including the ground floor; these are shown in Figure 3. The ground floor is at the bottom right, and is connected to the floor above it by the staircase shown in the centre of the view. The view at the top right is in turn connected by the central staircase to the view at the bottom left. On the upper floors there are a range of hotel bedrooms depicted, and a conference room on the top floor. On the ground floor various areas

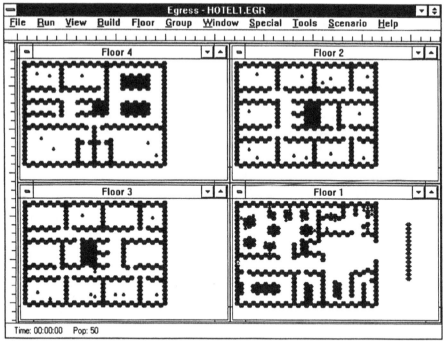

Fig. 3. Initial situation for hotel simulation

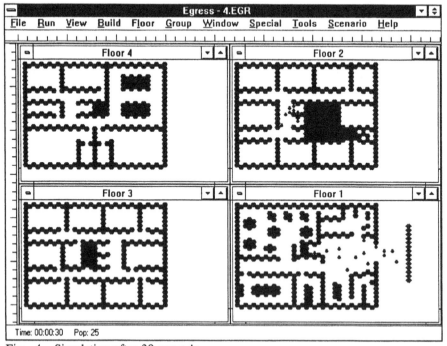

Fig. 4. Simulation after 30 seconds

are represented including reception, lobby, dining room and a small bar. The people are represented by small dots on the plan. There are a significant number in the bar and restaurant (top of Floor 1) who are at the tables. Outside the building on the ground floor an area has been defined which represents a place of safety. During the simulation people are removed when they reach this area.

For demonstration purposes a scenario has been included where a fire occurs in a store-room behind the staircase on Floor 2. This spreads throughout the hotel as time progresses. Figure 4 shows the simulated situation 30 seconds after the alarm has been sounded, and people have begun to evacuate. Some crowding occurs around the staircase on Floor 1, and people are still leaving the lobby area, but this simulation results in a successful evacuation.

This example has included some very simple assumptions, and is for a very small building only for simplicity. AEA EGRESS is more usually employed for larger cases, including different groups of people, with varying response times, and a variety of objectives, dictated by their expected behavioural response. Once the plan has been drawn up it becomes relatively easy to examine the effects of both structural and procedural changes in relation to a variety of scenarios. However, it is hoped that this simple example has given a feel for AEA EGRESS and has illustrated some of the flexibility of these techniques.

9   Acknowledgements

This work has been sponsored under a Joint Industry Project by AEA, Exxon, HSE, Shell, and Texaco, for offshore studies, and under the framework of Major Industrial Hazards, with support of the CEC, HSE and AEA. We would like to acknowledge additional support and interest from London Underground Ltd.

References

1. "British Standard BS5588: Fire precautions in the design and construction of buildings", British Standards Institution.

2. Fruin, J. J., "Pedestrian Planning and Design", Metropolitan Association of Urban Designers and Environmental Planners, New York, 1971.

3. Pauls, J., "Movement of People", SFPE Handbook, Section 1, Chapter 16, pp246-268, 1st Ed., National Fire Protection Association (1990)

4. Predtechenskii, V. M., and Milinskii, A. I., "Planning for Foot Traffic Flow in Buildings", Amerind Publishing, New Delhi (1978)

5. Ketchell, N., Hiorns, N., Webber, D.M. and Marriott, C.A., 'When and How Will People Muster? A Simulation Approach'. In 'Response to Offshore Incidents' Conference Proceedings, Marriott Hotel, Aberdeen, June 1993.

19 INTEGRATED EMERGENCY MANAGEMENT

J.R. STEALEY
John Stealey & Associates, Bickley, Kent, UK
C.F. PAYNE
Christopher Payne & Associates, Hutton Rudby, Cleveland, UK

Abstract
Emergency planning is a structured, framework designed and tested to deal with and manage unusual or extreme events in exceptional circumstances. Development of a plan requires both quantitative and qualitative techniques recognising that human activity and decision making in conditions of uncertainty are as much a contribution to the initiation of an emergency as to its successful resolution. Planning also extends beyond the immediate response and damage limitation capability. It is a powerful management tool which looks to identify how and why emergencies might arise, suggests options for reducing hazards and consequences, sets out procedures for dealing with initial events and provides a system for maximising recovery and business continuity. Its strengths lie in flexibility to respond to a wide range of unusual events in extraordinary circumstances, outside normal management and operational experience. At its best emergency planning should fit seamlessly into an organisation's policy and culture of good working practice. Its integration into the total management system is as much in the interests of the outside public and customer base as for the internal organisational use; for this reason it should be considered as a responsibility by everyone working in an organisation and as open and transparent as possible.
Keywords: Emergency planning, integration, management, policy, risk assessment.

1. Introduction

Organisations, whether large or small, simple of complex, have to deal with a continual series of minor disruptions and breakdowns caused by a variety of internal and external influences. People working at all levels in the organisation manage these disruptions as part of their day-to-day responsibilities.

However, once disruption reaches a level where the system has failed or ceased to function either as designed, or specified, then the future of the organisation as a recognised and effective entity is threatened. This vulnerability, where normal

Fire Engineering and Emergency Planning. Edited by R. Barham.
Published in 1996 by E & FN Spon. ISBN 0 419 20180 7.

experience and practice are insufficient to cope with restoration, requires special arrangements to manage the situation:- should disruption continue unchecked the organisation may never recover its former competence. Procedures and arrangements for this particular management problem have developed over many years and consequently, are referred to in various ways including disaster limitation, business recovery and continuation, crisis management and emergency planning; a diversity of terms which reflect differences in priorities rather than concept or content.

This paper uses the term <u>emergency planning</u> for simplicity but is applicable across any other of the procedures including the integrated management functions.

2. The planning concept

Emergency planning should not be simply a reactive management tool brought out when all else fails. As a proactive function it has four main aims, Table 1, which, whilst addressing different time scale priorities, such as immediate response or long term business restoration, are nevertheless interconnected.

Table 1. Aims of emergency planning.

1. To identify and describe potential hazards and recommend effective solutions to remove or minimise those hazards.
2. To provide a tried and tested management framework which can respond to emergencies and deal effectively with the immediate response.
3. To identify the requirements for longer term recovery and business continuity.
4. To designate management responsibilities for resource allocation to meet both the immediate response and continuity needs.

The implementation of any or all of these aims does not usually generate any direct income for the organisation and for this reason activity is sometimes trimmed to the minimum requirement, commensurate with say a statutory responsibility. However, the process can provide other benefits which are not always readily apparent.

For example, it contributes to good and caring management practice both for the on-site workforce and to the surrounding off-site location. It offers a method for minimising loss control and maximising operational efficiency; furthermore, in the event of a disaster or emergency occuring, it links directly into business recovery and continuity to bring an organisation back onto a full operational basis as quickly and cost effectively as possible and retain the customer and market base.

Disruption which results in activation of an emergency plan frequently has damage and loss components. In extreme cases it can include injury and death and contamination to the surrounding environment. However, it should be stressed that it is not only those events which make headline news that might be termed disasters or emergencies and for which the emergency planning process is relevant. Any organisation trying to manage its operations but outside its normal experience is facing an emergency. For example, the loss of financial trading through breakdown of information technology hardware, the evacuation of staff from premises for security reasons or the identification of potentially dangerous operations forcing closure of manufacturing, production or distribution can all lead to an uncertain future.

Thus, emergency planning is a conceptual framework such that it becomes part of an organisation's policy and culture with a recognised commitment from the "top of the office", be that the board, the chief executive, the general manager or the owner, downwards.   Operational recovery and business continuity are likely to be the touchstones and as a consequence the planning should be as open and transparent as possible: it becomes the hallmark of the quality of the organisation giving confidence to the outside world and the customer base.   The planning and response capability are as much to the benefit of the organisation's clients as to the organisation itself and applicable to a very wide range of operational systems, from chemical process industries and City financial institutions, to local government practice, food retailing companies and small businesses.

Equally, it is important to recognise that the view that "planning, once carried out will suffice forever" is inappropriate.   In addition, the response by the organisation to emergencies needs to be managed in a manner outside normal experience but unless thought and commitment have been given to planning, training and evaluation of this particular capability then the efficiency and effectiveness of the response will be low in quality.

The development of this argument sets emergency planning into a framework of an integrated system with inputs, outputs, feedback and interactions.   In this way the systems becomes a dynamic entity, moving and responding to accommodate new ideas and experiences.   But this integration is not solely the recognition of ensuring cross relationships or correlation across service delivery heads for, unlike many other management functions which are designated to a particular working level in a hierarchy, an input to emergency planning is required from every person working in the operational system recognising their contribution and rôle should the need arise in the event of an emergency.

Certainly, there is a need to encourage those who are responsible for delivering a service also take responsibility for planning how that service must be given in extraordinary circumstances.   And, should a plan ever need activating then there is a even greater need to look at what happened, and why, and feed this information back into the improvement of the system.

However, there is something perhaps a little more subtle than looking at emergencies (real or simulated) as the only source of data to the planning function, its evaluation and modification.   An understanding of the way people perform and manage daily uncertainties are a genuine input to the emergency planning process for there may not be a discrete step between what is day to day management under stress of uncertainty to where this situation becomes a crisis or emergency.

Whilst this may already exist in some organisations it is by no means the universal practice, and more importantly it is not carried through in a methodical manner. Therefore, what one groups thinks is important to collect and analyse does not always relate to that by another.   This leads to difficulties in sharing or pooling ideas and evaluating the real quality of the output.

Thus, the conceptual framework requires a structured and methodical process which can be recognised and implemented by all those working in the operational system.

These general principles of integrated emergency management might be expressed in an alternative manner as: Generality: Devolved delegation: Co-operation.

Generality recognises that there are many similarities in dealing with emergencies and coping with consequences. In this way plans are built on more day to day routine arrangements in a graduated, incremental manner responding appropriately to their scale and nature of need. Furthermore, there is a recognition that the response itself will be made up on a number of elements, each in themselves capable of use in different circumstances, such that the task of management reflects dealing with effects and not with cause.

Devolved delegation of responsibility to the most appropriate level of management for control, direction and use of resources is a widely accepted and agreed principle in management today. This same philosophy must apply to emergency management such that responsibility for planning the response lies where the resources are normally controlled. But, as importantly, delegation lays stress on the need for co-ordination.

No one style of management practice can be recommended; it is something which very much reflects the culture of the organisation. Some prefer strong centralised decision making whereas others tend towards a collegiate style, perhaps operating in a pivotal or facilitating rôle.

However, if the goal of integration is to be reached then co-ordination of planning requires the strongest co-operation between various managers to whom responsibility for planning has been delegated.

Consequently, whilst it may be possible to separate the organisation's policy on emergency management from the operational command levels responsible for delivering that service, maximisation of effectiveness will occur only if and when these two management structures interact and co-operate.

3. Planning process and method

There is an important issue to be addressed in looking at the emergency planning process: that is, the process offers an approach to problem solving. There are no prescribed rules or procedures; neither is there a best solution although some ways of tackling the problem may be better than others however that comparative judgement is made.

However, it should address the four interactive factors described in Table 2:

Table 2 Emergency Planning Objectives

Prevention
Preparedness
Response
Recovery

The prevention phase encompasses measures which are adopted in advance of an emergency and which seek to prevent that emergency occurring. Obviously this phase links closely to preparedness which addresses how systems might fail, the range of possible hazards which might ensue and the consequences which might arise. This information can be used to calculate relative risks, perceived or actual, and would assist in identifying where best to direct resources or arrange stock piles as part of the prevention phase. The analysis also should identify options to prevent or reduce system fragility by increasing the reliability and performance characteristics.

At the same time, the process recognises that disruption may still occur and there is a need to be prepared and able to respond to the extraordinary circumstances. Finally, there is a requirement to recover from the disruption as quickly, efficiently and effectively as possible.

The concept of integrated emergency management process extends back over a number of years. In 1983 The US federal Emergency Management Agency (FEMA) published a paper [1] setting out their conceptual view of the process. This view stated that the most effective way to achieve emergency management preparedness was through "increased emphasis on developing the common and unique capabilities required to perform specific functions across the full spectrum of hazards rather than focusing on the requirements of specific hazards".

The systematic approach developed by FEMA contained a number of detailed steps set out in Table 3 upon which reasonable and justifiable plans could be made and effective action taken to increase emergency management activities.

Table 3 US FEMA Emergency planning process overview

Hazard analysis
Current capability assessment
Emergency operations plans
Maintenance
Consequence management
Emergency operations
Evaluation

This framework has been used by others to develop the concept of integrated planning further such that hazard analysis contains elements for scenario modelling and risk assessment. Knowing what might happen, the likelihood of it happening and an assessment of the magnitude of the problems which might occur are essential first ingredients to emergency planning.

The next step would then be to assess the current capability for dealing with a range of hazards identified in the first step. However, this assessment must be made against pre-determined standards or criteria and include other related functions; for example alerting and warning populations, evacuation methods and plans, emergency communications, monitoring the effects of the hazard, consequence management and decision making all influence the current capability.

Once emergency operations plans have been developed the ability to take appropriate and effective action against hazards must be maintained continually or the value of those plans will diminish with time. Attention to maintenance of planning through simulated activities may be of great importance to those areas or organisations which do not experience frequent activation of plans.

Consequence management is as important as the "planning to prevent" phase of integrated emergency management. Significant effort applied to resource allocation and utilisation is likely to affect markedly the level of loss or suffering.

Finally, should emergency operations ever need activating then the response should use current and existing plans. Attempts to incorporate proposed upgrades and revisions are likely to lead to failure since not all personnel involved in the system may be aware of the changes. At the same time, output from dealing with the emergency provides first class data for feeding back into existing capabilities under real operational conditions. Thus, any plan should have, as part of its content, a capacity to record and log information not pertinent to management of the emergency itself but as back up for subsequent analysis; this capacity should not be at the expense of handling the emergency itself in the first instance.

The outcome of emergencies must be analysed and assessed in terms of actual versus required and expected capabilities and results fed back into the reformation and structure of the emergency plan and the management capability. This evaluation process should be extended to include the lessons and experiences gained from exercises, training, rôle plays and simulations.

4. Preparing a plan

Contents of individual emergency plans are likely to differ, reflecting the particular circumstances for which they were written. However, a outline plan might include the general elements shown in Table 4.

Table 4. An emergency plan outline structure

Hazard assessment
Scenario evaluation,
Consequence modelling
Risk analysis
Technical data gathering
Command, control and co-ordination arrangements
Emergency management information
Other agencies
Public relations
Business recovery strategy

The development of some or all of these elements can be supported by a range of techniques and methods designed to examine and evaluate the added value acquired from carrying out that development.

Quantitative techniques, such as problem analysis, hazard assessment, command and control operational design, cost/benefit evaluation and risk management have a strong rôle to play. In addition, as systems become more complex, and the use of automated processes increases, so the opportunities for human involvement are likely to occur only at critical control and management steps when automated functions may have broken down or key decisions have to be taken. Human response, behaviour and decision making, perhaps under great uncertainty, whilst very much subjective and qualitative in content, may dominate the analysis and evaluation at these stages.

Analytical techniques, including quantified risk assessment (QRA), cost:benefit analysis (CBA) environmental risk assessment (ERA) have come to the fore over recent years as regulatory authorities have responded to and formulated legislation to deal with potentially hazardous industries. But these have been tempered by the need not to over prescribe and inhibit industry through excessive costs of regulation. Therefore, concepts to minimise hazardous activity related to cost have been devised including use of "best available techniques not entailing excessive cost" (BATNEEC), "as low as reasonably possible" (ALARP) and "best practical environmental option" (BPEO).

5. Validating a plan

Once prepared a plan must be tested, understood and validated.. Lessons learnt are fed back into the planning process and the plan revised. This requires training and exercising to ensure that all parties recognise and have practised their responsibilities. Table 5 offers a number of ways in which a plan might be validated and tested through exercises.

Table 5 Emergency plan validation

Awareness exercise
Discussion seminar
Table top exercise
Command, communication and co-ordination exercise
Full scale exercise

Techniques range from simple awareness and discussion seminars and exercises where participants talk about options in response to certain challenges. Table top and command exercises involve a certain level of rôle play against structured scenarios with external directing staff influencing and modifying the stages in further action according to output of previous activity. Full scale exercises offer the potential of investigating the capability of the system to respond under the most realistic conditions, but equally they are very costly both to prepare and produce as well as tying up staff time throughout the play of the exercise. Consequently, their use could be very limiting and should only be considered after the other techniques have been exhausted and detailed cost:benefit analysis has been carried out.

6. Managing the complexity

Emergency management is a complex, dynamic function which could lend itself to a systems study generally described in Table 6.

Table 6 Integrated emergency management; a systems approach

Make clear statement of objectives
Define outcome measures and state in terms of objectives
Define system to be studied
Identify alternative plans and evaluate against outcome measures - sensitivity analysis
Choose "best" plans - validate and revise system

This methodology brings together the management functions (making clear the policy, objectives and requirements and stating the outcome measures) with the operational functions (defining the system, identifying alternative plans, investigating uncertainty and validation) in an integrated manner.

This might be more readily seen in Figure 1 which shows that integrated emergency planning has many layers any of which might have flaws or uncertainty in their aims, objectives, operational readiness or response capability. At any time various internal and external factors contrive to bring these faults together in such a way that disaster occurs.

Figure 1 The layers of integrated emergency management

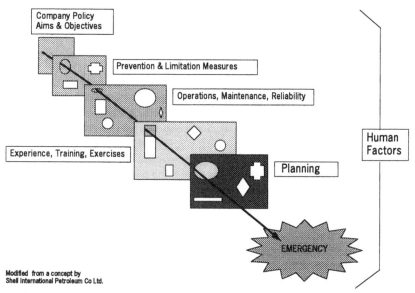

Emergency Management

Company Policy
Aims & Objectives

Prevention & Limitation Measures

Operations, Maintenance, Reliability

Experience, Training, Exercises

Human Factors

Planning

EMERGENCY

Modified from a concept by
Shell International Petroleum Co Ltd.

Summary
An emergency can occur at any time in an organisation's working life and should that happen then special measures will be required to restore normality. Thus, a responsibility is placed on everyone working in that organisation to consider their own working practices and identify where experience can be used as input to produce the integrated emergency planning and management process.

Quantitative techniques are available to help in the planning process; however, the qualitative factors involving human response and decision making are likely to be as equally important.

Human activity may well cause the emergency in the first instance but at the end of the day human resourcefulness will probably prove to be the salvation in managing the emergency to a proper conclusion.

Reference
1 US Federal Emergency Management Agency: *An Integrated Emergency Management System*, Washington DC 1983.

20 FIRE COVER COMPUTER MODEL

C. REYNOLDS
Fire Research Development Group, Fire and Emergency
Planning Department, The Home Office, UK

Abstract

The Fire Research and Development Group of the Home Office have developed
a computer model to assist local authority fire brigades with the task of fire cover
planning. The model was first written in the 1970's as an ambulance routing program
and over the years has developed to be used to assess risk categorisation. The model
has recently been further developed to use a Geographical Information System.

The model takes into account the main aspects of a fire brigade which can affect its fire
cover. These include:

* the geography of the area and the road network of the brigade
* the number of incidents and their location
* travel times of appliances attending incidents
* turn out times of appliances
* risk categorisation within the brigade
* brigades resources (the number, siting and crewing of stations)

and,

* the standards of fire cover which the brigade should meet.

The model uses tree spanning algorithms and probability theory to predict attendance at
incidents, using information on road networks and incidents. It provides on-screen
maps of the brigade area, visual representation of fire cover, performance indicators
and detailed information in a series of output screens and tables.

This paper describes the principles behind the model and provides some examples of
use.

Fire Engineering and Emergency Planning. Edited by R. Barham.
Published in 1996 by E & FN Spon. ISBN 0 419 20180 7.

Introduction
The Fire Research and Development Group of the Home Office undertakes fire research in support of UK local authority fire brigades. The group is based at two separate sites: Horseferry House in London and on the site of the Fire Service College at Moreton-in-Marsh, where the Fire Experimental Unit is located.

The work undertaken by the group ranges from practical tests of equipment and tactics, through ergonomics and human factors studies to the application of management sciences and operational research. The group generally has approximately 40 projects active at any one time.

The Fire Cover Model
The fire cover computer model was written in the 1970s as a tool for routing of ambulances. Over the years it has developed into a management decision tool for achieving the optimum disposition of operational resources within a fire service.

Planning Fire Cover
The task of planning fire cover is a relatively complex one which must take into account the national standards and also all of the factors listed here:

Figure 1: Planning Fire Cover

National Standards	The national requirement for fire cover
Brigade Resources	The number, type, crewing and location of appliances
Road Network	The roads most used by appliances and the time taken to travel along them.
Workloads	How busy appliances are, where and when incidents occur.
Brigade Policy	Any local policies which may affect fire cover, eg remote rural classification of risk.

Let us look at each of these considerations in turn.

National Standards of Fire Cover

The UK currently has national guidelines for the weight and timeliness of responses to fire calls. These guidelines depend largely upon the amount of risk of fire spread within the area. The risk of an area is assessed and divided into four broad categories; 'A', 'B', 'C' and 'D'.

Risk	1st appliance	2nd appliance	3rd appliance
A	5 mins	5 mins	8 mins
B	5 mins	8 mins	
C	10 mins		
D	20 mins		

Brigade Resources

The main considerations in assessing resource availability are:

Station siting	Where stations are located in relation to the road network and the incidents they attend.
Type of staffing	Fire appliances can be crewed by wholetime, day staffed or retained personnel. The type of crewing affects the time taken to mobilise an appliance on receiving a call.
Number of appliances	The number of appliances located at a station can vary and also some specialist appliances may be available.
Local policy	Local policy may affect how the appliances are mobilised, for example, some brigades move appliances between stations (standby moves) during a large incident.

The Road Network

The road network is a vital consideration when planning fire cover. Response time estimates are based upon the type of roads to be used and expected travel speeds. The following average travel speeds for different road types are assumed, although these can be changed locally and globally.

Type	Speed (mph)
Motorway	40
Class A road	30
Class B road	27
Class C road	25
Unclassified	20

The road network is characterised by the road types and intersections, but for the purposes of modelling, it can be reduced to a number of main and minor roads which are regularly used by appliances when attending incidents.

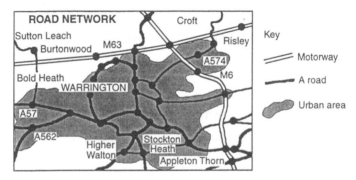

Figure 2: The Road Network

The road network is modelled using a series of road links and junctions (nodes). Figure 3 depicts nodes placed on each of the major road junctions in the area and then numbered sequentially. Fire stations have also been located on the map on appropriate nodes.

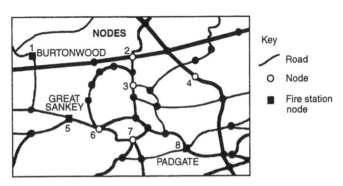

Figure 3: Modelling the Road Network

A model of the road network can now be built using these nodes as reference points and using travel times between the nodes as direct links.

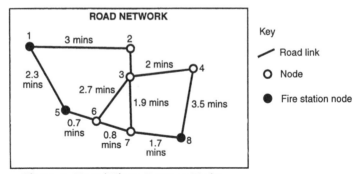

Figure 4: Travel Times Between Nodes

The travel time between adjoining nodes is calculated from the scale of maps employed by the user. From this a matrix of minimum travel times from each station to all nodes can be built, using Djikstra's spanning tree algorithm.

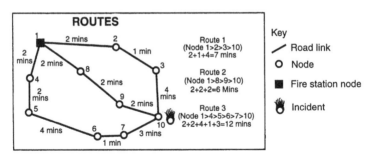

Figure 5: Choosing Shortest Routes

Workloads: Modelling Incidents
The following characteristics of incidents are used to model the requirements within an area:

* location,
* time of day,
* number and type of appliances attending,
* average time taken to deal with an incident,
* incident type.

In order to model the often large numbers of incidents effectively, the area under study is divided into small areas, or zones. These zones surround a single node and are assigned a single risk. Figure 6 shows each node in the area has a corresponding zone, which has been coloured to indicate risk category.

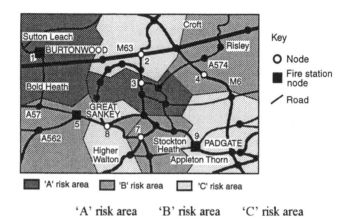

'A' risk area   'B' risk area   'C' risk area

Incidents can then be plotted onto this zonal map and all incidents falling within each zone are assigned to their corresponding node. In this way, very complex incident patterns can be effectively modelled.

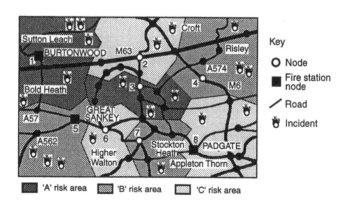

Workloads and Average Attendance Times
How busy appliances are attending incidents has a direct effect upon the average attendance times. Where appliances are not busy, the majority of incidents will be

attended by appliances from the nearest station. However, when appliances are busy then the nearest station to an incident may be empty and appliances need to travel from the next nearest station to attend incidents. This has the effect of increasing the average travel time to all incidents.

This effect can be calculated by the following iterative process. Firstly, assume the probability that an appliance is available is 1 and therefore it has no initial workload. Then, repeat steps 1 and 2 below until the change in the probability that an appliance is available is negligible:

1. Calculate expected attendance times for each appliance
2. Use this to refine the estimate that the appliance is available

Consider first the expected attendance time calculation (step 1).

Let

P_i =   the current estimate of the probability that the ith nearest pump is available

$IncidentRate_n$ =   the number of incidents at the node n

AttendTime   =   the average length of attendance at each incident

The probability that the first appliance is available is P_1. Therefore, the initial workload for the first appliance in attending single appliance incidents at node 1 should be incremented by

$$P_1 * AttendTime * IncidentRate_1$$

However, there is a probability $(1-P_1)$ that the incidents at this node are answered by the second nearest appliance, providing it is available. The probability that the appliance will have to come from the second nearest location is

$$(1-P_1) * P_2$$

and the corresponding additional workload for the second nearest appliance is

$$AttendTime * IncidentRate_1 * (1-P_1) * P_2$$

But again, there is a probability that incidents remain to be attended by the third nearest appliance, providing it is available, which is given by $(1-P_1) * (1-P_2)$

Hence, the corresponding increase in workload for the third nearest appliance will be

$$(1 - P_1) * (1 - P_2) * P_3 * AttendTime * IncidentRate_1$$

This allocation of work at each node continues in this way until the 10th nearest appliance has been considered, although in practice the probability yet to be allocated usually becomes negligible before the 10th nearest appliance is reached.

Where the requirement is for two or more appliances to each incident, then the allocation can be generalised such that $(P_1 * P_2)$ is the probability that two appliances will come from the nearest and next nearest locations.

So

$$P_1 * P_2 * AttendTime * IncidentRate_1$$

should be added to the workload for these appliances. However, there is a probability

$$(1 - (P_1 * P_2))$$

that the incidents at this node cannot be serviced by this combination of appliances, so the next alternative is to consider attendance from the nearest and third nearest appliances.

The probability that the second appliance will have to come from the third nearest location is

$$(1 - (P_1 * P_2)) * P_1 * P_3$$

so

$$(1 - (P_1 * P_2)) * P_1 * P_3 * AttendTime * IncidentRate_1$$

should be added to the workload of the first and third nearest appliances.

This process continues for all appliance pairs in the area to provide steady state availability probabilities for all appliances.

WORKLOAD EFFECT

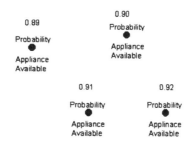

Figure 8: Steady State Appliance Availability

These can then be used to calculate which appliances attend incidents and, therefore, average attendance times.

Uses of the Model

When all the relevant data has been input, attendance times can be calculated, and the model can be used to study a variety of fire cover scenarios. These scenarios may involve moving stations from one site to another, examining the effect of traffic calming measures, planning emergency cover during major incidents, or planning for future incident patterns and densities.

> **The model can predict the level and extent of fire cover within an area for different fire cover scenarios, quickly and easily, without carrying out the changes in real life.**

Results from the Model

The model provides information on the level and extent of fire cover in a series of results using fire cover terms such as:

*   attendance time,
*   areas where standards are not met
*   overall percentage of time when standards were not met

The model also provides two indicators in order to compare varying scenarios.

The **fire cover failure index** is based upon the number of incidents attended where standards were not met, and by how much time the standards were failed. This indicator will, therefore, rise in scenarios where fire cover is worsening and fall where changes in resource disposition improves fire cover.

The **overall performance measure** also considers incidents where fire cover standards have been met and by how much time the attendance was inside the standard. This index, therefore, gives an overall measure of performance for fire cover within the area.

In summary, the fire cover model is a management decision tool for use by fire brigade officers, designed to answer 'what would happen if ...... ?' questions.

21 'LINCE': COMPUTERISED EMERGENCY MANAGEMENT

J.L. ROMAN MONZO
Instituto Tecnologico de Seguridad Mapfre, Madrid, Spain

Abstract: The "LINCE" (Logical Informático para el Control de Emergencias) programme, is an advanced computerised application, designed to establish specific recommendations for intervention, resource management and other vital information needed during an emergency at industrial installations, harbours and other areas where hazardous materials could be involved.

This management is performed in real time, presenting the user graphic information of the accident scene over the user's own drawings and plans.

The programme provides the management structure for a multitude of emergencies including conventional fire situations, Haz-Mat spills with or without fire, toxic or inflammable vapour clouds, BLEVE, solid explosive detonations, hydrocarbon land or sea spills and a host of other potential disaster situations.

Keywords: Lince, computerised, emergency, management, haz-mat, accident, resources.

Fire Engineering and Emergency Planning. Edited by R. Barham.
Published in 1996 by E & FN Spon. ISBN 0 419 20180 7.

1. PREFACE.

The "LINCE" (Logical Informático para el Control de Emergencias) programme, is an advanced computerised application, designed to establish specific recommendations for intervention, resource management and other vital information needed during an emergency at industrial installations, harbours and other areas where hazardous materials could be involved.

This management is performed in real time, presenting the user graphic information of the accident scene over the user's own drawings and plans.

The programme provides the management structure for a multitude of emergencies including conventional fire situations, Haz-Mat spills with or without fire, toxic or inflammable vapour clouds, BLEVE, solid explosive detonations, hydrocarbon land or sea spills and a host of other potential disaster situations.

Needed advises for management of the emergency are presented in real time conditions, offering the user the necessary guidelines and data (including graphics) which facilitate the labours of intervention, human and material resource activation, and virtually eliminates delays or errors as it permits total recall of all the possible response variables contemplated in the Emergency Plan.

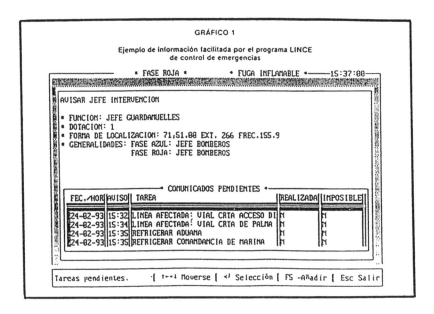

Fig. 1. Type of information supplied by the Emergency Control Software LINCE.

```
┌──────────────────────────────────────────────────────────────────────┐
│                              GRAPHIC 1                                 │
│  ┌──────────────────────────────────────────────────────────────────┐ │
│  │  ADVISE RESPONSE CHIEF                                             │ │
│  │                                                                    │ │
│  │    ■      POST:            FURNITURE DEPOSIT CHIEF                  │ │
│  │    ■      STAFF:   1                                               │ │
│  │    ■      ADVICE:  PHONE 71 51 00 - EXT 226 - FREQ 155,9           │ │
│  │    ■      GENERAL:         BLUE STAGE:    CHIEF OF FIRE BRIGADE     │ │
│  │                            RED STAGE:     CHIEF OF FIRE BRIGADE     │ │
│  │  ────────────────────────────────────────────────────────────────│ │
│  │                     COMMUNICATIONS PENDIG                          │ │
│  │                                                                    │ │
│  │  DATE/TIME   ADVISED JOB                           COMP   NOTPOSS   │ │
│  │  02-24-93          15:32    AFFECTED LINE: ACCESS WAY XX   N     N  │ │
│  │  02-24-93          15:34    AFFECTED LINE: ACCESS ROAD X   N     N  │ │
│  │  02-24-93          15:35    COOL CUSTOMS OFFICE           N     N  │ │
│  │  02-24-93          15:35    COOL NAVAL COMMAND OFFICE     N     N  │ │
│  └──────────────────────────────────────────────────────────────────┘ │
│  ┌──────────────────────────────────────────────────────────────────┐ │
│  │  Pending tasks    ↑──↓ Move    Enter: Select   F5-Add   Esc-Exit   │ │
│  └──────────────────────────────────────────────────────────────────┘ │
└──────────────────────────────────────────────────────────────────────┘
```

The programme provides graphic information on the areas affected by the emergency, drawing the effects of the accident over that area (in the form of concentration of the toxic cloud, radius of BLEVE, heat radiation levels, etc.) identifying the specific units or sectors affected by the emergency (buildings, process areas, reservoirs, streets, drains and sewage systems, underground wells, etc.).

LINCE, indicates the actions to be taken in each case depending on:

1. The accident type
2. Vulnerability of affected units
3. Occupation of the plant
4. Availability of resources
5. Other aspects of influence

The programme also indicates who must do what, when and where during every possible situation such as shifts, holidays, absence, etc. ; giving the operator relevant information (phone number, radio frequency, extension, etc.) to locate whatever resource needed.

The LINCE programme provides also a vast data base on dangerous substances which permits accurate estimates of the consequences of technical accidents (including but not limited to fires, explosions, Haz- Mat incidents) which could occur in the plant or installation being considered. This particular application makes it an extraordinary tool for use in the phase of risk evaluation as well as a training aid for the people who must manage the emergency and for the development of drill scenarios in all phases.

The data base of substances contains relevant information for intervention and first aids in emergencies with some 1,300 dangerous substances, with a rapid recall system which locates the required file by UN number, substance name, hazardous properties and other specific means.

The resources data bank includes the categories and characteristics of all human and material resources, including location, status, amounts available and all other pertinent and needed information.

LINCE, can be operated in any of three modes:

1. Reference consulting and updating
2. Emergency
3. Utilities (auxiliary mode)

Information is presented via windows and Pop-up menus, in a real user friendly way.

2. REFERENCE CONSULTING AND UPDATING MODE

This is the part of the programme which allows for initial definition, up-dating and consultation of any and all information concerning risks, resources, installations, hazardous materials, transport of substances, historical background of incidents, as well as the estimation of consequences of potential incidents in order to create pre-incident scenarios and preplanning.

Access is protected by soft (password) and hardware which avoids entry by unauthorised persons, thus eliminating the possibility of accidental or intentionally un-wanted modifications.

3. EMERGENCY MODE

This is the emergency management part of the programme which facilitates, in real time sequence, all of the possible variables in the course of emergency situations, from initial response to resource management, avoiding errors or delays, as the programme analyses all of the possible solutions to every accident suggesting the best of them, according with the guidelines defined in the Emergency Plan.

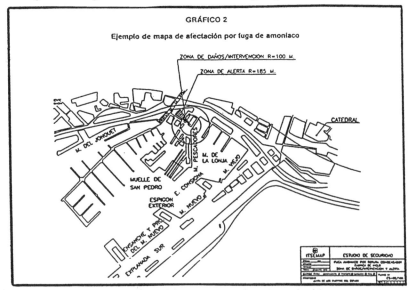

Fig. 2. Site Plan: Area affected by an ammonia release incident.

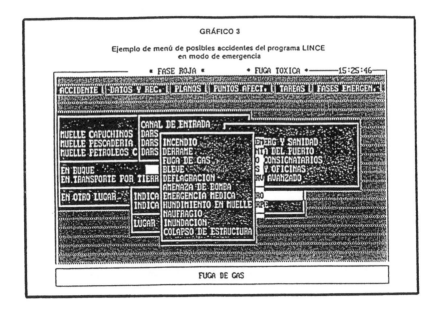

Fig. 3. Software LINCE'S Menu of Possible Accidents in Emergency Mode.

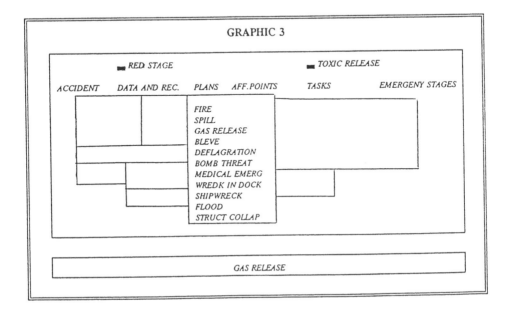

5. TYPES OF EMERGENCIES

LINCE, can manage all of the emergency situations contemplated in Emergency and Disaster Plans (fires, chemical spills, bomb threats, explosions, sinking, etc.), providing, in the cases that this is possible, the estimation of consequences based on mathematical models, if hazardous materials are involved in the incident.

Authorised user (with password) can also define up to six additional types of emergencies over and above those contemplated in the programme.

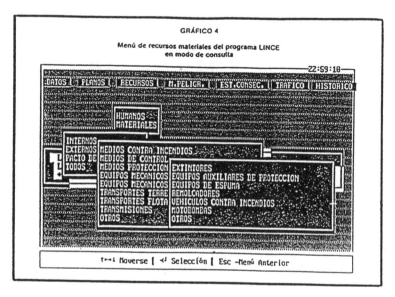

Fig. 4. LINCE's Software: Material Resources Menu
QUERY Mode

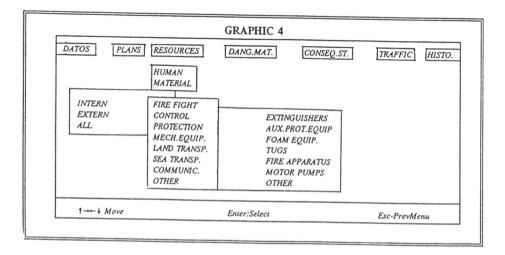

6. HAZ MAT FILES

LINCE has a built-in data base of the intervention instructions index of Haz-Mat incidents according to the IMDG codes, complemented by instructions of other internationally recognised entities for some 1,300 dangerous substances, by means of a rapid search system which presents the required data using UN numbers, product names, dangerous nature of the products, etc. .

The LINCE is delivered also with a data base which contains the physical-chemical properties of a vast number of dangerous substances in order to estimate the consequences of technological accidents (fire, explosion, gas escapes, spills).

The incorporation of information and data about news substances, including properties and response information can be easily performed by the operator.

7. ESTIMATION OF CONSEQUENCES

The programme incorporates several models for estimating the consequences of different incidents involving hazardous materials:

1. Stationary fire (solid or pool fire)
2. Jet fire
3. Vapour cloud explosions
4. Reservoir explosion
5. BLEVE
6. Boil-over
7. Ignition or explosion of solids.
8. Gas cloud dispersion (toxic or flammable)
9. Toxic or contaminating marine spills.
10. Toxic or contaminating spills on land.

Accident scenarios can be defined for any type of physical situation (storage tanks, on-board ship, pipe lines, process units, etc.) and for different meteorological conditions.

With the aid of these models the operator can analyse and define by himself the possible consequences of accidents postulated for his installations. These results can be stored (prerecorded scenarios) for future consultation.

7.1. PREPARATION OF EMERGENCY SCENARIOS

This package includes the possibility of prerecording emergency scenarios (in Consult Mode) for different physical-chemical-meteorological characteristics, so that they can be used at any time (in Emergency mode) by any one, eliminating the need to reintroduce data.

This option substantially reduces response time during an emergency as it eliminates the need to introduce information which has been previously recorded. This also permits advanced estimations of damages, and recommendations of actions, even when uncertainties still exist concerning some information (for example meteorological information) and which can be corrected later if necessary.

8. GRAPHICS

The LINCE, programme incorporates the cartography of the zones so that the projections of the consequences (shape and spread of clouds, spills, explosions, etc.) are made on these maps and plans, giving a "true picture" of the nature and gravity of the emergencies.

The cartography can be easily modified by the user via a CAD package, as an interface for DXF files is incorporated.

9. UTILISATION

Once installed, LINCE can be used in a number of different applications:

9.1. Direct assistance during emergencies

Through interaction with the user. He will feed the necessary information concerning the incident to the programme, which will give him in return concrete recommendations and/or relevant information (layouts, electrical switch gear, etc.) to manage the emergency in a real time condition.

9.2. Emergency simulator

Which permits the user through periodic use, to become more completely familiar with hazards and the possible solutions to emergency situations which could affect his plan to this application mode is particularly useful in a variety of ways:

1. To identify and correct defects in the safety systems and the Emergency Plan.
2. Evaluate possible improvements in these systems, and
3. Anticipate problems which could occur during a real emergency.

9.3. Training tool.

By an extensive and periodical use of the above mentioned capabilities of the programme.

22 ANALYSING EVACUATION MODELLING TECHNIQUES OF MIXED-ABILITY POPULATIONS

L. RUBADIRI
Department of Built Environment,
University of Central Lancashire, UK

Abstract
With an average of over eight hundred persons dying in fires in the UK each year, and several more suffering injuries as a result of fire, there is a pressing need to investigate the critical aspects affecting fire safety in different occupancy settings. This paper discusses the basic principles of evacuation models that investigate occupant movement in fire emergencies and discusses, in particular, their limitations; the most prominent being the omission of disabled people from the simulated populations. The notable effects of disability on occupant movement are revealed by using a novel concept that provides a mechanism for measuring evacuation capabilities of various classes of disabled people. This concept is developed through the use of an *evacuation performance index* (EPI) which is the relative ease of evacuating a disabled person compared to evacuating an able-bodied person. The application and use of the concept in research in Fire Safety Engineering is described in the final section of the paper.
Keywords: Evacuation modelling, *evacuation performance index*, fires

1 Introduction

Fire has significant potential to cause costly destruction. A vivid reminder of the large number of lives lost and the extent of damage in fires is illustrated in major disasters such as the Bradford [1], Hillsborough [2] and Woolworth's fires [3]. Both structural design and occupant behaviour were seen to be at fault in these scenarios due to the lack of early warning and the delayed occupant response in attempting to evacuate. A strategic approach towards understanding the relationship between the design of fire warning systems and occupant response in fires is therefore essential.

Fire Engineering and Emergency Planning. Edited by R. Barham.
Published in 1996 by E & FN Spon. ISBN 0 419 20180 7.

Evacuation modelling techniques are generally used for this purpose and some current ones are described in subsequent sections.

The basic principle of evacuation modelling is defined by the following equation which should be satisfied if occupants are to evacuate safely during an emergency:$t_{TL} \geq t$, where t_{TL} specifies the time for tenability levels to exceed their allowable limits for different classes of fires, and t specifies the maximum likely evacuation time of the occupants to leave a building.

For the purposes of this paper the emphasis is on t the available evacuation time and the sub-components that affect this value.

2 Background of basic model categories

Most of the current evacuation models can be grouped into two basic categories: psychological [4] [5] [6] and mathematical [7] [8] [9]. In both categories, models can be probabilistic, in that a range of possible occurrences is allowed for, or they may be deterministic which means that they are scenario specific and predict single possible outcomes. Most models are predominantly mathematical in that their underlying relationships are described by mathematical functions. Alternatively, mathematical models can be described as phenomenological in that they simulate the actual physical phenomena that affect safety. They can be further divided into several groups which include: analogue, empiric, systemic and knowledge-based models.

Psychological models focus on human behavioural aspects and can be sub-divided into two groups, namely: those with a time-based approach describing specific occupant behavioural stages and those with a time-based approach defined by discrete time frames. The fundamental difference between the approaches lies in the fact that the former emphasises the stages in behaviour that occupants experience as a fire develops and environmental conditions change. These stages are described as occurring within a series time frames of no specific duration. In the latter group, at every discrete time frame of a specific duration the action of an occupant can be determined as a result of analysing the surrounding environmental conditions. In this group the development of stages in behaviour are not discrete because they are governed by momentary changes in the surrounding conditions and are not necessarily manifested in clearly defined stages.

3 Objectives

The objective of this paper is to highlight the relevance of evacuation modelling in fire safety engineering, from a research viewpoint. Oftentimes greater emphasis is given to fire modelling and the time components affecting t_{TL}. The paper identifies the areas in which there is a significant need for further research into occupant movement and behaviour in fire emergencies. Some notable limitations are highlighted, particularly the need for investigative studies into the evacuation of occupants with disabilities and the need for a coherent framework relating current research findings on evacuation movement, which at present tend to be somewhat isolated. In attempting to find a solution to the above mentioned problem, the author describes a system for measuring evacuation capability of a mixed-ability population using an *evacuation performance index* (EPI) which relates three aspects of fire safety, namely; individual characteristics of disabled occupants, the amount of assistance they require, and building design and environmental factors. The author believes that the *evacuation performance index* of a class of individuals is primarily dependent on these three categories. Use of the index enables the assessment of the relative effects on the evacuation capability of an evacuee of changes to the amount and type of assistance provided to the evacuee, and changes to the building design and environmental factors. The concept encourages the use of a coherent approach to analytical research methods in evacuation modelling [10].

The following section discusses some of the weaknesses in evacuation modelling. The author hopes that this overview will highlight key areas that warrant further research and where improvements to modelling techniques can be made.

4 Limitations in Evacuation Modelling

There are several limitations in evacuation modelling of which the most important are as follows:

• *Evacuation modelling does not lend itself to precise observation or analysis.*
As a result several assumptions are made during the use of simulation methods to study occupant movement and behaviour during fires. According to Kendik [11] who has analysed a number of major evacuation models, "... all of them appear to make several assumptions partially to overcome the gaps in technical literature which makes their validations against real-world events or fire drills necessary...". She continues by stating that only a few of these models are in fact calibrated in this manner and are able to provide quantitative results.
• *The use of complex of equations and the attaching of unjustified significance to numbers in these equations*
Basic principles are sometimes hidden behind sophisticated and complex notations. This limits the scope of understanding of those responsible for the application of model findings in the design of means of escape. According to the EGOLF research group [12] the quantification of levels of safety rely heavily on statistical information and techniques. The models are mainly computer-based and are criticised by some engineers as being too complicated. Frequently unjustified significance is attached to numbers resulting from model equations because they cannot be interpreted with adequate background knowledge.
• *Simulated populations are limited to able-bodied occupants*
Empirical studies on occupant movement are often limited to able-bodied populations. Occupants with disabilities that may their hinder movement and that of others, are rarely acknowledged. This is despite the fact that disabled people appear to be at the greatest risk in fire emergencies where movement speed is so important. However, both disability and unsympathetic design of buildings appear to be key negative

influences on occupant movement and the techniques used to measure movement are so diverse, often leading to inconsistent results.

• *Outdated validation studies*
The empirical data to test the validity of models is often outdated. It is unfortunate that the cost and the complexities associated with setting up validation studies for various models are prohibitive. However, they are essential in research to ascertain the credibility of these models.

• *Large numbers of parameters to analyse*
Some complexities arise from the development of models with a large number of parameters. In addition, the depth of analysis in these models is questionable. Hinks [13] states that these parameters will have varying levels of importance and influence on escape potentials of occupants. It is essential therefore to have a system that classifies these parameters in some hierarchial arrangement and that has provision for accurately measuring their effects.

• *Closed characteristic of most models*
The closed characteristic of most models limits their expansion when new research information becomes available. This limitation implies that additional information may be continually tagged on to existing information with the risk of losing track of the primary objective of a given model.

Summary
It is true to say that, 'fire safety is often considered in a fragmentary way' [14]. The elements which combine to produce fire and possible loss of life and property tend to be effectively regarded as independent of each other. There is a lack of coherency in the techniques used in evacuation modelling. What is lacking is a unifying philosophy that links the approaches to modelling in such a way that these approaches are recognised as extensions of a basic safety criterion.

5 A Coherent Approach to Evacuation Modelling - EPI Concept

In this section the author describes a coherent approach to evacuation modelling that provides a method for predicting evacuation times and for applying these results to the design of escape routes [12]. In order to predict the time an occupant would take to evacuate a building, it is necessary to have a quantifiable attribute which defines his/her evacuation capability and which is sensitive to variable external conditions, for example, building design. This measure of the intrinsic evacuation capability of an occupant is described as his/her *evacuation performance index* (EPI), which is defined as the unassisted speed of a person relative to that of an able-bodied person along a straight obstacle-free route of a pre-specified distance.

Basic EPIs of occupants are determined primarily by their respective disabilities and mobility aids. During evacuations this basic EPI is dynamically modified into an effective EPI which determines the actual evacuation capability of the occupant during an emergency. The modification of basic EPI into effective EPI is brought about by the following primary factors.

1) Individual characteristics: of occupants can be either physical and psychological. Physical characteristics include occupants' disability-mobility aid combinations while psychological characteristics include behavioural factors such as 'panic'.

2) Managerial aspects: can be measured in terms of the assistance provided to occupants. EPIs have been observed to undergo notable changes if additional help is provided [10] [15]. Whenever necessary assistance is provided to evacuating disabled people, this has the general effect of increasing their speeds or EPIs. Unnecessary assistance typically leaves the EPI unchanged or may even reduce it.

3) Environmental factors: generally expressed as crowd densities have been shown to influence occupant speeds [10] [16]. The authors observed that at low crowd densities, able-bodied occupants make purposeful attempts to avoid being an obstacle to disabled people. In some cases, they may even assist a disabled person by, for example, opening doors. In large crowd densities, however, occupant speeds tend to decrease.

4) Building design factors: such as the configuration or geometry of building layouts, number and positions of fire doors, dimensions of corridors, positions of lighting, floor coverings and textures also influence EPIs.

6 Application to Research

The actual process of measuring EPIs begins by deconstructing a given evacuation route into primary sections such as the rooms, corridors, stairwells etc. Having classified occupants according to their disability-mobility aid combinations, their EPIs could be measured along each section of the route. This could be carried out by using carefully monitored fire drills. The time taken to traverse each section could be recorded using strategically placed video cameras with a time-display mechanism. The simplest building layout would require not more than eight mounted cameras and some observers to monitor the evacuation.

Characteristic times could be calculated for each disability-mobility aid combination, made possible through the use of a design procedure incorporating measures of EPI. A first approximation of the proposed design procedure might proceed as follows: Consider that nominal EPIs have been determined for different disability-mobility aid combinations. Design in such a case may proceed along the lines of, for each evacuation route out of the building:

1) First, the designer computes the time t_{TL} for tenability levels to exceed their allowable levels, and the travel distance d through the escape route.

2) Second, the effective EPI of each disability-mobility aid combination along the evacuation route is calculated. This calculation will utilise expressions defined to take into account any assistance which will be available to disabled evacuees, and any building design and environmental factors of the evacuation route that might modify the nominal evacuation performance indices of the evacuees.

3) Next, the worst case evacuation time, t_W is computed from Equation (1)

$$t_W = \frac{d}{I_M \times v_O} \tag{1}$$

where, I_M is the smallest effective EPI computed from Step (2) above, and v_O is the nominal evacuation speed of an able-bodied person. Note that v_O itself varies with building design and environmental factors, although not with level and type of assistance.

4) Finally, the designer ensures that $t_{TL} \geq t_W \times SF$, where SF is a factor of safety.

If the inequality $t_{TL} \geq t_W \times SF$ cannot be satisfied, the designer has to modify the assistance to be made available to disabled evacuees and the building environmental factors such that Step (2) returns a high enough I_M. However, if the fire safety evaluation is being performed at the architectural design stage, then the layout of the building may be modified so that Step (1) returns a higher value for t_{TL} or a smaller value for d. Once the inequality has been satisfied for a particular combination of level and type of assistance to be made available to evacuees and building environmental factors, then managerial steps have to be taken to ensure that these conditions prevail in an emergency. A real design situation is more complex as the inter-relations between a number of different possible routes have to be analysed.

The design procedure described above computes a lower bound or ultra-safe evacuation time since t_W is based on I_M, the smallest effective EPI over the range of disability-mobility aid combinations. An alternative procedure which is perhaps more realistic and which returns an upper bound value of t_W is to use characteristic times in place of the above relation in Step (c) such that,

$$t_W = \max \sum_{i=1}^{n} \frac{d_i}{I_i \times v_i} \tag{2}$$

where, $i = 1,....,n$ subdivides the evacuation route into elemental sections for which EPIs are known and the summation in Equation (2) is computed for each disability-mobility aid combination.

7 The Value of the EPI Concept in Research

This novel concept defines a coherent procedure for fire safety engineering design founded on a measure of evacuation capability. The evident strengths in this concept from a research viewpoint include:

- *The use of simple, economical empirical exercises*
The suggested evacuation exercises to obtain EPIs are simple and inexpensive. Resources are limited to ordinary video cameras and measuring equipment such as tapes for measuring distances. In this way the difficulties experienced with complex experimental methods are avoided.

- *Flexibility in the number of parameters analysed in a given scenario*

Any number of parameters can be analysed in a given scenario. A wide range of parameters whose effects on evacuation capability are considered critical can be measured. The model is therefore flexible and grows with increasing knowledge from research.

- *Recognition and classification of disabled people*

The presence of disabled people is acknowledged using a classification system based on their disability-mobility aid combinations. Thus the differences in evacuation capabilities in an evacuating crowd are adequately represented.

- *The precise definition of evacuation capability and the flexibility of EPI*

EPI provides a precise definition of evacuation capability which is measurable. Although it shares some similarities with alternative measures of escape potential [17] [18], its uniqueness lies in its clear description of evacuation capability and its flexible nature which is manifested by its variation with changes in surrounding conditions.

- *The use of simple methods for determining evacuation capabilities*

The determination of EPIs and the use of the design procedure are not complicated and simple to understand.

- *EPI provides link between research findings and design*

There has often been some degree of conflict between research findings and their application in the codes of practice. EPI provides a gateway for channelling results from research into a format that can be easily adopted by the design codes.

- *The role of EPI as an assessment tool*

The efficiency of an evacuation route can be assessed by computing characteristic times using EPIs. If EPIs are below acceptable levels for any class of people then the corresponding times will be too high and modifications may be implemented at the drawing board stage.

- *The wide scope of application*

The transferability of the concept to various occupancy settings is an added advantage.

8 Conclusions and prospects for further work

The EPI concept provides a valuable stepping stone for future strategies in research in evacuation modelling of mixed-ability populations. It is evident that this concept links all approaches to research without necessarily favouring any particular one. It provides a framework that coherently links all approaches to modelling all the while ensuring that the scope for investigating all critical evacuation parameters is sufficiently broad. It is a useful assessment and design tool that is simple to understand and apply. Further work in developing the concept is encouraged by the author using as many ranges of disability-mobility aid combinations to provide a representative mixed-ability population. The results could be adopted by the codes of practice. While the EPI concept may not solve all the drawbacks of modelling it has attempted to address some of the primary limitations.

9 References

1. Popplewell (1986) **Committee of Inquiry into Crowd Safety and Control at Sports Grounds: Final Report.** HMSO.
2. Taylor (1990) **Inquiry into the Hillsborough Stadium Disaster: Final Report.** HMSO, London.
3. Joint Fire Prevention Committee (1980) **Report of the Planning/Legislation Sub-Committee on the Fire at Woolworth's' Piccadilly, Manchester on 8 May 1979.** HMSO. Fire Department.
4. Canter, D and Matthews, R. (1976) **The Behaviour of People in Fire Situations: Possibilities for Research.** CP 11/76, BRE.
5. Sime, J. (Private Communication).
6. Bryan, J.L (1988). **Behavioural response to fire and smoke.** SFPE Handbook of Fire Protection Engineering. Quincy, MA: National Fire Protection Association, section 1, ch. 16, pp. 269-285.
7. Hallberg, G and Nyberg, M. (1987) **Human Dimension and Interior Space.** Department of Building Function Analysis, Royal Institute of Technology, Stockholm.
8. Galea, E.R., Galparsoro J.M.P. and Pearce J. (1993) A brief description of the EXODUS evacuation model, **Proc. Int. Conf. on Fire Safety,** 18, pp.149-162.
9. Rubadiri L. (1992) Transfer report from MPhil to PhD, University of Central Lancashire.
10. Rubadiri, L., Roberts, J.P. and Ndumu, D.T. (1993) Towards a coherent approach to engineering fire safety for disabled people, **Proc. CIB W14 Seminar/Workshop on Fire Safety Engineering.** Ulster University, Northern Ireland.
11. Kendik, E. (1986) Methods of design for means of egress - Towards a quantitative comparison of national code requirements, **Fire Safety Science - Proceedings of the First International Symposium,** Grant, C and Pagni, P.J (ed.), pp. 497 - 511.
12. EGOLF (1994) **A Framework For Research in the Field of Fire Safety in Buildings by Design.**

13. Hinks, A (Private Communication)
14. Beard, A. (1986) Towards a Systemic Approach to Fire Safety, Fire Safety Science - **Proceedings of the First International Symposium**, Grant, C and Pagni, P.J (ed.), pp. 943 - 952.
15. Rubadiri L. and Roberts J.P. (1994) A review of the effects of mixed-ability populations on evacuation model predictions, submitted to **Safety and Health Practitioner Journal**.
16. Rubadiri L., Ndumu D.T. and Roberts J.P. (1994) Assessment of human and structural safety of sports grounds, to appear **IABSE Symp. on Places of Assembly and Long-Span Building Structures**, Birmingham, 7-9 September.
17. Marchant, E.W. and Finucane, M. (1978) Hospital fire safety - Non-attendance and patient mobility, **2nd International Seminar on Human Behaviour in Fire Emergencies**, Edinburgh, pp 115-144.
18. Hallberg, G. (1988) Evacuation safety in dwellings for the elderly, **Safety in the Built Environment**, Sime, J. (ed.) SPON.

PART TWO

APPLICATIONS

23 INTRODUCTION

S.-E. MAGNUSSON
Department of Fire Safety Engineering,
Institute of Science and Technology, Lund University, Sweden

There are many new and interesting developments taking place in industrial fire safety and I would just like to mention a few of these to introduce what's going to happen today. As you all know, industrial safety was started in the nuclear industry and in the airplane industry. It has now found a wide application in other areas such as the chemical process industry and the off-shore industry and what is happening now, I think, (and this is very important) is a number of developments.

The use of quantitative risk assessment is coming into force over a large number of sectors - at the same time really. So I think that what will be required in the next few years is the introduction of much more stringent procedures. This can be by the standardisation of calculation models; it can be the standardisation of the selection of scenarios, or it can be by the standardisation of input data and its presentation. Anyway, things are going to be much more strict and stringent in the next few years. It will not be a numbers game any longer, which it has been for the last ten years.

The management of quality, or "quality management", is becoming a very important issue in this area and, of course, you are all aware of the Euro-standards such as ISO 9000 and so on. These are going to have a large impact on the way we are doing these things.

There are other factors emerging, such as the emphasis on environment and, all in all, it points to the fact that procedures which have been widely diverging across a number of industrial sectors now are converging. This is a result, not the least, of the CEC Major Industrial Hazards Project. Environment protection is an area under quite rapid expansion and development; therefore, today's papers are very topical and they point to that fact.

The first subject that we will hear about raises some of these issues while addressing the subject of the use of foam and water for the protection of equipment engulfed in fire. Later we will hear a review of the experimental results from a major industrial research project on fires in chemical warehouses. This is a project which is still ongoing - and will hopefully be finished in 1996. However, a lot more research remains to be done in this area - but I do think that we are now seeing some progress.

We are also going to hear about some very interesting research with practical applications. The behaviour of glass in fire is an important subject and deserves much more study than has been possible up to now. We are, therefore, going to listen to what I know will be a very interesting paper on this topic. It is a very practical problem and I, for one, am looking forward to the development of more research in this area in the future. Also, today, we have a paper about the testing of axially loaded and restrained steel columns. This is a very complicated problem. I, myself, was doing research on this in the early 1970s and I must admit that the results were rather meagre. I am very pleased to see that progress is being made in this area.

Later in the day, the emphasis goes from looking at the applied research and development of the various fire safety aspects to a wider area entitled "E.C.

Fire Engineering and Emergency Planning. Edited by R. Barham.
Published in 1996 by E & FN Spon. ISBN 0 419 20180 7.

Perspectives". This will start with a review of the very large and complex area of European standards and then look at other aspects of laws and directives which affect fire safety. There has been a lot of European activity in these areas during the last few years. This wider perspectives theme then allows us to introduce some papers later in the day which take us into looking at several varied techniques - underlying principles and applications of various types of modelling, fault tree analysis, decision support systems, neural networks, knowledge-based systems, etc.

It is good to see that there is such a lot of challenging work going on and that we are now seeing the start of a steady flow of knowledge from research into actual applications in fire safety operations.

APPLIED RESEARCH AND DEVELOPMENT – INDUSTRIAL FIRE SAFETY

24 FOAM AND WATER FOR THE PROTECTION OF EQUIPMENT ENGULFED IN FIRE

J. CASAL and E. PLANAS
Department of Chemical Engineering,
Universitat Politècnica de Catalunya, Barcelona, Spain
L. BORDIGNON and A. LANCIA
TRI srl, Scanzorosciate (BG), Italy

Abstract
This communication presents some results of the work which is being developed at Barcelona to treat a part of the large amount of original data collected in the frame of the project "Foam and water deluge system for offshore oil platforms" CEC Project TH-15125/89-IT, developed by the Italian firms SABO SpA, Tecsa Spa and TRI-Tecsa Ricerca & Innovazione srl and the British company RM Consultants Ltd.
A set of 92 large-scale tests were effectuated, including pool-fires, liquid jet-fires and 3D-fires. Amongst them a group of 8 tests has been selected in which a tank was completely engulfed in pool-fires of hexane and kerosene (4 m^2 and 12 m^2); a foam and water deluge system was used to extinguish the fire. The evolution of tank temperatures and fire structure as a function of time is described, for the different operating conditions. Conclusions are derived on the efficiency of the system.
Keywords: cooling, equipment, large-scale tests, fire, foam & water

1 Introduction

In the event of a fire in an industrial installation with a high density of equipment and with a significant inventory of potentially hazardous materials -like, for example, an offshore oil platform or a process plant- the two objectives of an active fire protection system are:

1. Extinguishing or at least controlling the fire.
2. Cooling the structures and the equipment.

The second scope is often the most important one since further structural damages mean further loss and specially because the damage to certain objects can result in additional loss of containment of flammable substances and hence in the escalation of the fire scenario. This is particularly true for the offshore rigs, where a scenario scalation can result in the loss of a number of lives and even of the complete platform, with possible consequences on the environment.

Fire Engineering and Emergency Planning. Edited by R. Barham.
Published in 1996 by E & FN Spon. ISBN 0 419 20180 7.

In the conventional approach, cooling and extinguishing are addressed by two different sistems: water deluge systems and foam systems.

Several interesting features have made water the most used fire extinguishing agent: it has high specific heat and very high latent heat of vaporization, and furthermore it is usually available at a low cost. However, the use of water has as well several disadvantages, the main one being originated by its relatively high density, which makes it more dense than most hydrocarbon fuels. Furthermore, hydrocarbons are also immiscible with water; therefore, in the event of a pool-fire water will nor cover the burning fuel -thus extinguishing or reducing fire- neither mix and dilute it. The hydrocarbon will maintain a lighter burning layer, and even in some cases water may flash (if it enters a hot mass of fuel) and spread fuel.

If a cover for the burning surface is required -and this will help significantly in the control of fire- foam solutions must be used. However, foam agents are much more expensive than water, and on the other hand usually they are available in much more reduced amounts.

An interesting approach relies on using a single deluge system which discharges foaming solution for a given minimum time and then continues discharging water. Such systems, known as "Foam & Water Deluge Systems", should discharge a kind of foam which is fluid enough to behave like water for cooling purposes but stable enough to possess useful extinguishing properties in spill fires.

F&W deluge systems are quite attractive, specially for oil off-shore platforms and process plants, due to the possibility to achieve a fast fire control, to the simplification in the plant structure and in its operating procedures, and finally due to the resulting reduction in the costs of the system.

The actual "deluge" systems are the ones where the foam discharge heads are located over the fire area and apply the foam in the form of a snow fall. This paper is mostly adressing the use of F&W spray systems, i.e. based on a number of spray heads which are discharging the cooling fluid directly onto the surface of the parts to be protected.

In all these systems, calculation methods are available and widely used for one of the aspects involved, the fluid flow through piping. However, there are no calculation procedures available for the other aspects related specifically to the operation of sprinklers; this is due to the fact that very complex mechanisms play a role in the extintion of a fire with water or foam, with badly-known variables like, for example, the size and behaviour of droplets. Therefore, the design of sprinkler systems is rather based on empirical approaches. This is why an effort should be done -and this was the aim of this project- to contribute to the progress in designing this kind of fire protection systems.

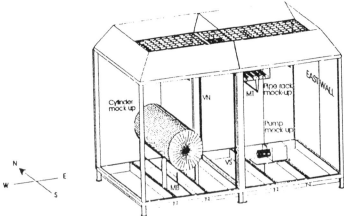

Fig. 1. Structure of the platform module mock-up used for the full scale tests series in Borås.

2 Experimental installation

A total of 93 full scale fire tests were conducted at the indoor fire test facility of SP (the national Swedish testing laboratory) in Borås [1].The tests were carried out within a specially constructed rig simulating a typical offshore or process plant module and comprising in particular the mock-up of a leaking pump, of a pipe rack and of a pressure vessel. Two sides were closed by steel bulkheads and a series of deluge manifolds were pre-installed (Fig 1).
   The mock-up was equiped with 90 thermocouples and 8 plate radiometers (Fig. 2). Total and convective heat release rates were measured by the "SP Industrial Calorimeter" which is able to measure up to 17 MW HRR. 3 radiometers were installed in the vicinity of the fire module. Pressure and flow rate of the deluge system were also logged. One Hughes thermographic camera and two camrecorders were used in all tests.

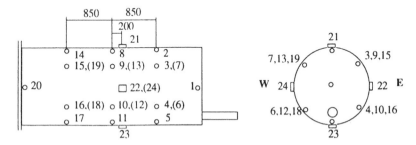

Fig. 2. Positioning of some thermocouples.

3 Description of the tests

8 tests were effectuated in which a horizontal cylindrical tank (ϕ = 1.2 m; L = 3 m) was completely engulfed in a pool-fire. During the test, temperature was measured at 20 points on the inner wall of the tank; furthermore, temperature was measured at 4 external points (gas temperature). 4 tests were effectuated with hexane (with a pool surface of 4 m²) and 4 with kerosene (with a surface of 12 m²).

Table 1. Description of the selected experiments.

Test number	Pool area	Fuel	Extintion with	Configuration	S.A.R.[2] (l min⁻¹ m²)	Deluge flow (l min⁻¹)
12	4 m²	Hexane	Foam	two SPK0/125°	14,2	246
13	4 m²	Hexane	Foam	two SPK0/125°+SPY0/90°	14,2+	296
14	4 m²	Hexane	Water	two SPK0/125°+SPY0/90°	14,2+	296
15	4 m²	Hexane	Water	two SPK0/125°	14,2	246
88	12 m²	Kerosene	Foam	two SPK0/125°+SPY0/90°	14,2+	296
89	12 m²	Kerosene	Foam	two SPK1/90°+SPY0/90°	14,2+	296
90	12 m²	Kerosene	Foam	two SPK0/125°	6,5	112
91	12 m²	Kerosene	Foam	two SPK1/120°	6,5	112

SPK0,SPK1 are medium velocity sprinklers , from two different manufacturers.
SPY0,SPY1 are "pig-tail" spiral shaped sprayers , from two different manufacturers.

The distribution of the 20 thermocouples and the 4 plate radiometers on the tank walls can be seen in Fig. 2. Fig. 3 shows the position of the three radiometers installed at a certain distance from the tank.

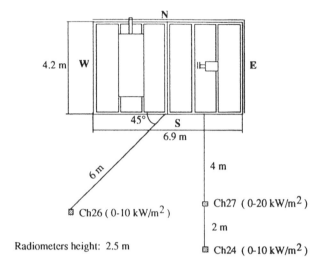

Fig. 3. Position of the radiometers with respect to the experimental module.

4 Reproducibility of tests

In order to verify the reliability and reproducibility of experimental measures, the evolution of temperature (during the first minute of the test) has been plotted for runs number 12, 13, 14 and 15, and 88, 89, 90 and 91 as a function of time for each thermocouple; the values should be identical for these tests for the time previous to the sprinklers operation.

Generally speaking, these results show a practically linear trend (Fig.4-a), essentially identical for the three first tests; run number 15 shows lightly higher values with respect to the rest of results, although the trend is similar. It should be noted that the thermocouple number 14 exhibited a different trend, rather parabolic than linear. For the 12 m² pools (Kerosene), the development of fire over the whole surface of the pool was much slower (fig.4-b) and less reproducible.

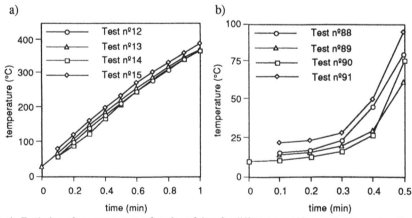

Fig. 4. Evolution of temperature as a function of time for different tests (thermocouple number 5).

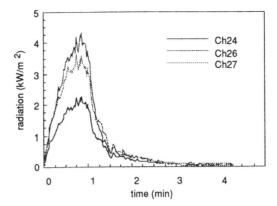

Fig. 5. Radiation intensity vs time at three different points external to the experimental module (test number 13).

On the other hand, the data obtained from the radiometers have been used to plot the variation of radiation intensity as a function of time (Fig. 5); as it can be observed, after the first minute the fire can be considered to be fully developed and with a regime practically stationary.

5 Discussion

The study of the diverse plots of temperature vs. time show that in most of them five different zones can be established, the transition from one to the next one being indicated by a change in the slope (Fig. 6). A careful analysis of the video films taken during the different tests helped in the interpretation of this behaviour.

First of all there is an initial step in which fire develops to steady state (corresponding approximately to the first minute after the ignition); in this stage the vessel temperature increases linearly with time, at a high rate. The slope of the line varies with the location of the thermocouple (depending on the degree of flame impingement). The slope changes significantly (decreases) in the moment in which the operation of sprinklers starts.

With the start-up of sprinklers the presence of important turbulences is observed, with a wind stream from the W wall of the module (see Fig. 7 (g) and (h)); this makes the flames tilt towards the E side. Thus,the W side of the vessel is less affected by flame impingement and the thermocouples located on this side measure lower temperatures, with the corresponding decrease in the slope of the plot; at the same time, in the E side the opposite phenomenon is observed, with an increase in the slope.

At a certain moment, however, a maximum temperature is reached and the slope becomes negative; this situation corresponds to the moment in which the cooling effect of water starts. Nevertheless, as flame impingement continues, the temperature decreases slowly.

Later on, as the pool has been progressively covered by the foam, the flames become smaller and there is no more impingement on the tank wall; the cooling action of water is now very strong and the slope changes significantly, with a dramatic decrease of wall temperature. Finally, when the flow from the sprinklers is stopped, a light increase of temperature is observed again due to the radiation from the module hot walls surrounding the tank.

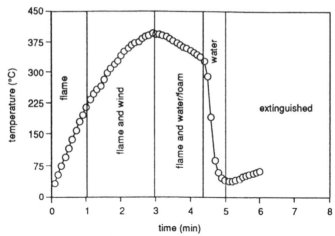

Fig. 6. Variation of temperature as a function of time for Ch6 (test n°12).

Fig. 7. Flame shape at different moments during the test (test n°12).

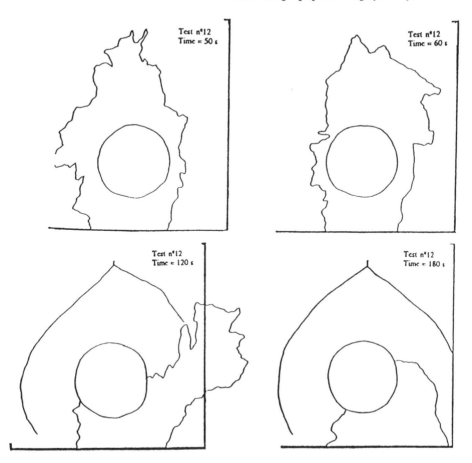

Fig. 7. (Cont.)

The different heat transfer mechanisms actuating in each step can be summarized as follows:

Step 1	Convection from hot gases Radiation from hot gases Conduction through the steel
Step 2	Convection from hot gases Radiation from hot gases Conduction through the steel
Step 3	Convection from water Radiation from hot gases Conduction through the steel
Step 4	Convection from water Conduction through the steel
Step 5	Radiation from surrounding walls convection to air conduction through the steel

Heat conduction through the wall of the tank is not important, as the heat conductivity of steel is very high, the prevailing mechanisms being therefore those actuating on the external surface (convection and radiation).

These comments concern to the more complete plots (Fig. 6). In some cases, one of the intermediate steps did not appear; for example, if flame impingement finished in the same moment in which water reached the measuring point, then step 3 did not exist: just after the maximum a sudden decrease of temperature was observed (Fig. 8-a).

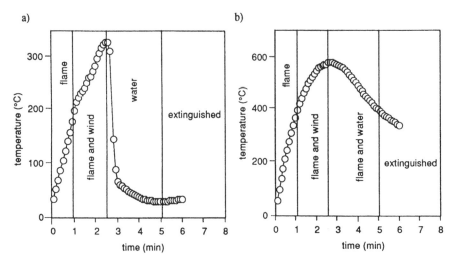

Fig. 8. Variation of temperature as a function of time   a) for Ch12   b) for Ch5.

A different behaviour was observed in the bottom zone of the tank. Here, there was flame impingement during practically all the test and cooling water could not reach this zone; therefore the change in the slope was smoother, with much lower cooling rates (Fig.8-b).

All these results concern to the tests in which foam and water mixtures were used. When only water was used as extinguishing agent, a similar behaviour was observed, although the existence of the different steps was not so clearly notticeable; this was due to the fact that in most runs the extintion of fire was not reached and even the control of flames was difficult. Therefore, only the three first steps were observed and in many cases the temperature increased again at the end of the test.

In tests 13 and 14 the extinction system was provided as well with one sprayer in the central zone of the tank bottom (besides the two aforementioned sprinklers). This originated a much faster cooling of tank wall in this zone. The effect of this sprayer was practically the same both when only water or foam and water mixtures were used; however, foam was useful to cool the wall and simultaneously it contributed to extinguish the fire, while water had only a cooling effect.

In the case of tests 88 to 91 with 12 m² pool, the temperatures observed were usually higher than for the other cases; nevertheless the trend of the plots of temperature vs. time were very similar to previous tests. In tests 88 and 89 the fire was easy to control, probably because the preburn time was established on 30 seconds and the fire was not totally developed. In tests 90 and 91 the extinction was difficult, and even impossible when only water was used as extinguishing agent.

Conclusions

The use of foam and water mixtures applied directly on the equipment engulfed in fire has proved to be useful both for the protection of this equipment - cooling - and for extinguishing the fire. The action of foam is essential - specially during an initial time - to reduce the fire by the formation of a covering layer on the pool. As this decreases progressively the impingement of flame on the equipment surface, the foam/water mixture or only water in a subsequent step leads to an intense cooling of equipment; this process is clearly shown in the temperature vs. time plots, where the changes in slope show the transition from one step to the next one. Water, very useful for the cooling purposes, is much less efficient in the first step, the control of fire being essentially achieved by foam.

The presence of a vessel shadowing a pool fire area from the direct deluge application has a negative effect on fire control performance by the deluge system. This problem can be solved when upward sprayers are mounted below the vessel for cooling reasons. In such a case an increase of deluge rate is not required.

For the 12 m^2 pool fires, longer preburn time has a negative effect, but not very strong. The request for limiting the system actuation time to 30s can be justified more in terms of fire effect on plant than on the ability to control the fire.

References

1.   "Foam and water deluge systems for off-shore oil platforms", Project Summary Report. EC Contract: TH-15.125/89-IT.

2.   "Fixed water spray and deluge protection for oil and chemical plants", IRI Information IM.12.2.1.2, June 3, 1991.

25 DEPLOYMENT OF FIRE PREVENTION EQUIPMENT: THEORY AND EXPERIENCE – FUTURE COMPUTERISATION AND THE EXAMPLE OF THE BOUCHES-DU-RHÔNE

J.-C. DROUET
University Technology Institut of Aix-en-Provence,
Health & Safety Department, Marseille, France

**Can the number of fire appliances
that will be needed to deal with a particular
risk situation prédicted?
If so, how and when?**

**The paper that follows is the answer to
this question from the Fire and Emergency Department
of the French county ("Département") of Bouches-du-Rhône.**

PRESENT SITUATION.

On every summer evening, just before 6 p.m., the French weather bureau ("Météo France") broadcasts a weather forecast for the afternoon of the following day which includes predictions of fire risk bases upon available data. The bulletin is updated ans broadcast again just before 10 a.m. the next morning.

The information is routinely telexed to the different services concerned. It is also accessible on the French videotex system ("MINITEL") by using a particular password, where it is accompanied by the hourly or 3-hourly observations from the automatic weather stations together with the calculations.

At the levels of the regions, this prediction of risk is one of the factors taken into account in deciding whether the "water bomber" aircraft should take off as a preventive measure. This decision is taken by the Civil Defence Interregional Centre for Operational

Fire Engineering and Emergency Planning. Edited by R. Barham.
Published in 1996 by E & FN Spon. ISBN 0 419 20180 7.

Coordination (CIRCOSC) located at Valabre near Aix-en-Provence. A similar decision has to be taken in each county concerning the deployment of land-based teams by the county Fire and Emergency Operational Centre (CODIS).

The decision-making process can be very precise because each county is divided into "meteorological zones" and forecasts are issued for each (there are five in the Bouches-du-Rhône). The following information is broadcast for each zone :

-shade temperature;
-wind speed and direction;
-dew point and relative humidity;
-cloud cover;
-the ground temperature

This value can now be calculated, it is measured and given on request by the Marignane station only. This reading was originally introduced for research purposes. From these data, the ground moisture index is calculated using Thornthwaite's formula.

The following aids to decision-making are also calculated and broadcast :

-the speed of propagation of fires;
-the possible fire "threshold" (the contribution of the wind to the probability of a fire starting);
-the numerical risk;
-the final risk.

These will be described in the order given.

Speed of fire propagation on the wind, and other factors used

to determine the fire configuration.

The method used for calculating fire propagation speed comes from the thesis by J.C. Drouet, published in its final version in 1972. The problem originally put (1966) to the author by Captain Maret (now Inspecting Colonel in the Civil Defence), then Director of the Valabre test centre, was to devise a géometrical model for the fire. The idea was that by knowing the position of the fire in advance it is possible to take decisions about deploying appliances on the ground in order to put down a chemical barrier.

The elliptical configuration was chosen primarily because the shape of a free fire in its early stages, seen from above, resembles thet of an ellipse and, secondly, because to a first approximation the envelope curve obtained for two circles of the same radius whose centres are very slightly apart is an ellipse. This very low eccentricity is obtained with a very light wind for a fire propagating in a isotropic environment. The mode of propagation will be described later in the paper. The functions and variables were determined by logical reasoning or from information known elsewhere and the values assigned to the different coefficients were deduced from observations made on the ground.

The modele therefore uses what is now called a semi-empirical approach; it was produced in 1972 and observations made *in situ* since then have confirmed its validity.

The formula used by Météo France is the following :

$$VPmto = 180*Exp(TE*1.714)*Tgh((100-Res)/150*(1*2*(0.843*Tgh(V/30-1.25))))$$

This is the "reduced" version of the formula : it is possible to add the possibility of knowing the influence of sunlight on the fire propagation speed on the wind :

replacing exp $(T*1.714)$ by $Exp(TE*(1+(XSE/1.4))*0.035)$

The width of the ellipse is obtained using the formula :

$$VPE = VSA*SIN((3.14/2)*(((-V/140)+1)/((V/140)+)))$$

where VSA is the propagation speed disregarding the effect or the wind.

The values obtained correspond to a fire propagating in homogeneous and isotropic vagatation on horizontal ground with kermes oaks and argeiras, the correction factors (multiplying coefficients) are for pines $(1-V/250)$ and holm oaks $(1-V/250)*Exp(-XSE*0.11)$.

Where :
-VPmto = propagation speed of the fire on the wind for horizontal terrain and homogeneous vegetation -mixed vegetation dominated by kermes oaks- broadcast by Météo France;
-exp = exponential;
-TE = temperature un degrees Celsius;
-Tgh = hyperbolic tangent;
-Res = ground moisture index (saturation to 150 mm);
-V = wind speed;

-XSE = sunlight factor (focal number of a camera divised by 10 wich gives 10 in full sunlight);

-VPE = speed perpendicular to the wind, semi-latus rectum of the ellipse.

Various correction factors exist and are put in "by the hand" :

-the geological nature of the ground (for types of vegetation on an unusual subsoil leading to changes in propagation speed);

-other plant combinations (indicated above);

-correction for the moisture content of vegetation (during the growing season it is richer in water for a given ground moisture index that when mature - thesis by Ollivier U3 Marseille 1975).

-an unusual drought (1989 with impact on 1990) leads to a 5 mm moisture index with an increase in speed of 25%.

-when the vegetation dies down during the winter the moisture index is fixed at 45 mm because the vegetation tested manually burns in the same way as when the reserve is 45 mm in the summer.

The ground relief is taken into account as follows :

The boundary of the fire is taken to be the intersection of a cone penetrating point downwards into the ground with the boundary of the fire on horizontal ground, the apex of the cone being vertically below the starting point of the fire.

The author's task is to inform CODIS 13 in advance (since vehicle registration numbers in the Bouches-du-Rhône county include a "13" the practice is to give this number to the county's services) of the status of the different correction factors and to observe events to see whether the different theories do in fact apply on the ground. The objective is to ensure that knowledge and experience acquired by the service can be applied in the future as automatically as possible by the computer system now being set up. It is important to note that the computer's function will be to suggest solutions to the CODIS 13 staff but the actual deployment of aquipment towards the fire will always require manual confirmation by an operator.

At present the entire operation of controlling forest fires relies upon early detection of the fires and the immediate "reflex" despatch of large numbers of personnel and equipment for rapid and effective action. It is a matter of anticipating needs so as not to be controlled by events.

The objective of computerisation is therefore to determine in advance the personnel and equipment requirements necessary for effective firefighting. The practical result should be that the person responsible for directing operations will find sufficient ressources when he reaches the location to make it unnecessary to call for reinforcements.

The notion of the influence of the wind on the possible starting of fires.

(originaly known as the "possible fire thresholds") comes from the propagation model for the fire front described below. The principle is that if radiation from the fire directed forward and downwards into the ground cannot vaporise the moisture present in the ground debris and raise its temperature to the ignition point, the fire is unable to propagate. Radiation represents about 10% of the energy produced by the fire (conduction 1%) and the radiation used for propagating the fire varies from 25% of the total in zero wind to a little over 30% when the wind is flattening the flames. The height of the flames can be calculated fairly precisely by assuming that the combustion gases rise in the horizontal air flow of the wind. This height is the greater the more the situation is stoichiometric because the Oxygen Limit Index (OLI) must be taken into account. For light debris burning in these flames the index is about 14.75%, meaning that only a quarter of the oxygen in the air is used (50% in the embers).

In practice this means that a burning brand falling to the ground cannot cause a new fire if the moisture content of the ground debris is above a certain value.

As the wind increases, the flames are flattened, increasing the quantity of radiation reaching the ground. The fire moving up a slope produces the same effect. The heat taken by the convection gases, which is about 90% of the heat given off by the fire, serve to dry and ignite the upper parts of the vegetation but has little effect on propagation (excluding fire propagating through tree tops).

The moisture content of the ground debris is obtained from tables as a function of the water vapour content of the air (the dew point is used to quantify the phenomenon) and the ground temperature. This can be measured using a suitable thermometer or calculated using a formula developed by Météo France (Bernard Sol) utilising conventional meteorological data. The result is expressed as the difference between the wind speed and the speed theorecally necessary for the fire to advance.

A favourable slope "increases" the wind by 10 km/h as does the presence of dry grass standing clear of the ground.

The numerical risk

The numerical risk was developed in 1988 by Météo France (Bernard Sol) at the request of the county fire services who wanted information on the "forest fire" risk expressed as a single number.

The result obtained reflects the risk of a fire starting and the risk of propagation. Current work is concerned with developing an index for the fire initiation risk alone. At present the propagation speed can be taken as a propagation index and the effect of the wind as the initiation index. The initiation-index correlation which appears capable of improvement explains that research is in progress on this topic.

The numerical risk was determined as follows :

In 1987-88 Météo France was doing statistical tests on the various risk indices proposed at the time using criteria suggested by CODIS 13.

The proposals made included one from Pierre Carréga, Senior Lecturer at the Nice Institut of Geography. In order to improve the results, J.C. Drouet suggested that the functions proposed in this index should be replaced by those used in the calculation of propagation speed, while retaining the same structure. The outcome was an appreciable improvement in the results. Finally, purely mathematical work was done to improve the results.

The final risk

The final risk as broadcast has preserved the appearance of the old index developed in 1962 by Mr Orieux, then an engineer with the National Meteorological Office. In practice the final risk can now be modified by Météo France engineer present at CIRCOSC according to the other risk indices and recent events (rain for example) while still retaining the appearance of the "Orieux Risk". This is given in one of four categories : Low, Normal, Severe and very Severe.

Manual plot of the fire configuration

Using the propagation speeds of the fire parallel and perpendicular to the wind it is possible to plot the fire boundary by hand on a map. In order to allow for changes in the fire boundary caused by slopes that favour or hinder propagation, the method mentioned above involving the cone was originaly proposed. For the purposes of the computer system now being installed, consideration is being given to using a multiplying coefficient obtained from the ratio of the angles of the radiation directed towards the ground with and without slope (ratio between the quantities of radiation received by the ground).

The initial process, by applying other rules, could also give an indication to the risk that the fire could leap an obstacle.

It will be noted that this method could not easily handle changes in the boundary related to changes in vegetation or wind direction. With the computer system, the boundary of the fire will be plotted by considering each point on the boundary is a new initiation point (using points located less than 150 m apart) and that the fire boundary is the envelope curve of all the ellipses obtained. In each stage of the calculation, the distance travelled by the fire is limited to a particular distance (about 150 m at the most).

For the convenient calculatio of propagation speeds in the field, a circular slide rule was used first, then programs were tranfered to a pocket calculator (CASIO FX850). Originaly (1970 onwards) such a calculation required a fixed computer taking up more than 1 cubic metre, wich illustrates the improvements in computing and the resulting possibilities for managing fires. The first plots of the fire boundary (homofocal ellipses) were made on transparencies and showed the successive boundaries of the fire at half-hour intervals on a 1/25,000 scale for a propagation speed of 1 000 m/h.

Equipment in service

The basic appliance used for fighting forest fires in the Bouches-du-Rhône is a 4-wheel drive vehicule carrying 2 000 litres of water and four men. The current tendency is to increase this water capacity to a maximum of 3 000 litres. For reasons of efficiency concerning their capacity, wich is increasing, and control, wich has been simplified, these vehicles normally attend fires in groups or four constituting an "intervention group".

Each group is commanded by an NCO or officer who has had special training and who travels in a light all-terrain vehicle with a driver. The group therefore consists of 18 people with an increasing proportion of women (2 to 3 per group in 1993).

There are also similar vehicles carrying four people and 4 000 litres of water (maximum 6 000). These vehicles account for a small proportion of the fleet in the Bouches-du-Rhône but are present in largers numbers in certain counties.

The vehicules carrying 2 000 litres have a 30 m3/h pump, a reel with 80 metres of 23 mm semi-rigid hose and another reel carrying 80 metres of 45 mm flexible hose.

The current trend is to provide protection for the personnel; they now travel in a cab and there are no more exposed seats. The cab is provided with compressed air supplies so that the occupents do not breathe in smoke if the vehicle is halted by the fire. Individual protective gear (hoods, capes, etc.) is also carried. Finally, the more recent vehicles carrying 4 000 litres and more are provided with spray systems for self-dousing if necessary. The county has about two hundred vehicles, half of wich are formed into groups on days when the risk so justifies.

In the conventional way the fleet includes by road tankers (4x2) carrying about 10,000 litres of water which replenish the other vehicles as close as possible to the fire. They collect their water either from natural resources (very scarce), from artifical reserves (buried tanks holding 60 to 120 m3, artificial lakes of various sizes), canals and piped water supplies. The Bouches-du-Rhône has 40 tankers at present.

Over the last two or three years, new vehicles with very different characteristics have appeared.

First are the "mist generators" which direct a very fine spray of "mist" using a special lance. The water is fed through nozzles and entrained by the flow air from a fan. The water flow reaches 21,000 litres an hour and the air flow 55,000 m3/h. In the absence of wind the jet carries about 30 m. These units are extremely useful for industrial fires (the building are filled with air that is not propitious to sustaining combustion, the surfaces are wetted and cooled and smoke or air-mixed gases are reduced).

In the forest they appear particularly useful for putting down chemical barriers because they can be used to place a very fine film of fire-repellant chemical on the vegetation with no run-off. Moreover the wind, by helping to carry the water, improves their performance on the days whenit also encourages the propagation of the fire.

Heavy duty vehicles have also appeared (6x6) which carry, together with a 3-man crew, 10,000 litres of water, 350 litres of emulsifier for forest firesand 650 litres of multi-purpose emulsifier for liquid fuel fires (hydrocarbons and polar solvents). These vehicles are fitted with a water/foam cannon that can deliver 180 m3/h with a useful range of 65 m in the absence of wind. This range is increased to over 80 m with a following wind blowing at 60 km/h and is still 30 to 35 m against the wind. This is extremely important because such a distance is over twice the width of the flame band at the head of a fire propagating at 1 800 m/h. The equipment for hydrocarbon fires means that these vehicles are used throughout the year which improves their viability.

NEW CALCULATIONS TO BE INTRODUCED AFTER COMPUTERISATION.

To beable to determine what equipment is necessary to deal with a fire just reported but for which detailled information is lacking, reasoning and calculational facilities are needed for predicting how the fire will develop.

The first assumption made is that the fire wille develop in the best way it can. In fact it is better to start sending aquipment towards the fire and then call a halt, than to delay and have to deal with a free moving fire of great intensity.

Confirmation of the actual growth of the fire usaually comes fairly quickly because the network of lookout posts (32) gives good coverage of the area. This is supplemented by the surveillance patrol vehicles (50) wich head for every reported fire (the working area assigned to a vehicle is such that it can reach any point from its location at the time of the alert within about 5 minutes). These patrol vehicles carry two men and 600 litres of water; they are fitted with a 6 m3/h pump with 80 metres of 23 mm semi-rigid hose.

The following calculations (detailed in the example at the end of the text) involve approximations that all err in the direction of safety, i.e. the calculated amounts of water are the largest possible while being compatible with the resources available to the fire brigades.

Thes calculations are the mathematical interpretation of the knowledge and experience built up by the personnel of the Bouches-du-Rhône county and more particularly those of CODIS 13.

The methods to be used for the computer calculations are as follows :

-the boundary of the fire is regarded as a rectangle of width equal to the width of the theoretical ellipse as regards the "following wind" part of the fire. As regards the edges, the length of the rectangle will be the distance between the point at which the fire started and the leading edge; the upwind part is neglected.

-the water needed for extinguishing the fire consists of two parts. First, the water used to "douse the flames", i.e., it is assumed that the water is projected in a suitable form into the flame and that its vaporisation during a given period absorbs all the heat given off by the fire during this time in the area treated. The paradox according to which the faster the lance is moved the less water is needed per metre of fire front is confirmed by experience. The other part concerns the cooling of the ground. This is a metter of vaporising water using the heat accumulated by ground in the area covered by the flames.

These data are used to calculate an amount of water theoretically needed to extinguish the fire. From this value it is possible to calculate the number of firefighting groups needed to carry out the operation having regard to the time they required to arrive and the time it will take to empty their water tanks.

The numerical data used for obtaining these results are as follows :

-arrival times : there are programs which give this type of answer once the distances between the various intersections in the sector of study and the average speed at which vehicles travel on the linking roads are entered into the machine.

-heat given off by the flames : the Byram formula is applied only to the lighter parts of vegetation which burn in the flames. In fact the band of fire consists of fine parts which burn in about 30 seconds in the "flames" regime. The parts which burn behind the flames do not affect the propagation of the fire. The values to be taken into account are nearly always the same because leaves constitute a perfect solar trap of constant surface area (about 1.8 times the surface area of the ground with thicknesses for the leaves fairly close by), and as regards the fuel on the ground which can be mobilised in this mode of combustion it is considered to consist of fallen leaves from the previous year which have not yet been biologically destroyed.

Pines can be an exception because the needles are slow to decompose but there is then a lack of oxygen which slows down combustion and also because fine dry needles remain on the trunks and branches and form an excellent fuel.

-this calculation is limited by the fact that the lances cannot be moved as far and as quickly as would be desirable; thus the leading edge of the fire a displacement of 25 cm/s is taken and for the sides a length of 300 m at the most as being capable of being treated by one group, while the water carried would make it possible to treat much more. To fix our ideas, the water requirements calculated in this way are as follows for a fire propagating at 1 800 m/h (fifty centimetres a second).

-for onr metre of fire front at its leading edge with the heat given off being 18.81 kJ per gramme of material and a water absorption between ambient temperature and 100 ° C of 2.5 kJ per gramme of vaporised water we have : 500*18.81/2.5 = 3.762 g/s.

-the mass of fuel is taken as 1 kg/m2 or 400 g (very high value) for fine parts ("solar trap"), the same value for ground debris and a margin of safety for the dry grass that might be present.

-for the heat absorbed by the ground, assuming that the flames constitute a perfect black body parallel to the ground and that this absorbs without re-radiating, for a flame temperature of 1 250 K (by definition this is reached only during a period lesse than the total time of passage of the flame - 1 000 K would appear more reasonable-).

$(30*31)/2*(5.67*10^{-8})*(1\ 250)^4/2*(1/2.5) = 25{,}668$ g of water.

In total therefore, if the water can be brought to the fire front in a suitable form, less than 30 litres of water is needed per metre of front to stop the fire. In fact the movement of the lance is more likely to be 25 cm/s rather than 100, wich multiplies the initial amount of water by four. In wich case the water requirement is 40 litres a metre, an amount which only the new appliances carrying 10 000 litres of water can provide.

It must be noted here that the speed of propagation of a fire which the firefighting system can now control with virtual certainty (meaning that the fire is burning less 100 hectares) comes out at 1 660 m/h for 1994 with an improvement of 30 m a year (in practice 100 m every three years).

It must be admited that if the extinguishing operation is not successful, it is not through lack of water but essentially because when the appliqnces arrive they discover a fire too far from the nearest point they can reach and that the hoses they have available cannot be laid on hot ground.

In fact the condition laid down is that they arrive within 10 minutes at the most, which in our example corrresponds to a maximum propagation of 300 m. The operation then means moving the appliances in order to reach the flames, a manoeuvre which is not always successful.

It must be added that many fires do not propagate, because they are under trees which slow down the wind such that the propagation speeds observed are much less than those calculated for a free fire. This considerably facilitates firefighting.

Mention may be made of other methods of calculating the water necessary to stop a fire.

The first involves calculating the amount of heat necessary to bring materials on the ground to their ignition temperature, on the assumption that the fire cannot propagate if the ground is "wet" enough over an adequate width. This is equivalent to assuming that high vagetation burns under the effect of pilot flames which are those produced by the combustion of fine material on the ground.

The second method involves using statistics about the amounts of water used on actual fires in the past. This approach gives a value of 20 litres per metre of fire front. This value lies between that calculated above for the leading point of the fire and that for the sides of the fires for wich the same method of calculation gives 10 litres per metre or fire front.

The third method involves using statistics about the number of appliances present at a fire at the time it stops advancing. The values are found to lie on a curve shaped like an equilateral hyperbola passing through the following points :

-at 10 minutes there is one vehicle per 20 metres of fire front,
-at 30 minutes to one hour there is one vehicle every 50 metres
-and over a very long period of time the result tends asymptotically to the value of one vehicle every 100 metres.

RESULTS GIVEN BY THE COMPUTER SYSTEM

At the moment these are predictions because the computerisation process should be completed during October 1994.

Within a very short period -of the order of 15 to 30 seconds- an image will form on the operator's screen known as the "reflex screen" (at present the initial phase of work after a fire has been reported is known as the "reflex" phase) showing the following information :

-a map plot showing the boundaries or the fire predicted at 10 minutes intervals for the first hour of propagation. To permit rapid calculations, the boundaries of the fire will ellipses corrected only for any difference in altitude between the point at which the fire started and the upwinf end of the ellipse. The wind direction and the type of vegetation will be those at the point and time at which the fire began.

-the water requirements curve calculated using the simplified rectangle method and that of the amounts of water available according to the predicted arrival of appliances and groups at the site of the fire. The curves will be plotted until the time at which the available water overtakes that of the water requirements, plus 15 to 30 minutes.

-the highlighting of any sensitive points located inside the ellipse, with a resulting proposed increase in water requirements.

The orders to deploy appliances and groups can be issued on the basis of these data. It must be pointed out that for a first report it will be difficult to determine the starting point of the fire within the 75 metre mesh of the map. Since the theoretical ellipses are wider than the real ellipses a safety margin will have to be found in this difference.

The fact that data are displayed so quickly means that it is possible to deal with several fires that might occur at about the same time.

Next comes the program which draws the fire boundary as precisely as possible, with each point on the boundary at any time being regarded as a new fire initiation point with the data known for the nearest point on the map (75-metre mesh-). The programming will be such that the "normal" points used will never more than a certain distance (between 100 and 200 m, to be specified) apart.

The results displayed will not be very different from those of the first calculation but will be more presice and will take into account factors such as obstacles o the fire. It will also be possible to interven manually, for example using the mouse, to intoduce corections to incorporate observations from the site or to feed in experience acquired by CODIS operators.

FIRE AT MARTIGUES-PONTEAU ON 10/8/93

The following is a description of an an actual fire and of what should happen when the situation is dealt with by the computer.

This fire occured in the commune of Martigues near the Ponteau power station. It was reported at 4.58 p.m. (14 58 h UT) as suspicious smoke and identified as the beginnings of a fire a few minutes later.

The starting point of the fire was on the line on the maps linking the meteorological stations of Istres and La Couronne which are 21.5 km apart, and 3.5 km to the north of La Couronne. The wind direction was virtually parallel to this line.

The propagation speeds broadcast on the Minitel at 5p.m. local time (15 h UT - Universal Time, Greenwich meridian-) were 2 100 m/h for La Couronne and 1 200 m/h for Istres. Because of rounding, the figure of 2 100 should be taken as meaning "between 2 100 and 2 200" and 1 200 to mean "between 1 200 and 1 300".

Interpolation of the propagation speeds, xhich can be programmed on the computers, gives 200 m/h. Interpolation of the wind speeds would give a similar result.

The vegetation was known and at the starting point of the fire ans for the first 300 metres consisted of dense Aleppo pines, about 6 m hight. The propagation speed is lower in this environment at 1 616 m/h.

Apart from the fact that the fire was tackled very quickly, there are two causes of deceleration which must be taken into account (not done in this paper), i.e., a descending slope over these 300 metres followed by a rise where because of the very low density of vegetation which may equally well slow down the propagation of the fire through lack of fuel or speed it up by facilitating its transmission by burning brands.

The fire started along a road at a point exposed to the wind. Hence the fire gained speed immediately as indicated by the fact that thepines burned over their full height from the outset of the fire.

The speed of propagation observed was 875 m in 36 to 37 minutes (this uncertainty of one minute is used to demonstrate that ther are uncetainties on the observations made) giving a speed of 1 418 m/h. Ignoring the deceleration of the fire owing to the low vegetation density on the rising slope, which introduces a safety margin, the difference between the calculation and reality is 198 m, or 12. 2%.

At the time the fire was stopped, after 45 minutes of progress, the water requirements were as follows :

-dimensions used :

length 2x1 100 m for the sides and 550 m for the "leading edge" using rounded-up values.

-to "douse" the flames the requirements were : 1 000 x 4.5 x 4.18 x550 x (1 418/3 600)/2.5 = 1 626 litres of water for a lance jet moving at 1 metre per second.

The water requirement for the sides is the same. In fact one cannot expect the lance to move at more than O.25 m/s, which multiplies the water requirement by four if no account is taken of the distance the fire moves during the watering time.

-for "soaking" the needs wers : at the leading edge (138/2.5) x 30 x30.5/2.5 = 20.203 g of water per metre of fire front.

at the side : (138/10° X 30 x 03.5/2.5 =5 050 g of water per metre of fire front.

In total therefore, the requirement is 10 l/m for the sides and 30 l/m for the leading edge.

The necessary amount of water is therefore : (10 x 2 200) + (30 x 550) = 38 500 l.

This would be true if the appliances had free access to the fire, which is far from being the case.

For the leading edge therefore, assuming it is accessible, provision must be made for two groups which represents 20 000 litres for a calculated requirement of 16 500 litres, and for the sides thres groups for each side (1 100 m compared with 1 200 m of possible attack). In fact as far as the sides are concerned, it is not the amount of water available which limits the work that can be expected of an intervention group, but the mobility of the jets.

We reach a total ot 10 groups while in fact 12 were deployed, and our reasoning does not take into account the appliances which had to work at the starting point of the fire to protect dwellings.

The statistical method, for extinction after 45 minutes and a fire front of 2 800 m, would lead to 2 800/50 = 56 appliances approximately, in other words very close to the number of vehicles deployed (12 attack groups, 2 isolated appliances and 2 carrying 11 000 litres).

The result is that the calculations give values that are very close to reality and in a number of very different ways.

It may be pointed out that if helicopters with cannons had been available to bring to the fire only the necessary amounts of water, the total water consumptionhad been lower.

The question of aircraft may also be considered since they would have very good access to the sides of the fire.

There was a problem with the particular fire in question, since it developped under power lines which largely prevented the use of aircraft in the operation.

CONCLUSION

The experience gained in fighting forest fires and the organisation of preparations for these operations in the Bouches-du-Rhône, based upon theoretical studies confirmed on the ground, can be transposed to a computer system. Once this system is up and working, further iprovements can be made since it will then be possible to maintain recors that will be easy to consult. With experience it schould become possible to improve the operation of the system.

The computerisation of the Bouches-du-Rhône fire service will be realised by "INTERGRAPH", french part of the american company "INTERGRAPH".

Bibliography :

Publications of the author in "Revue Technique du Feu"; "Revue Générale de Sécurité" F 38100 Grenoble.

26 A CRITICAL INSIGHT INTO THE BEHAVIOUR OF WINDOWS IN FIRE

S.K.S. HASSANI, G.W.H. SILCOCK
and T.J. SHIELDS
Fire SERT Centre, University of Ulster, Jordanstown, UK

Abstract

Windows constituting part of enclosure walls are important features in that they depict the maximin potential ventilation opening should the glazing fail. Thus in the development of fire safety engineering solutions, it is necessary to be able to accurately predict the behaviour of glazing systems subjected to different thermal fields. This paper reviews the current level of knowledge gained through both analytical and experimental work reported in literature relevant to this phenomenon. It is also intended to highlight the limitations in the current knowledge base and point out areas where future research is required.
Keywords:

1 Introduction

It is well understood that the growth and development of a fire in an enclosure is influenced by the performance of the glazing system. Until recently it has been assumed that glazing always fails catastrophically when the temperature difference between the shaded region and exposed section of the glass reaches a predetermined value [6,7,10]. Such an event will turn the window into a vent through which a fresh supply of oxygen is transported into the fire enclosure which will then increase or decrease the level of fire severity within the enclosure. The cause of the initial edge cracking is known to be due to the thermal expansion of the central exposed region of the glass which is heated directly by radiation from flames and hot surfaces and by convection from the hot gases. Since the edges of the glass are shaded by the frame they remain at much lower temperature. Compared to the rest of the glass, this mismatch in thermal expansion in the glass puts the edges of the brittle glass pane in tension and ultimately causes cracking at stress concentration sites along the edge.

In the development of fire safety engineering models it has long been desirable to predict the time when a glazing system fails during the development of an enclosure fire. In the context of fire modelling the definition of the glazing failure is still ambiguous, for example, in the majority of cases, it seems that the time at which the first crack occurs is taken as the time of failure. However, recent experiments carried

Fire Engineering and Emergency Planning. Edited by R. Barham.
Published in 1996 by E & FN Spon. ISBN 0 419 20180 7.

out in the Fire SERT Centre of University of Ulster [17,18] have shown that the pane of glass may remain in place for a long time after the initiation of the first crack. The aim of this paper is to review current Knowledge and understanding of this phenomenon and suggest areas of further in depth research.

2 Literature Review

Following an initial literature review of glass cracking as the result of thermal insult, it was concluded that the investigations carried out in solar induced thermal cracking [1,2,3,4] would also serve in the further development of a general understanding of behaviour of glazing in enclosure fires. This is due to the fact that in both cases of solar and fire induced failure of glazing system there is a common mechanism prevailing and that the stress σ in either cases can be represented by Hookes Law, ie,

$$\sigma = \text{constant}.E\beta\Delta T \quad \text{(from solar literature)} \tag{1}$$

or

$$\sigma_y = E\beta \ (T_\infty - T_o) \quad \text{(from fire literature)} \tag{2}$$

when:

E	=	Youngs modulus for glass
β	=	coefficient of linear thermal expansion
ΔT	=	temperature difference between shaded and unshaded part of glass
T_0	=	initial temperature
T_∞	=	temperature of central exposed section of glass

In this regard a matrix was generated in which relevant conclusions and findings from research carried out on solar induced glazing failure were reported, table (1).

From a closer examination of this information, several important findings were obtained. For example, the thermally induced stresses were constant over the edge length with the exception of the corners where the stress was reduced to zero when the exposed section of glazing was heated uniformly. However, when a shadow was cast over a part of the pane of glass this altered the stress profile at the edges. The effect of non-uniform heating as the result of window shading was thought to be analogous to non-uniform heating due to two zone gas layer in an enclosure fire.

It must be noted that recent literatuare in fire engineering domain [table 2] has not fully dealt with the effect of two zone gas layer environment on the performance of glazing systems.

3 Development of Experimental Research Programme

Because the phenomenon of glass breakage in real fire is influenced by many factors the current analytical models are unable to predict the behaviour of glazing in all the situations. This is in part due to the oversimplification of boundary conditions which do not always reflect the two layer gas zone condition imposed by a typical enclosure fire. Thus, in order to gain a better understanding of glass cracking and breakage in fire necessary for the development of analytical models a sequence of tests involving glazing systems in configurations typical of domestic dwellings were commenced and are currently in progress.   To date, the data from these tests have yielded the information necessary to enhance our knowledge base in relation to the stress fields generated with respect to timeby the thermal impact of the descending hot as layer [17].

It is anticipated that the resulting bifurcation patterns that occur after cracking can be related in an empirical manner to the evaluated stress fields.

From observations made in these and other tests [10,13,18] it is probable that with certain types of bifurcation patterns the glazing will remain in place for periods in excess of 20 minutes.

This outcome will have an important bearing on the use of current and future fire models to predict the severity of enclosure fires.

4 Analysis and Discussion of Preliminary Results

To explore the effect of two zone fire a series of tests were carried out at University of Ulster at Jordanstown in which a large window was employed in a half scale room [17,18]. From the gas temperature profile over the height of glazing it was evident that the window is subjected to a descending hot gas layer. This is ignored by the analytical models [see table 2] in which it is presumed that glass is uniformly heated over the exposed area. The implication of such non-uniform heating was explored further by investigating the in-situ stresses in the part of glass as the fire developed employing strain gauges[17]. The resulting thermal and stress fields in the glazing were evaluated in conjunction with the bifurcation patterns. It was concluded that in the case of tall windows in an enclosure fire the cracking will occur in the top section of the glass which is impinged by hot gases during the initial stages of an enclosure fire where there exists a well defined hot upper gas layer. It was also noted that the crack bifurcations did not propagate instantaneously and that the lower edge section of the glass tended to remain stress free or in a state of compression. This may explain why the glazing remained in place for periods in excess of 20 minutes. Therefore, if the glass can remain in place for 20 minutes iis it valid to assume that the glass has failed when the first crack appears?

5 Conclusions

1 A comprehensive literature review is essential when developing a programme of research in this or any other fire related topic.

2 Glazing systems will in the main be subjected to the effects of a descending hot gas layer during an enclosure fire.

3 The edge cracks developed due to the thermal insult from the descending hot gas layer generate in turn bifurcation patterns which will ultimately determine whether or not the glass will fall out.

4 The long held assumption that glazing will fail when the upper gas layer temperature exceeds 600°C is no longer valid. Also, the time to first crack may not be considered feasible as a criterion for the glass failure.

No.	Author & Date	Thermal Environment	Executive Summary
1.	Blight, G.E. 1974	Solar Experimental	- glass was heated by insertion into a hot cabinet. - heat absorption of clear and reflective glass was compared; clear glass absorbs less, ∴ remains cooler, ∴ less susceptible to thermal fracture. - demac (demountable) gauges were used for strain measurement. **output:** a. glass is isotropic but β increases with increasing temperature for T >60°C. b. stress in shaded edge is uniaxial tension parallel to edge. In the heated area stress is uniaxial compression parallel to edge. in the corners stress approximates to an isotropic compression. c. by using Griffith criterion ε_t = const. $(\gamma/EC)^{\frac{1}{2}}$ explains why cracks initiate at edge and are controlled by defects. d. stress in shaded edge increases with increasing edge width. e. there is a size effect with induced edge strains increasing with increasing glass size. f. recommends the use of a conducting strip to direct heat into the shaded region to minimise thermal stress.
2.	Stahn, D. 1980	Solar Analytical	- predicts thermal stresses for various temperature distributions imposed by different shadow geometry's by finite element technique using solid SAP (Static Analysis Prog.). **output:** a. thermal stress is constant over majority of edge length and reduces to zero at corners. b. tensile stresses at the edges are equal to principal stresses except at corners ∴ these can be directly used for safety assessments.

Table 1 Literature dealing with the thermal fracture of solar control glazing .

No.	Author & Date	Thermal environment	Executive Summary
3.	Mai, Y.W. Jacob, L.J.S. 1980	Solar Analytical and Experimental	- use of stress analysis (partial differential equations determined numerically to predict stress from $\sigma = const. E\beta\Delta T$. - use of fracture mechanics, concepts to predict stress from $\sigma_r = 2.24\,\dfrac{K_{ic}}{\sqrt{r}}$. where; K_{ic} = stress intensity factor, r = mirror radius of fracture - experimental evaluation of stress in the glass utilising strain gauges. **output:** a. experimental and predicted values of stress agreed well. b. shadows increase thermal stresses by 10%. c. the use of thermal conductive sealant is suggested.
4.	Pilette, C.F. Taylor, D.A. 1988	Solar Analytical	- uses finite element analysis to investigate the effect of varying parameters on thermal stresses in double glazed windows . - *model* : three dimensional solid elements for sealant and gasket and thin shell element for glass. *parameters investigated*: window size, frame absorption, outdoor air temperature, gasket, solar heat flux, sealant stiffness, exterior air film conductance, influence of vertical and horizontal shadows. **output:** a. maximum thermal stress occurs along the edge where shadow line intersects the edge. b. three dimensional effect in a double glazed window are negligible, i.e., in plane forces are not transmitted to other pane.

Table 1 (cont.)

No.	Author & Date	Thermal Environment	Executive Summary
5.	Emmons, H.W. 1986	Fire Review	- Addresses those areas of fire engineering which has not been researched and those areas which needs further attention. - One of such areas thought of as important in the growth and development of enclosure fire was the breakage of window glass in an enclosure fire. This paper highlighted the need for scientific study on the behaviour of glazing in fire.
6.	Joshi, A.A. Pagni, P.J. 1990	Fire Analytical	- make use of transient, one dimensional (into the glass normal to the pane), inhomogeneous (in-depth radiation absorption) energy equation to model temperature profile in the glass. output : a. surface temperature history $T(0, \tau)$ of the glass. b. temperature at breaking, ie, when $\{T(0,\tau) - T_i\} . \alpha . E = \sigma_b$. c. suggest $\Delta T = 50°C - 100°C$. d. results are presented for a set of varying parameters. e. in this work it is assumed uniform flux imposed on glass - no effective hot and cold zone in fire compartment is considered.
7.	O. Keski-Rahkonan 1988	Fire Analytical	- uses two dimensional heat equation to get temperature and stress distribution in the glass subjected to fire. Output : a. $T(x, \tau)$ - temperature profile in terms of x and time, assuming $dT/dZ = 0$ where Z is in thickness direction, $dT/dY = 0$, i.e., no temperature variation along the height of glass and $Bi < 0.1$. b. $\sigma_y(x, \tau)$ - stress in y direction in terms of x and time. c. suggests that distinction must be made between cracking and loss of integrity. For integrity two levels must be identified; 1. cracks at t_1 making fire spread through a barrier possible. 2. breaking of glass pane at t_2 where large open areas of window (fall outs) allow gas flow.

Table 2 Literature dealing with the thermal fracture of glazing in enclosure fire.

No.	Author & Date	Thermal Environment	Executive Summary
8.	Joshi, A.A. Pagni, P.J. 1991	Fire Computer Model	- instructions on use of BREAK computer programme. output : a. escribes temperature profile T(x,t) where x is distance into thickness. b. T(x,t) is inserted in $\Delta T = \frac{1}{L_0}\int\limits_0^L T(x,t)dx - T_i = g\frac{\sigma b}{E\beta}$ to calculate the time to glass breakage. (g - geometry factor of order 1).
9.	Pagni, P.J. Joshi, A.A. 1991	Fire Analytical	- extends the analysis described in ref.6 above to obtain temperature profile and stress profile in terms of x,y and time, taking into account the effect of heat dissipation into the shaded area of the glass. - assumption is made that glass is subjected to a uniform hot gas at the inner surface
10.	Skelly, M.J. Roby, R.J. Beyler, C.L. 1991	Fire Experimental	- experimentally investigated the window glass breakage in enclosure fires. - fire compartment ; 1.5 x 1.2 x 1.0m. - glass pane 0.28 x 0.5 x 0.024m. - fire source ; liquid hexane. - window was placed in fire compartment in such a way that in all tests entire glass was in hot zone within 10 secs. - glass was cut by hand and edges were not prepared in any way. output: a. time-temperature profile for shaded and unshaded glass for various fire sizes. b. tabulated temperature difference with time of crack initiation. c. bifurcation patterns. d. compared experimental results with theoretical work of Pagni (ref. 6) and Rahkonen's (ref. 7).

Table 2 (cont.)

No.	Author & Date	Thermal Environment	Executive Summary
11.	Silcock, G. W. Shields, T.J. 1993	Fire Experimental	- fire compartment ; a half scale fire room. - single and double glazing with wooden frame. - fire source ; wooden cribs. <u>output:</u> a. time to first crack. b. shaded/unshaded temperature difference. c. crack bifurcation patterns. d. effect of two zone fire environment on the breakage of glass.
12.	Joshi, A.A. Pagni, P.J. 1994	Fire Analytical	- Glass thermal field obtained from model in ref. 9 are examined. <u>output:</u> a. glass surface temperature increasing with decreasing decay length of flame radiation. b. glass surface temperature increasing with decreasing flame radiation heat flux. c. breaking time decreasing with increasing shaded width. d. breaking time decreasing with increasing decay length. f. most of imposed heat influx is stored in the glass, increasing its temperature.
13.	N. A. McArthur 1991	Fire Experimental	- an investigation into the performance of windows with aluminium and wooden frames subjected to bush fire. - furnace and standard time-temperature curve was used as the means of heating. <u>output:</u> a. time to fracture. b. crack and bifurcation patterns. c. fracture face observation. d. comparison of performance of aluminium and wooden window frames subjected to bush fire conditions.
14.	Joshi, A.A. Pagni, P.J. 1994	Fire Experimental	- use of four point bend test method and Weibull analysis to determine the breaking stress of glass. <u>output:</u> a. $\sigma_f \approx 40$ MPa. b. Weibull parameters ; $m = 1.21$, $\sigma_0 = 33$ MPa , $\sigma_u = 35.8$ MPa c. the results of Skelly's experiments (Ref. 10) are compared with Pagni's BREAK programme (Ref. 8) and found good agreement.

Table 2 (cont.)

No.	Author & Date	Thermal Environment	Executive Summary
15.	Cuzzilo, B. Pagni, P.J. 1993	Fire Analytical	- describe Joshi and Pagni's work (Ref.6). - applies the method to a double paned window subjected to a wild fire. output: a. the pane farther from fire stays cool, if the pane facing fire were to break and fall out then the cool pane begin to heat u break. b. fire facing pane with low-E coating will stay cool enough to avoid breaking.
16.	Emmons, H.W. 1988	Fire Analytical Review	- review of O. keskie-Rahkonan's work (Ref. 7) to predict glass crack initiation. - draws attention to distinction of "crack growth" from "Crack Initiations". - suggests that $\sigma_y = E. \beta (T_\infty - T_0)$ be experimentally validated, especially stress measurement at the glass edge is request - puts forward an explanation for crack bifurcation based on beam theory.
17.	Hassani, S.K.S. Shields, T.J. Silcock, G.W. 1995	Fire Experimental	- Experimental evaluation of thermal and stress field in glazing units subjected to real fires. - Thermal strains were measured in-situ using strain gauges. output: a. experimental determination of dynamic stress development in glazing subjected to real fires.
18.	Shields, T.J. Silcock, G.W. Braniff, J. 1992	Fire Experimental	- experimental investigation of performance of double glazed window units in enclosure fires. The glazing units were spe constructed to incorporate thermocouples to monitor surface temperatures on all of the glazing surfaces. Output : a. Data in graphical form depicting the temporal variation of surface temperatures, associated gas temperature variations respect to time in both in the descending hot gas layer and lower cool zone.

Table 2 (cont.)

Notation

Bi	(=hd/k) Biot number
E	Young's modulus
T_o	initial temperature
T_∞	asymptoric value for gas temperature
X	= x/d
b	half width of glass
d	glass thickness
h	surface heat transfer coefficient
k	thermal conductivity
Δ	= δ/d
δ	width of shaded edge
β	glass thermal expansion coefficient
σ_y	stress in y direction
λ	fracture surface energy of the glass
C	equivalent dimension of an initial flow in the glass
ϵ_f	strain at failure
ΔT	temperature difference
σ_f	stress at failure
L	glass thickness

References

1. G. E. Blight, *Thermal strain and fracture of building glass*. First Australian Conference on Engineering Materials, NSW University, NSW, pp. 685-700, (1974).

2. D. Stahn, *Thermal stresses in heat-absorbing building glass subjected to solar radiation*. Proceedings, International conference on thermal stresses in materials and structures in severe thermal environment, Virginia Polytechnic and State University, Blacksburg, VA, March , pp. 305-323, (1980).

3. Y. W. Mai, L. J. S. Jacob, *Thermal stress fracture of solar control window panes caused by shading of incident radiation,* Materiaux et Constructions, 13, no. 76, pp. 283-288, (1980)

4. C. F. Pilette, D. A. Taylor, *Thermal stresses in double-glazed windows.* Construction Engineering, 15, No. 5, pp. 807-814 (1988)

5. H. W. Emmons, *The Needed Fire Science.* C. E. Grant, P.J. Pagni (eds), Fire Safety Science-Proceedings of the First International Symposium, Hemisphere Washington, D.C., pp. 33-53, (1986).

6. A. Joshi, P.J. Pagni, (1990) *Thermal analysis of a compartment fire on window glass.* Report No. NIST-GCR-90-576, (June 1990).

7. O. Keski-Rahkonen, *Breaking of glass close to fire, I.* Fire and Materials, 12, pp. 61-69 (1988).

8. A. A. Joshi, P. J. Pagni, *Users Guide to Break, The Berkeley Algorithm for*

Breaking Window Glass in a Compartment Fire. Report No. NIST-GCR-91-596, (October 1991).

9.   P. J. Pagni, A. A. Joshi, *Glass Breaking in Fires.* Proceedings of Third International Symposium on Fire Safety Science, pp. 791-802, (1991).

10. M. J. Shelly, R. J. Roby, C. L. Beyler, *An experimental investigation of glass breakage in compartment fires.* Journal of Fire Protection Engr., 3, No. 1, pp. 25-34, (1991).

11. G.W. H. Silcock, T. J. Shields, *An experimental evaluation of glazing in compartment fires.* Interflam 93, Proceedings of the sixth International Interflam Conference, pp. 747-756, *1993.

12. A. A. Joshi, P. J. Pagni, *Fire-Induced Thermal Fields in Window Glass, I-Theory.* Fire Safety Journal, 22, pp. 25-43, (1994).

13. N. A. McArthur, *The performance of aluminium building products in bushfires.* Fire and Materials. 15, pp. 117-125, (1991).

14. A. A. Joshi, P. J. Pagni, *Fire-Induced Thermal Fields in window Glass, II-Experiments.* Fire Safety Journal, 22, pp. 45-65, (1994).

15. B. Cuzzillo, P. J. Pagni, *Windows in wild fires.* Fire and Materials, 12, pp. 61-69

16. H. W. Emmons, *Window glass breakage by fire.* Home Fire Project Technical Report No.77, Harvard University Cambridge, Massachusetts, (1988).

17. S.K.S. Hassani, T.J. Shields, G.W.Silcock. *Experimental Evaluation of Temperature and Stress Field in Glass Subjected to an enclosure fire.* (Due to be published)

18. T. J. Shields, G. W. H. Silcock, J. M. Braniff, Building Regulation Interaction-Report. 1, Fire Research Centre, University of Ulster, September 1992

27 TESTING OF AXIALLY LOADED AND RESTRAINED STEEL COLUMNS

W.I. SIMMS
School of the Built Environment, University of Ulster, Jordanstown, UK

Abstract

This paper presents details of the design and commissioning of test apparatus, the test methodology and initial test results which will form the basis of an extensive study into the effect of axial restraint on the performance of steel columns in fire. The elevated temperature tests described are conducted using half scale I-section samples subjected to a constant value of applied load and a known degree of axial restraint. Theory is introduced, in the form of a simple equation, which successfully predicts the restraint force generated in actual fire tests.

1 Introduction

The project described in this paper investigates experimentally the affect of axial restraint on steel columns exposed to fire.

This investigation has been initially inspired by the structural report on the Broadgate Phase 8 fire [1] which highlighted evidence of a thermally induced restraint force contributing to column failure. As real fires are likely to affect a single compartment or isolated area of a building heating of individual structural members is likely to occur. A column within a building frame which is heated while the surrounding structure remains cool will experience additional axial forces due to the restraining effect of the surrounding structure on its thermal expansion. This has been reported by Furumura and Shinohara [2] as early as 1976. This early analytical

Fire Engineering and Emergency Planning. Edited by R. Barham.
Published in 1996 by E & FN Spon. ISBN 0 419 20180 7.

investigation attempted to understand the behaviour of columns in structural scenarios which were found in real building structures. This was different from the more usual situation where columns were treated as isolated elements in standard fire resistance tests. Furumura and Shinohara's analytical study made qualitative evaluation of the affect of various end restraints and quantitative evaluation of axial restraint.

The complexity of steel columns in frames has been analysed more fully in recent finite element analysis studies. Computer programs have been developed by Arbed and the University of Liège (CEFICOSS), University of Sheffield (3DFIRE and NARR2) and by the BRE (FLAMEFIRE) among many others. Other analysis methods have been developed such as Jeyarupalingam and Virdi [3] and Lie and Chabot [4] which are based on moment-thrust-curvature relationships. In Australia Bennetts et al, BHP [5] have developed a method for the numerical analysis of axial restraint forces, and most recently a more general method of analysis has been reported by Poh and Bennetts [6]. These methods of analysis cover the full range of structural and material properties and aim to provide an accurate prediction of structural behaviour.

However, until recently, testing has focused mainly on standard fire resistance tests, which normally have simple end conditions and concentric load application. As the purpose of such tests is to determine the fire resistance time of a structural member the data reported from such tests is limited. Consequently little experimental data is available which relates to the effect of axial and rotational restraint conditions which are commonly encountered in real structures. This has created a problem in the validation of computer models, as even for simple structural situations, few tests exist which have sufficiently detailed reporting of test results. Franssen et al [7] are currently attempting to create a database of test results and have experienced considerable difficulty in obtaining suitable data.

Bearing in mind this lack of appropriate experimental information this project is part of a five year combined testing and computational study on the behaviour of columns in fire influenced by axial and rotational restraint. The aim of the project is to provide test results from structural scenarios relevant to those encountered in real structures. The initial project described here considers axially loaded and restrained columns and forms the background to a second project which will consider rotational restraint. The project also aims to obtain experimental results which will provide sufficiently detailed data to assist validation of computational analysis methods.

2 Description of Test Apparatus

In order to fulfil the requirements of the test programme to be conducted on half scale samples a loading rig and furnace where developed.

The loading rig was required to fulfil two main objectives. These were namely that the rig would be capable of providing variable degrees of axial restraint to the specimen and secondly that the rig should be able to apply a constant value of axial load during the fire test. A rig has been designed which meets these objectives and is shown in Figure 1. The vertical members of the rig are constructed from 20 mm diameter stud bars attached to 80 x 80 SHS. The bottom of the sample is supported on one set of two 254 x 76 RSC fixed back to back to the 80 x 80 SHS. At the top of the sample load and restraint capabilities are provided using two sets of two 254 x 76 RSC also joined together in a back to back configuration. Load is applied to the sample using two 30 tonne Enerpac rams via the top upper channels. These rams jack against load cells which in turn are restrained by a steel plate and nut attached to the threaded bar. The load is transferred from the top upper channels via a stub column to the top lower channels which rest on top of the sample. The axial restraint of the specimen is achieved using the top lower channels. Upward movement of these channels is resisted by a system of springs and load cells which are restrained at their upper end by a plate and nut attached to the threaded bar. By varying the stiffness of these springs the degree of axial restraint can be adjusted to different levels.

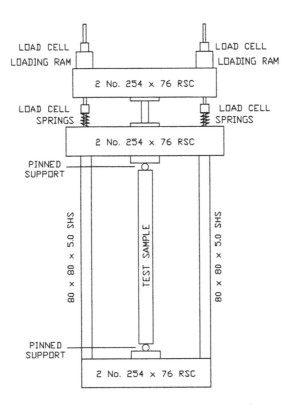

Figure 1: Test Apparatus

The design criteria for the furnace were as follows. Firstly the furnace had to obtain a maximum temperature of 800°C. Secondly the gas temperature within the furnace had to be reasonably uniform so that internal thermal gradients would not be created in the sample. Finally the dimensions had to be confined so that the loading rig was required to span only a short distance.

The basic structure of the furnace is a light steel frame made up of rolled steel angles and supported on a base of 50 x 50 SHS. This frame is then infilled with insulating material to form the walls, floor and roof of the furnace. The insulating material used is a rigid calcium silicate board lined internally with a calcium silicate fibre blanket.

The furnace dimensions are 1.6 m high and 600 mm square on plan. The whole front wall of the furnace can be removed to allow test specimens to be placed inside. The specimens fit through holes in the floor and roof of the furnace into the surrounding loading frame.

3 Commissioning of test apparatus

To insure the loading rig and furnace satisfied their design criteria a number of tests were conducted during the development to assess their performance.

The loading rig was tested at various stages of development to ensure that the loading and restraint characteristics performed their desired functions and also interacted in a satisfactory manner. The flexibility of the rig and the degree of axially restraint it provided, without springs insitu, was also assessed during these tests. The test devised consisted of replacing the test sample with a 100 x 100 SHS which would not have a large axial displacement at maximum load. Increments of load were then applied to the rig up to its maximum design load. A number of dial gauges were used to measure the vertical deflections of the top channels thus allowing the flexibility to be assessed. The final loading rig has been found to have a flexibility of 0.0315 mm/kN. This value of flexibility can be used in calculations when the full restraint capability of the rig is employed and when no springs are being used to generate additional flexibility.

Temperature tests where also conducted on the furnace to assess the uniformity of temperature produced. During these tests gas and steel temperatures were recorded and the temperature variation was found to be only 20°C over the full length of the steel specimen at 800°C.

4 Instrumentation

During the tests five parameters are recorded, these are applied load, restraint force, lateral deflection, axial deflection and steel temperature.

Referring again to Figure 1 it can be seen that four load cells are attached to the rig. The upper load cells associated with the loading rams record the applied load on the column which remains constant during the test. The lower load cells record the additional load on the column due to the restrained thermal expansion.

Lateral deflection is measured, via quartz rods, by two displacement transducers (LVDT) located at mid height of the test specimen outside the furnace. Lateral deflections are measured on either side of the weak axis of the specimen and therefore it is unimportant which direction deflection occurs.

Axial deflection is measured, at the centre of the top channels, using a LVDT attached to an independent reference framework.

Steel temperatures are recorded at the two quarter span positions on the sample by five thermocouples on the cross section at each location. A thermocouple is positioned either side of the web on each flange and the fifth thermocouple is located centrally in the web. All the thermocouple junctions are located in holes drilled half way through the material so true steel temperatures are recorded.

5 Test Procedure

The test procedure consists of two main stages namely the loading stage and the thermal programme. The application of load to the test specimen is conducted in incremental steps. As each increment of load is applied the lateral deflection of the test specimen at its mid height is recorded. This process enables additional information to be gathered on the initial curvature of the sample by utilisation of a Southwell plot.

Once the desired level of load has been applied the column is restrained. The thermal programme can now begin, in which the steel temperature is increased by 10 °C/min until failure occurs.

Failure is defined as the point at which axial deflection returns to its initial value or when lateral deflection reaches a runaway situation.

6 Analysis

The results presented in this paper are for a test conducted on an axially restrained column with an applied load. The follow equation has been developed to predict the magnitude of the restraint force generated as the column is heated. The general form of the equation is as follows:

$$R = (\alpha LT - \frac{R}{K_c} - \frac{P}{K_c}) K_r$$

where R is the restraint force, αLT is the thermal expansion, P is the initially applied load, K_c is the column stiffness and K_r is the stiffness of the rig. The term in the above equation involving the initial load, P can be ignored in this test as restraint is applied to the column after it has been loaded. The above equation can then be rearranged to give:

$$R = \alpha LT \frac{K_c K_r}{K_c + K_r}$$

or

$$R = e_{th} L \frac{K_c K_r}{K_c + K_r} \tag{1}$$

where ϵ_{th} is the thermal strain. In situations where $K_c \gg K_r$ equation 1 becomes:

$$R = \frac{K_r}{1 + \frac{K_r}{K_c}} e_{th} L$$

or where $K_r \gg K_c$

$$R = \frac{K_c}{1 + \frac{K_c}{K_r}} e_{th} L$$

Equation 1 relates to the particular case for the results presented in this paper. The slenderness ratio of the column is high and therefore material properties are not seriously effected by temperature. It should be noted however that for stocky columns where a higher restraint force is required to cause failure the temperature will have a more serious effect on the material properties. This reduction in material properties will cause an additional compressive strain to occur due to the initial loading. This additional strain, $\Delta\epsilon_P$, is given by:

$$\Delta e_P = \frac{P}{AE_{(t)}} - \frac{P}{AE_{20}}$$

where P is the initially applied load, A is the cross sectional area, $E_{(t)}$ is the Youngs modulus at elevated temperature and E_{20} is the Youngs modulus at ambient temperature. Hence Equation 1 becomes:

$$R = \frac{K_r K_c}{K_c + K_r} (e_{th} - \Delta e_P) L$$

7 Results and Discussion

The results of the fire test FT_3101 are presented. The section tested is a Special Light Section SLS 4" x 2.25" x 4.08 lbs whose overall dimensions are 101.6 mm x 57.15 mm and whose steel quality is equivalent to Grade 43A BS 4360 [8]. The sectional properties are presented in Table 1.

Table 1: Section Properties

Area cm^2	r_{xx} cm	r_{yy} cm	I_{xx} cm^4	I_{yy} cm^4	Length mm	λ
7.74	4.73	1.24	137.3	11.86	1600	129

Table 2: Ambient Temperature Design Loads

Youngs Modulus kN/mm^2	Yield Strength N/mm^2	Euler Load kN	BS 5950 Load kN	Initial Load kN
205	275	94	80	45

Table 2 presents the design load at ambient temperature, for the weak axis of the column, calculated from BS 5950: Part 1 [9] and from classical Euler theory given in Equation 3.

$$P_{crit} = \frac{\pi^2 EI}{L^2}$$ (3)

The Loading which was initially applied to the column before the fire test began was calculated as 60% of the BS 5950: part 1 design load.

The column tested failed at a steel temperature of 238°C and when a total load of 141 kN was reached. The mode of failure was overall sinusoidal buckling which occurred rapidly.

The increase in restraint force is shown in Figure 2, and can be seen to have a non uniform rate of increase which is accounted for by the variation in rig stiffness with load. As the rig stiffness has been measured in the commissioning tests actual values can be used in Equation 1 to calculate the restraint force. These calculated values are also shown in Figure 2 and display reasonable agreement with measured values.

As shown in Figure 4 no lateral deflection occurs until the last 10 minutes of the test, at which point a gradually increasing rate is recorded before a sudden runaway situation occurs indicating failure. It should be noted that the LVDT ran out of Travel hence the premature termination of lateral deflection which in fact had a magnitude of approximately 250 mm.

Referring to Figures 2 and 3 it can be seen that a corresponding reduction occurs in both restraint force and axial deflection as the column begins to buckle.

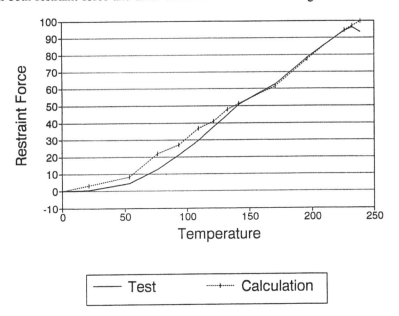

Figure 2: Restraint force versus temperature

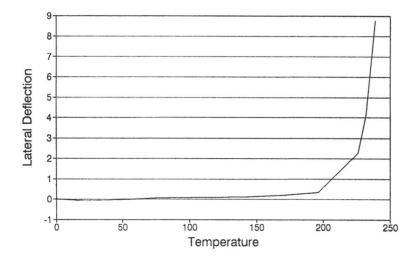

Figure 3: Axial deflection versus temperature

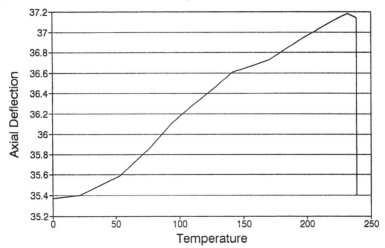

Figure 4: Lateral deflection versus temperature

References

1. Steel Construction Industry Forum Fire Engineering Group (1991) Structural Fire Engineering Investigation of Broadgate Phase 8 Fire. SCI, Ascot
2. Furumura,F. and Shinohara, Y. (1976) Inelastic Behaviour of Protected Steel Columns in Fire. Proc. IABSE 10TH cong., Tokyo, pp 193-198
3. Jeyarupalingam, N. and Virdi, K.S. (1991) Steel Beams and Columns Exposed to Fire Hazard. Structural Design for Hazardous Loads, (ed. J.L. Clarke, F.K. Garas and G.S.T. Armer), E & FN Spon, London, pp 429-438

4. Lie, T.T. and Chabot, M. (1993) Evaluation of the Fire Resistance of Compression Members using Mathematical Models. Fire Safety Journal, Vol. 20, pp 135-149

5. Bennetts, I.D., Goh, C.C., O'Meagher, A.J. and Thomas, I.R. (1989) Restraint of Compression Members in Fire. BHP Melb. Res. Lab. Rep. No. MRL/PS65/89/002

6. Poh, K.W. and Bennetts, I.D. (1994) Behaviour of Steel Columns at Elevated Temperatures. Submission for Publication ASCE Journal of Structural Engineering

7. Franssen, J-M., Schleich, M. J-B. and Cajot, M. L-G. (1994) Buckling Curves of Hot Rolled H Steel Sections in Case of Fire. CEC Agreement 7210-SA/515, 931, 316, 618

8. British Standards Institution (1986) BS 4360: Specification for Weldable Structural Steels. BSI, London.

9. British Standards Institution (1990) BS 5950: The Structural Use of Steelwork in Building Part 1: Code of Practice for Design in Simple and Continuous Construction. BSI, London.

28 TOXIC COMBUSTION PRODUCTS FROM PESTICIDE FIRES

L. SMITH-HANSEN
Risk Analysis Group, Systems Analysis Department,
Risø National Laboratory, Denmark

Abstract

The present paper gives results from combustion experiments with three pesticides using the DIN 53 436 furnace. These pesticides were dimethoate, dichlobenil and thiram. Two different experimental conditions were applied in order to simulate fully developed fires with high ventilation and non-flaming oxidative decompositions. The combustion gases (e.g. CO_2, CO, HCl, $COCl_2$, HCN, SO_2, NO_x) were quantified by means of the FTIR gas analysis technique. Dimethoate was burned as the pure compound and as a formulation in order to investigate the effects from the flammable solvent. Furthermore, combined flash pyrolysis/GC/MS experiments were carried out in order to obtain information on the type of organic products from thermal decomposition. Some of these products might survive the flames and be of importance, in particular in case of fires under oxygen deficient conditions. Such fires may represent the worst case situation with respect to the toxicity of the fire plume.

Keywords: Combustion products, pesticide fire, pesticide storage, toxicity, warehouse.

1 Introduction

A large number of storages containing hazardous chemicals are in operation in most countries. These chemicals are e.g. pesticides, fertilizers and solvents. A specific storage will often contain a large number of different pesticides. These pesticides could be pure chemicals or in the form of formulations containing inert materials or flammable solvents. The types and amounts of the pesticides will normally vary in accordance with the season.

Chemical fire is the most important hazard related to the chemical storage facilities. Large quantities of chemicals may be involved with the formation of significant amounts of toxic fire effluents. These products will be dispersed with the fire plume

Fire Engineering and Emergency Planning. Edited by R. Barham.
Published in 1996 by E & FN Spon. ISBN 0 419 20180 7.

threatening or causing harm to humans working at the chemical storage facility, living or staying in the vicinity or taking part in the fire fighting. Also environmental consequences to the soil, groundwater, rivers, lakes etc. from release of contaminated fire fighting water and deposition of soot particles are important problems in some of these fires.

It is, however, difficult to assess the consequences of a pesticide warehouse fires, mainly because only limited data is available concerning the identity and the amounts of the components in the fire effluent.

This paper presents results from pyrolysis/GC/MS and DIN 53 436 combustion experiments with three selected pesticides. The organic pyrolysis products were identified by means of GC/MS and the combustion gases from DIN 53 436 experiments (CO_2, CO, HCl, $COCl_2$, NO_x, HCN, SO_2 etc.) were quantified using the FTIR technique.

2 Pesticides

The investigated pesticides were: dimethoate, dichlobenil and thiram. Figure 1 shows the chemical structure of the three compounds.

- Dimethoate was investigated as the pure compound (technical grade) and as formulation. Dimethoate was chosen as a representative for the sulphur containing organophosphorous insecticides. The formulation consisted of 408 g dimethoate/l, 50 g Berol 946/l (emulgàtor), 455 g cyclohexanone/l and 138 g xylene/l.
- Dichlobenil (a herbicide) was chosen as it contains a cyanide group and a chlorine group. It was therefore expected that it would generate significant concentrations of HCN, HCl, $COCl_2$ and NO_x.
- Thiram (a fungicide) was chosen due to the high sulphur and nitrogen content in the compound. The distribution between SO_2, H_2S, COS, CS_2 and other sulphur containing compounds could then be studied as well as the production of HCN, NO_x etc.

3 Pyrolysis/GC/MS

Combined pyrolysis/GC/MS experiments were carried out in order to obtain information on the type and amounts of products from thermal decomposition (the chemical fingerprint). The basic idea is that some of these products are expected to survive a flame and will therefore be of importance, particularly in a fire under oxygen-deficient conditions. Such fires are expected to represent the worst case situation with respect to production of toxic organic decomposition products.

3.1 Experimental

A flash pyrolysis technique was applied. A PYROLA-85 foil pulse pyrolyser was combined with a Varian Saturn II GC/MS equipment. The compound subjected to pyrolysis was placed on a platinum foil in the pyrolysis chamber in a helium flow. It was left there for about 5 min. prior to the pyrolysis to ensure complete exclusion of

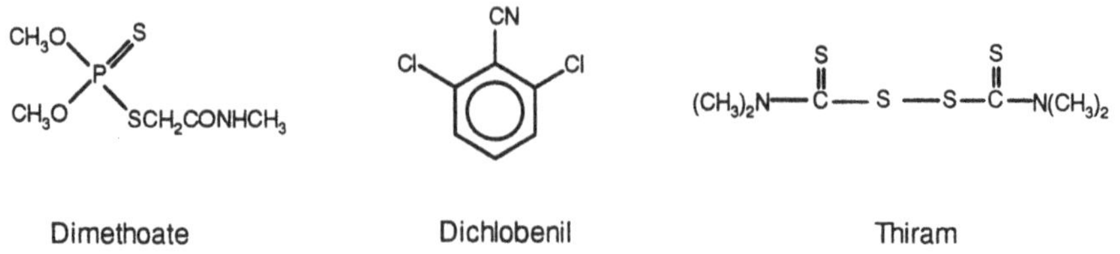

Dimethoate   Dichlobenil   Thiram

Fig. 1. Chemical structure of the investigated pesticides

air. The sample was pyrolysed in 2 sec. The pyrolysis temperature was about 900 °C in all experiments. The pyrolysis products were then separated on a GC column. The GC conditions were: column: CP-sil 5 CB, 25 m x 0.25 mm, df = 0,25 µm, Chrompack. Oven: 100 °C - 10 °C -> 285 °C/19.5 min. Carrier gas: helium. Finally, the pyrolysis products were identified in an ion-trap mass spectrometer. A WILEY 5 library of standard mass spectra was used for identification. All experiments were carried out twice to ensure repeatability. The pyrolysis/GC/MS technique is described in [1].

Not all products could be unequivocally identified, as mass spectra alone do not allow a distinction to be made between closely related compounds (such as isomers). Comparisons with reference compounds were not possible since such compounds in most cases were not available. Furthermore, no library of mass spectra containing all types of pyrolysis products from all substances is available. No real quantification of the pyrolysis products could be made. However, based on the peak heights in the chromatograms the compounds present in the largest amounts could be noted. Finally, it should be noted that the lowest molecular weight substances would not be detected (e.g. SO_2, HCl, HCHO).

3.2 Results
Below follow the lists of decomposition products from the three pesticides identified from the combined flash pyrolysis/GC/MS experiments.

3.2.1 Pyrolysis products from dimethoate

- dimethyl disulphide
- isocyanatomethane
- phosphoric acid trimethylester
- dimethyl trisulphide
- phosphorothioic acid, O,O,O-trimethylester
- N-methyl thioacetamide
- trithiolane
- phosphorothioic acid, O,O,S-trimethylester
- phosphorodithioic acid, O,O,S-trimethylester (major product)
- phosphorodithioic acid, O,S,S-trimethylester (major product)

Important products from pyrolysis of dimethoate are the phosphoric/phosphoro-thioic/phosphorodithioic acid trimethyl esters. These compounds are expected to be produced by many sulphur-containing organophosphorous insecticides.

3.2.2 Pyrolysis products from dichlobenil

- 1,3-dichlorobenzene
- benzonitrile
- dichloromethylbenzene (major product)
- 2-chlorobenzonitrile
- trichlorobenzene

The major product from pyrolysis of dichlobenil is dichloromethylbenzene. 1,3-dichlorobenzene, 2-chlorobenzonitrile and benzonitrile are all pyrolysis products formed by simple losses of substituents from dichlobenil. Trichlorobenzene may be produced by a secondary chlorination of 1,3-dichlorobenzene. [2] describes thermal analysis of dichlobenil in air followed by mass spectrometry. It was concluded in [2] that the compound evaporated without decomposition. However, the experimental conditions differed somewhat from those reported here.

3.2.3 Pyrolysis products from thiram

- isothiocyanatomethane *
- dimethyldisulphide
- dimethyltrisulphide
- tetramethylurea
- 1,2,4-trithiolane
- N,N-dimethyl-ethanethioamide
- tetramethylthiurea
- dimethyl-carbamodithioic acid methylester
- Sulphur (major product) *
- 1,2,3,5,6-pentathiepane
- hexathiepane
- tetramethylthiurammonosulfide (major product) *

A number of sulphur compounds from pyrolysis of thiram were found. The major products were elemental sulphur and tetramethylthiurammonosulfide. The results can be compared with those reported in [3]. This reference contains a pyrolysis scheme for thiram. The products which were also described in [3] are marked with "*" in the list above. In addition to these products [3] also found tetramethylthiuramtetrasulfide and various low molecular weight products which were probably also produced in the experiments reported here but not detected.

4 DIN 53 436 combustion experiments

Combustion experiments were carried out using a furnace in accordance with DIN 53 436 [4]. The method is considered to be relevant for the simulation of non-flaming oxidative decomposition and fully developed fires [5].

4.1 Experimental

The apparatus (Figure 2) consists of a quartz tube (diameter 4 cm, length 1 m) enclosed by a movable annular oven. According to the standard, the substance subjected to combustion should be placed in a 40-cm quartz boat inside the tube. However, in order to approach steady-state conditions the pesticide (1-3 g) was placed in 24 quartz vessels. The vessels were then placed in the boat. A similar procedure is described in the literature [6]. During the experiment the oven moves with a constant speed (1 cm/min) along the tube. An airflow (100 l/h) is maintained through the tube in the direction opposite to that of the oven.

The combustion experiments were carried out at 500 and 900 °C furnace temperatures in order to simulate the two relevant scenarios and a Bomem 100 FTIR gas analysis instrument was used for quantification of the combustion gases.

4.2 Results

Tables 1-3 show the results of the DIN furnace combustion experiments.

Table 1. Dimethoate combustion gas yields

Temp. °C	CO_2 g/g	CO g/g	HCN g/g	SO_2 g/g	NO g/g	CH_4 g/g	COS g/g	N_2O g/g
900 (pure)	0.844	~0.06	0.011	~0.52	0.006	~0.01	~0.008	~0.03
900 (formulation)	1.822	~0.00	0.006	0.225	~0.01	-	-	~0.03
Theoretical max.	0.960	-	-	0.558	0.131	-	-	-

Formaldehyde, acrolein, CS_2, H_2S, NO_2 and NH_3 were not detected
- Not detected

The combustion gas yields for dimethoate are given in table 1. The theoretical max. values are given only for the pure compound, since the exact composition of the formulation is not known.

The conversion of sulphur in dimethoate to SO_2 by combustion is almost quantitative. However, for the pure compound a small amount of COS is also produced. Concerning the pure compound as well as the formulation it can be concluded that dimethoate produces significant concentrations of HCN, and the conversion rates are almost the same in both cases. Finally, it should be noted that the pure compound produces substantial concentrations of CH_4 by combustion. In the combustion

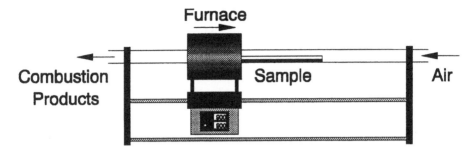

Fig. 2. Furnace in accordance with DIN 53 436

experiment using the formulation, the combustion is more effective and no production of CH_4 was observed.

Table 2. Dichlobenil combustion gas yields

Temp.	CO_2	CO	$COCl_2$	HCl	HCN	NO	N_2O
°C	g/g	g/g	g/g	g/g	g/g	g/g	g/g
900	~1.2	~0.46	0.008	~0.35	0.012	0.004	~0.03
900	1.581	~0.48	0.005	~0.36	0.009	0.003	~0.03
Theoretical max.	1.791	-	-	0.424	-	0.174	-

Formaldehyde, acrolein, CH_4, NO_2 and NH_3 were not detected

The combustion gas yields for dichlobenil are given in table 2. Since dichlobenil is a chlorinated aromatic compound, the concentrations of CO and HCl from combustion are high, as could be expected. The concentration of HCN and $COCl_2$ are rather low, but toxicologically significant. Finally, at the low temperature no decomposition took place at all.

Table 3. Thiram combustion gas yields

Exp.	CO2	CO	HCN	SO_2	NO	CS_2	COS	N_2O
	g/g	g/g	g/g	g/g	g/g	g/g	g/g	g/g
900	0.964	~0.003	0.009	~0.9	~0.02	-	-	~0.03
500	0.085	~0.5	0.027	0.579	0.003	~0.1	~0.1	-
Theoretical max.	1.098	-	-	1.065	0.250	-	-	-

Formaldehyde, acrolein, CH_4, H_2S, NO_2 and NH_3 were not detected
- Not detected

The combustion gas yields for thiram are given in table 3. In the high-temperature experiment an almost quantitative conversion of the sulphur in thiram to SO_2 was observed. In the low-temperature experiment a substantial amount of sulphur was converted to CS_2 and COS. Also rather high concentrations of HCN were seen, particularly in the low-temperature experiment. Finally, the concentration of CO was very high in the low-temperature experiment.

5 Conclusions

Conclusions can be drawn from the results of the experiments described in the present paper:

- Chlorinated compounds seems to produce high concentrations of HCl.
- Chlorinated compounds may also produce significant amounts of $COCl_2$.
- Chlorinated compounds produce higher amounts of CO than the non-chlorinated compounds.
- Nitrogen compounds are converted to NO, HCN and N_2O. A 100% conversion of nitrogen to these gases cannot be expected. NO_2 was not observed in any of the experiments reported here. However, it is expected that a compound such as NH_4NO_3 will produce NO_2 and NH_3 by combustion.
- Substantial concentrations of HCN are often seen.
- Acrolein and formaldehyde have not been observed in any of the experiments with combustion of pesticides reported here.
- An almost quantitative conversion of the sulphur content in the pesticides to SO_2 can be expected, particularly at high temperatures. However, important amounts of the combustion gases COS and CS_2 can also be produced. H_2S was not observed in any of the experiments reported here.
- A great variety of organic compounds, will be produced by combustion. The highest amounts are expected at low temperature and/or low ventilation.
- Pesticides produce highly toxic fire effluents. The main contributors to the toxicity are HCl, $COCl_2$, HCN, SO_2, COS, CS_2 and CO.
- The survival fractions of the pesticides are important factors with respect to the toxicity of the fire effluent, particulary for the highly toxic pesticides such as some of the organophosphorous insecticides.
- Organic decomposition products may also contribute significantly to the toxicity.

The work described in the present paper provides guidance on how to assess the hazards from chemical warehouses. Information on the types and amounts of the various toxic products generated from fires in different chemicals has been obtained.

However, if very precise concentrations of the individual combustion products are needed in a risk analysis study it is recommended that experimental work be carried out in order to make an exact quantification, since the quantity of the individual components produced in the fire depends on the individual pesticide burned and the fire scenario. The DIN 53 436 method supplemented with combined pyrolysis/GC/MS experiments has been shown to be a useful procedure.

6 Acknowledgements

The work described in the paper was partly sponsored by the CEC STEP programme (project: combustion of chemical substances and the impact on the environment of the fire products) and the CEC ENVIRONMENT programme (project: TOXFIRE - guidelines for management of fires in chemical warehouses).

Dimethoate was supplied by Cheminova Agro, A/S and thiram was supplied by KVK AGRO A/S.

7 References

1. Christiansen, J.V.; Feldthus, A.; Egsgaard, H.; Carlsen, L. (1993). Flash pyrolysis of coal sub-structures adsorbed on a carbosieve. *J. Anal. Appl. Pyrol.*, **24**, 311-323.

2. Matuschek, G.; Ohrbach, K.-H.; Kettrup, A. (1991). Thermal analysis of commercial herbicides. *Thermochimica Acta*, **190**, 111-123.

3. Staudner, E.; Beniska, J.; Kysela, G. (1976). Study of thermal decomposition of tetramethylthiuram disulfide. *Chem. Zvesti*, **30**, 336-341.

4. Deutsche Normen (1981). DIN 53 436. Erzeugung thermischen Zersetzungsprodukte von werkstoffen unter Luftzufuhr und ihre toxicologische Prüfung, Teil 1, 2, 3. (In German).

5. Fardell, P.J.; Woolley, W.D. (1988). State of the art of combustion toxicity. International Conference: Fire: Control the Heat....Reduce the Hazard. 24-25 Oct. 1988, London, UK. Organized by QMC Fire & Materials Centre, 12.1-12.12.

6. Einbrodt, H.J.; Hupfeldt, J.; Prager, F.H.; Sand, H. (1989). The suitability of the DIN 53 436 test apparatus for the simulation of a fire risk situation with flaming combustion. In: Advances in combustion toxicology, vol. 1, Hartzell, G.E. (ed). Technomic Publishing Company, Inc., Lancaster, Pennsylvania, USA, 240-251.

LEGISLATION AND REGULATION – TOWARDS EC STANDARDS AND CODES OF PRACTICE

29 EUROPEAN STANDARDS FOR FIRE SAFETY: A SUMMARY OF THE CURRENT POSITION

P.R. WARREN
Fire Research Station, BRE, Garston, UK

Abstract
The development of European standards related to fire safety is being undertaken, principally, under the auspices of the European Committee for Standardisation (CEN). A brief review is given of those CEN Technical Committees currently covering aspects of fire safety. While the majority of current interest relates to the development of test methods and systems concerned with fire safety in buildings, other fields, such as transport, are included. There is considerable interest in the development of the European Construction Products Directive. This has provided the driving force for the harmonisation of fire test standards through the work of CEN/TC127. The current position with regard to fire resistance tests is summarised. Some areas where standards are required, most particularly the reaction to fire performance of construction products and of upholstered furniture, are the subject of ongoing prenormative research programmes supported by the European Commission and outlines of the programmes are given.
Keywords: Fire, standards, harmonisation

1. Introduction

The requirement to reduce barriers to trade within the European Union has led to the need to develop harmonised technical standards for common use within all member states, not least in connection with test methods and systems related to fire protection. This paper reviews the background to the harmonisation process, setting out briefly the way that standardisation has developed in general and in relation to fire, in particular. The way in which the principal standardisation body, CEN, works is summarised and illustrated by reference to the work of those Technical Committees dealing with fire aspects, in particular CEN/TC127 -Fire Safety in Buildings.

Fire Engineering and Emergency Planning. Edited by R. Barham.

2. Trends in fire safety requirements

The trend in regulating for fire safety in most European countries has generally followed a common path. Initially, from the earliest times, the main concern was with preventing fire spread from one building to another and with controlling possible sources of ignition such as cooking fires. This led to regulations for the spacing of buildings and for the fire resisting construction of common walls. As prevention of fire spread within buildings became of concern, regulatory emphasis was placed on compartmentation and the size of uncompartmented spaces. It was recognised that buildings used for different purposes would need different approaches to fire safety, leading to classification according to use (warehouses, shops, theatres, dwellings etc.). In the latter part of the nineteenth century, as buildings became larger, taller and more complex fire safety regulations began to involve the provision of means of escape for occupants. This led to concern not only with containing fire but also with preventing or limiting its development, leading to restrictions on lining materials and, more recently, on contents. While, at first, fire safety legislation was prescriptive, setting out requirements in terms of materials, dimensions and methods of construction, the move towards performance-based regulations led to the need to develop test methods for fire resistance and later, for the reaction to fire performance of materials. These, in turn, led to the need for standardisation.

3. Standards making bodies

Of course, fire safety was not the only field in which the need for standardisation is of concern. Driven by the need for consistency, consumer expectations and regulation, official standards bodies, such as the British Standards Institution in the United Kingdom, DIN in Germany and AFNOR in France, came into being in many European countries during the inter-war period. The development of international trade and activities which crossed national boundaries, such as shipping, stimulated the need for common international standards leading to the formation of the International Standardisation Organisation, based in Geneva, and the International Maritime Organisation which sets standards under the international agreements such as Safety of Life at Sea (SOLAS).

The development of the European Community and the European Free Trade Area led, in 1961, to the creation of the European Committee for Standardisation (CEN) with its headquarters in Brussels. While CEN is the leading European standardisation organisation, standards, including those related to fire safety, are also produced by bodies dealing with specialist sectors, such as CENELEC (electrotechnical) and AECMA (aerospace). These bodies work closely with CEN.

4. European Committee for Standardisation (CEN)

4.1 Background
CEN activities were relatively limited for the first twenty or so years of its existence This was due both to the lack of pressure on industry to use other than national or ISO standards and, also, to the fact that European legislation tended to attempt to contain

detailed technical specifications, where these were needed, rather than to refer to external standards. However, the publication in 1985 of a White Paper on the completion of the internal market led to the so-called 'New Approach' directives. With the aim of speeding up the removal of technical barriers to trade, these New Approach directives have only limited technical content. They state essential requirements, principally addressing health and safety matters. Compliance with the directives is through European Technical Specifications which can include CEN or CENELEC standards and European Technical Approvals, prepared by another pan-European organisation, the European Organisation for Technical Approvals (EOTA). The advantage of New Approach Directives is that basic legislation can be agreed first and supporting details, such as standards and technical approvals, produced later.

The introduction of the new approach resulted in a considerable increase in the work of CEN which now has over 300 Technical Committees working on a wide range of topics. Work specifically related to the policies of the European Union is generally covered by a formal mandate from the European Commission. Approximately 20% of current CEN work items are covered by such mandates.

Where work is covered by a mandate the Commission provides support. A 'standstill' procedure ensures that the national standards making bodies of European member states do not draw up or introduce standards in areas where European standards are being prepared. This has resulted in a substantial decline in the preparation of new, or substantially revised, national standards.

4.2 CEN organisation

The overall direction of CEN is undertaken by the General Assembly on which all contributing national standards bodies of European states are represented, together with associate organisations. The principal body controlling the production of standards is the Technical Board (CEN/BT). It is responsible for all matters concerning the organisation, working procedures, coordination and planning of standards work and organises technical liaison with other relevant European organisations. To assist it to do this a number of Technical Sector Boards (BTSs) and Programming Committees have been set up. The most important of the BTSs in relation to fire-related standards are:

CEN/BTS1 -   Building and construction

CEN/BTS4 -   Health and safety of the workplace

While these provide coordination for standardisation within a sector, there are some topics, in particular fire safety, which cross sectorial boundaries. In this context a small Working Group, WG50, was set up in 1991, to advise on fire-related matters and to ensure consistency of approach in any standards dealing directly or indirectly with fire matters. The Working Group is currently in abeyance but it can be reconvened, as and when required.

4.3 Production of standards

Proposals for new a new standards project may originate from any CEN member, the European Commission, or relevant European trade or professional organisation.

Once agreed in principle, by CEN national members, through CEN/BT, there are three possible routes to the preparation of a CEN standard.

(a) Questionnaire procedure

If an appropriate document, for instance an existing national standard, is considered as likely to satisfy CEN requirements, then this procedure enables CEN/BT to assess its acceptability as a formal standard.

(b) ISO

In June 1991, an agreement was signed between CEN and ISO to ensure cooperation and the exchange of information between the two organisations. Known as the Vienna Agreement, this commits CEN to use ISO standards, wherever possible, as the basis for its work.

(c) Technical Committee

In the absence of a suitable ISO or other appropriate document, a CEN Technical Committee (CEN/TC) may be set up to prepare a draft standard (or prEN). A CEN/TC is composed of expert delegations representing individual national interests and may include observers from pan-European organisations with a particular interest in the topic concerned. It is common for TCs to set up a series of representative Working Groups (WGs) to cover particular aspects of their tasks. Technical Committees are responsible to CEN/BT and work to an agreed programme with a clear scope and timetable for the critical stages for a particular project.

Once a draft standard has been prepared, following any of the above routes, it is subject to public enquiry to allow wider consultation and the opportunity for technical comments at a national level. After the results of this stage have implemented, formal approval is sought from CEN members using a weighted voting procedure.

Once a European standard has been formally agreed, it must be implemented by CEN member states and given the status of a national standard. Existing conflicting national standards must be withdrawn.

5. CEN fire-related activities

5.1 Technical committees

The work of CEN necessarily presents a continuously changing pattern. However, an analysis[1] of CEN activities being undertaken in 1993, identified 30 active Technical Committees and 114 individual projects had a component related to fire. Principle areas of interest within these were fire resistance tests, reaction to fire tests and the standardisation of design fires and fire exposures. To illustrate the range of CEN concern with fire, Table 1 lists a number of the principal Technical Committees whose work has a significant fire-related component. These cover areas such as construction, furniture, fire-fighting and transport.

Table 1.   Selected list of CEN Technical Committees dealing with aspects of fire
            (fire safety plays a major role in those indicated by emboldened type)

Topic	Number	Title
Construction	TC33	Doors, windows, shutters and building hardware
	TC72	**Automatic fire detection systems**
	TC127	**Fire safety in buildings**
	TC128	Roof covering products and products for wall cladding
	TC191	**Fixed fire-fighting systems**
	TC250	**Structural Eurocodes**
	TC277	Suspended ceilings
Furniture &	**TC207**	**Furniture**
contents	**TC248**	**Textiles and textile products**
	TC263	Secure storage of cash etc.
Fire-fighting	TC70	Manual means of fire-fighting equipment
	TC79	Respiratory protective devices
	TC158	Head protection
	TC192	**Fire service equipment**
Transport	**TC256**	**Railway applications**

5.2 CEN/TC127   Fire Safety in Buildings
The principal Technical Committee covering fire safety is CEN/TC127 - Fire Safety in
Buildings. This TC has been in place since 1990 and its principal role is in response to
the requirements of the European Construction Products Directive, through mandates
from the European Commission.

5.2.1 Fire resistance tests
Considerable progress has been made in relation to the standardisation of fire resistance
tests, covering the following principal topic areas:

(a) General aspects of fire resistance testing
(b) Non-loadbearing elements
(c) Loadbearing elements
(d) Service installations
(e) Contribution to fire resistance of structural members
(f) Door and shutter assemblies

(g) Fire performance of roofs exposed to external fire

As at February 1995, 15 work items had reached stage of completion of the six month enquiry period; 2 work items were in the process of the six month enquiry; 11 work items had completed draft documents which were in the process of consideration by the main TC and another 6 work items were still at the drafting stage within working groups.

5.2.2 Reaction to fire tests

Whereas work on fire resistance has progressed steadily, work within CEN/TC127 on reaction to fire tests is now only just beginning. This has arisen principally because of the difficulties faced by the European Commission in defining the necessary mandates under the Construction Products Directive. However, as discussed below, a mandate covering the evaluation of construction products in respect of their reaction to fire was issued in September 1994, following a Commission Decision[2] of 9 September 1994. Work is now starting on the development of tests for non-combustibility.

6. Prenormative research on reaction to fire

6.1 Construction products

As noted above the European harmonisation of reaction to fire tests has been faced with a number of difficulties. Not least of these is the very large number of different national tests currently in use and the lack of correlation between these. The latter was illustrated by a comparison of the performance of 24 wall lining materials European countries undertaken by ISO/TC 92 (Fire Tests on Building Materials, Components and Structures) in the mid-60s and reported by Emmons[3]. This showed a very poor degree of correlation in the ranking of the various materials between the different national approaches.

In order to overcome this problem the European Commission has published a Commission Decision[2] setting out a classification system for the reaction to fire performance of construction products, based upon six classes determined by a combination of three test methods, as shown in Table 2 below. Of the three test methods, listed in Table 3, two are based upon existing ISO standards but the third, the so-called 'Single Burning item' (SBI) method, which covers the key central range of performance, has yet to be developed and is currently the subject of a short-term prenormative research project. In parallel with this, a long-term research programme is proposed to develop test methods more closely based upon the improved understanding of the physical processes of surface fire spread and related theoretical models that are now becoming available.

Table 2   Proposed European reaction to fire classification system

Euroclass	Class of products	Test Method	Fire Scenario
A	No contribution to fire	I	*Fully developed room fire* (Exposure > 60kW/m²)
B	Very limited contribution to fire	I + II	
C	Limited contribution to fire	II + III	*Single burning item in a room (Exposure approx. = 40kW/m²)*
D	Acceptable contribution to fire	II + III	
E	Acceptable reaction to fire	III	*Small fire attack on limited area of product (flame height 20mm)*
F	No performance determined		

Table 3   Test methods for use in the proposed European reaction to fire classification system.

Number	Description
I	Small furnace non-combustibility test and/or 'bomb' calorimeter (cf ISO1182 and ISO1716)
II	NEW METHOD - provisionally called the 'Single Burning Item' test.
III	Small flame test. (cf ISO DIS 11925-2)

It is anticipated that the long-term programme will take a number of years and, in view of the need to establish a system quickly, current emphasis is placed on developing the SBI method. Much preliminary work has been undertaken and work on the construction of a prototype test apparatus is about to start. The work is steered by an advisory group of fire regulators from EU member states, set up by the European Commission.

6.2 Upholstered furniture

Some European member states, notably the United Kingdom and Ireland, have in place regulations relating to the flammability of furniture. Consideration is being given by the Commission to the possibility of a directive dealing with the fire behaviour of upholstered furniture. Such a directive would be likely to require tests for ignitability and post-ignition behaviour. The work of CEN/TC207/WG6 (Furniture - Test Methods for Fire Behaviour) has provided the basis for harmonised ignitability tests but there is, at present, no basis for assessing post-ignition behaviour. In consequence, a comprehensive prenormative research programme has been undertaken, under the auspices of the Measurements and Testing Programme of CEC/DGXII (Science, Research and Development). Starting in early 1993 and involving 11 laboratories representing 8 European countries, the work has been completed and publication of the final report is imminent. The programme has sought to relate the hazard produced by burning furniture in a full-scale room scenario to the results of calorimeter tests both on full-scale furniture items and bench tests on composite and component materials. This approach may provide a useful guide to the long-term programme proposed for construction materials, mentioned above.

7. Conclusion

The development of harmonised European standards related to fire safety is a wide-ranging and complex process. It has only been possible in this paper to give a brief and incomplete overview. The preparation of standards involves considerable effort by experts from all of the countries of the European Economic Area, much of which is unseen but which, nevertheless makes a substantial contribution to securing improved fire safety and reduced property losses.

8. References

1. Becker W.   Unpublished communication.

2. Commission Decision of 9 September 1994 implementing Article 20 of Directive 89/106/EEC on construction products.

3. Emmons, H W. (1974). *Scientific American,* Vol 231, No. 7, p21ff.

30 A COMMENTARY ON THE FIRE RESEARCH/BUILDING DESIGN APPLICATIONS INTERFACE

J.C. ANGELL
Department of Built Environment,
University of Central Lancashire, UK
E.L. ANGELL
Chartered Architect, Preston, UK

Abstract

The paper explores the problem of communicating fire research findings and advice to those whose task it is to design new buildings. If progress is to be made, and safer buildings to be built, then research findings must be presented to Architects in a form which relates to the structure of the design process. Since all design involves compromise, advice on matters relevant to fire safety need to be presented in a way which ensures that, where compromises must be made, they are not made in ways which affect the safety of the building and its occupants.
Keywords: communication, conflicting criteria, compromise, change, cost

1 Introduction

If expertise in minimising the risk of major fire disasters is to be successfully applied to the design of new buildings and city complexes, then it must be recognised that the designers of buildings have many diverse criteria calling for their attention. Experts in all disciplines must address the problem of communication, and, since the central role of design co-ordinator is traditionally taken by the Architect, must ensure that the Architect in particular is aware of which of their criteria are essential and which are merely desirable. The form which that communication takes should have the object of ensuring the widest future application of the latest thinking in fire engineering.

2 The Brief

The design of any construction project begins with the client's brief, and the brief, if it is to result in a satisfactory building, must touch upon every aspect of the proposed works. Many basic criteria, of which fire safety is but one, are already covered by

Fire Engineering and Emergency Planning. Edited by R. Barham.
Published in 1996 by E & FN Spon. ISBN 0 419 20180 7.

legislation, but the responsible Architect will always query whether the client prefers a higher standard than the minimum provided by legislation. This may have a bearing, for example, on insurance premiums, which will have to be considered as part of future running costs. Thus it is easy to ask whether a client wants a higher standard of energy conservation, or greater freedom of access for the disabled, or a higher standard of fire safety than current legislation stipulates. Similarly, it is easy to ask these questions in a way which prompts the preferred answer. Would you like lower energy bills...? Would your insurers like a safer building...?

A client's instructions may include his wish that the building's appearance reflect his importance and its expense. He may be more concerned with a prestige appearance than with safety - you can't see safety. If he is an industrialist he may regard the building as a necessary part of the manufacturing process it houses, and it's cost is added to his products in fractions of a penny, for his legal duty as a director of the company is to maximise the profits for his shareholders. Fire safety regulations and recommendations are the ammunition the Architect needs with which to convince a reluctant client.

On the other hand, the brief for a public building enables a politician to emphasise criteria which give him a politically acceptable image regardless of his personal expense, which may or may not include fire safety, while avoiding any reference to economy of structure or practicality of construction. If a developer client is building speculatively, for sale, he will frequently emphasise built volume and economy of construction over other considerations. In short, regulations and recommendations for fire safety must recognise that a building project originates with the client, and the client's brief will often contain a hidden agenda which the Architect must satisfy if the client is to approve the construction of the design.

3 Compromise

Having established the initial criteria which make up the client's brief, the Architect begins to look for points of conflict. This is part of the process of analysis. Different criteria may easily result in physical conflict - ease of access and exit may conflict with security, as in a reference library or a jeweller's shop - and many design criteria will conflict with the available construction budget. If the building is actually to be built, it must be made possible for the client to afford it. So, if the final building is to be successful, all the conflicts in the brief must be resolved, and this implies compromise. This is the process of synthesis. Indeed, all design is compromise, and thus the requirements for fire safety are as likely to be compromised in practice as any other design criteria, whether at the design stage, during construction, or in the course of future maintenance.

From this arise the questions of what constitutes a successful design, and what amounts to a reasonable degree of compromise, which further leads to the question of how the criteria for fire safety may be so presented to the designer that they are successfully adopted in practice. On the first of these, successful design, it is not necessary here to expand at length about buildability, usefulness, or aesthetics, except to point out that if people actually like a building, if they think it is beautiful, if they enjoy living and working in it, then it is much more likely to be cared for and properly maintained. It seems to be often overlooked that buildings are usually required to simply endure through time - preferably with minimum maintenance costs. If maintenance costs are designed to be low, it is that much more likely that changing economic and social circumstances will not inhibit future maintenance. Naturally, fire safety items, both hardware and software, both installations and policies, require maintenance and updating as much as any other aspects of a building.

4 Concepts

How may buildings be designed to endure through time while incurring minimum maintenance costs, and thus with the minimum degradation in their fire safety regimes? Clearly, there are a number of building types for which the simplest design concepts are appropriate. For example, a nursing home for the elderly or disabled should be single storey. Every bedroom should have a french window (porte-fenêtre) or patio door with a terrace outside, and every bed should have wheels. Similarly, all infant and junior schools should be single storey, and every classroom should have a door to the playground. By avoiding a problem, rather than by creating a difficult situation which must then be solved, it seems obvious that such designs will be inherently "safe" and easy to escape from, even before considering additional hi-tech safety systems, although the risk of intruders, theft and arson may be increased.

The specialist knows more and more about less and less, until, as the saying goes, he knows everything about nothing. The Architect, on the other hand, is both by training and inclination a generalist. Since most areas of design require compromise, it is the Architect's job to discover or invent acceptable compromises between the conflicting requirements of the team of specialist engineers. Since many designers don't know how they do "it" - design - (Many intuitive designers will tell you that their ideas simply crawl out of the end of a pencil...) a basic requirement that a given building type be single storey is the sort of design criterion which, once adopted, is unlikely to be compromised in the finished building.

5 Criteria

How then may the criteria for fire safety be so presented to the designer that they are successfully adopted in practice? The first answer is that Architects need simple rules of thumb in the early stages of a project. So does the client, who invariably queries the cost of things he cannot see. So much for sophistication, and sophisticated design! There is much emphasis nowadays on sophistication in construction. Hi-tech architecture, which emulates NASA space hardware or offshore oil rigs are a current fashion adopted by a number of prominent Architects. Masts, tents, metal skins, external bracing and external services abound. But is this anything more than passing fashion? The basic characteristics of hi-tech artefacts are that they are intended to be mass-produced after extensive research and development, that they consume energy, require servicing, and are quickly obsolete. Few building projects have a budget for extensive research, and many people today will have thrown away a calculator, replaced an old computer, or scrapped a car. Society, however, has different standards for buildings.

The simple fact is that the brief for most buildings, if properly understood, requires that the building should sit on the site, enduring through time whilst needing the minimum of maintenance - at least until the mortgage is paid off. In other words, the brief for most building types implies a low-tech solution, simple, unspectacular, and relatively inexpensive to build and to maintain. The most significant difference between the vernacular low-tech buildings of today and those of the past is in the concept of the services, and these are largely able to be fitted retrospectively.

Innovative solutions adopted without extensive testing are a sure recipe for disaster, just like the Local Authority tower blocks built in the United Kingdom in the 1960s. In that case, the innovative form of the development was a requirement of the brief - the Local Authority clients demanded blocks of flats, and directed tenants to live in them. The results were profoundly unpopular and very difficult to maintain or up-grade, and the majority of these buildings have had to be prematurely demolished, leaving the Local Authorities still paying the original mortgages.

If innovative fire safety solutions are imposed on the built environment without adequate research and testing then a similar fate is a distinct possibility. Conversely, if there is a call for simple, tried and tested fire risk solutions to be applied to new projects, the chances are that the detailed proposals will be understood, intelligently incorporated in a design, and maintained in practice. Just as complex tax laws result in tax dodging, so complex fire safety rules will result in cheating.

6 Formulae

Once basic fire safety rules of thumb have been incorporated in a sketch scheme, and inappropriate alternatives rejected, then more subtle criteria can be worked into a design. Even so, if these criteria can be reduced to a handful of easily-remembered formulae, there is a greater chance that they will be used successfully. Simplicity here may indeed result in over-provision of safety features, but as long as the budget is not compromised this is not a bad thing. For example, a basic decision to incorporate sprinklers, or electronic alarm systems, or both, can be readily discussed with and understood by the client, and economic provision made accordingly. Of course, some overkill may be inappropriate, or even dangerous. I once spent some time persuading a Fire Prevention Officer that one-hour fire doors with hydraulic closers were not a good idea for the classrooms of an infant school. Either they would have been permanently wedged open, or many small fingers would have been pinched, or children trapped by doors too heavy for them to open.

7 Headings

The simple provision of fire safety features in a building is traditionally understood under the straightforward headings of:
    combustibility and resistance
    compartmentation
    means of escape
    access and facilities for fire fighters
    automatic systems: e.g. alarms, sprinklers, ventilation

By and large, Architects are comfortable using such simple concepts in terms of the problems of initial design, and future developments in fire safety will hopefully take place within these or similar headings. However, these future developments must be adopted with more account being taken of the cost of implementation than perhaps has been the case in the past.

8 Costs

Cost is a vital consideration, and it is only too easy for unforseen costs to invalidate a commercial proposal. No-one in the construction industry likes to put a price on human life, least of all the Fire Prevention Officer and the Architect, although, of course, insurance actuaries do it all the time. The client will have his own view of the value of human life. A little more realism regarding the cost of fire safety would often be helpful. Similarly, for the sake of cost control, there should be a cut-off point for fire safety design input. Changes of mind and extra fire safety requirements once the working drawings stage is reached makes cost control very difficult, and changes once construction has begun on site put all hope of financial control beyond reach. This is of vital concern to Architects, who risk being sued for professional negligence if they are seen to take an irresponsible attitude to the task of spending their client's money. Back in the early 1970s we had a large public building actually under construction when *Summerland* burnt down in the Isle of Man. The Fire Prevention Officer was on site within a week with an extensive list of expensive alterations, and all the additional money had to be found from within the existing budget. Naturally, it was the final finishes and fittings which suffered. Had it been a commercial or industrial building, the whole project could have been abandoned.

9 Maintenance

The problem of maintenance has already been mentioned. The fire safety systems and concepts of a building must be maintained throughout the life of that building. Here we must distinguish between maintenance to ensure that all systems are working as the designer intended, and future maintenance or refurbishment to upgrade a building to comply with future standards. Legislation with retrospective effect is never easy to implement, although in the United Kingdom we have the recent experience of raised fire standards in the hotel industry to guide us. A thoughtful designer, aware that the brief for a building stipulates an intended life span, will try to make allowances for unspecified future upgrading of the building's performance. The inverse corollary is that a building which cannot be upgraded is likely to be demolished.

How then can we design a building for long life? Firstly, by using materials which are naturally long-lasting, fire-resisting, and which require little maintenance. Secondly, by a generous over-provision of space, so that future alterations may be more readily incorporated. Thirdly, by designing systems which require a minimum input of energy. For example, smoke vent systems which are designed to be fan assisted are more likely to fail that those which function naturally. These enhanced design standards would in the past have failed to meet cost yardsticks, but life cycle costing techniques give more opportunity today for thoughtful design.

10 Presentation

How might all this information be presented to the Architect? Basic philosophies of fire safety should be presented as simple rules of thumb, and detailed fire safety criteria as straightforward formulae. A clear distinction should be made between what is essential and what is desirable, and areas of potential conflict with other criteria should be highlighted. Tables, graphs and diagrams should be used where possible, because designers tend to think visually. With an eye to future maintenance, simple systems are preferable to complex ones. If all this essential information can then be presented on one side of one piece of paper, so much the better. More helpfully, as computer aided design becomes more common, the information could be made available on disc.

11 Conclusion

Clearly, the objective is to design and construct better buildings, but if better buildings in terms of fire safety result in poorer buildings in other senses, the fire safety advice will be ignored. There is no logical reason why a fire-safe building should be ugly, or inconvenient, or unrealistically expensive, or depressing to work or live in. If such conflicts appear during the development of a design, the fire safety advice will be held in contempt, and will be less likely to be put into practice.

31 FIRE LEGISLATION: A UK VIEW OF EUROPEAN FIRE SAFETY REGULATION

R. BARHAM
Department of Built Environment,
University of Central Lancashire, UK

Abstract

This paper reviews the move within the European Union towards "risk-based, goal-oriented" health and safety legislation and towards a harmonisation of legislations (subject to considerations of "subsidiarity") in member states. It considers some of the main points raised in a consequential review of fire safety legislation recently carried out by the British government's Department of Trade and Industry and the author comments on some areas of common concern for European Union member states.

Introduction

In the past, the domestic policies of each of the member states of the European Union have each been separately pursued and, in general, it can be noted that in many states the Fire Safety Legislation has developed in a very ad hoc manner. However, with the development of, first, the common market and, latterly, the development of the European Union as more than just a trading block, the formerly fully independent states of the Union are now subjected to laws, or "Directives", by the European Commission which impinge either directly or indirectly on the domestic laws of each of the member states.

An avowed intention of the European Commission is to attempt to harmonise, wherever possible, legislation and regulations across member states where such is in the best interests of the citizens of the European Union. Fire Safety Legislation, also becomes subjected to such intentions and the purpose of this paper is to draw attention to the fact that, currently, little information is available on a pan European basis to enable the declared objectives of the Commission to be achieved in areas which affect Fire Legislation. It also becomes clear that the pursuit of Fire Engineering, Fire Engineering Management or Fire Science without a knowledge of the legislative framework within which it is to operate could mean that what might be an acceptable academic solution could be inappropriate in some member states but accepted in others.

Research is presently in hand to try to supply the deficiency in the availability of a European overview of legislation and regulation. It consists of a programme of reviews of domestic policies affecting member states' Fire Safety Legislations and it intended that, ultimately, it will be possible to identify where changes could, or should, be made to accommodate any proposals for European harmonisation. In the

Fire Engineering and Emergency Planning. Edited by R. Barham.
Published in 1996 by E & FN Spon. ISBN 0 419 20180 7.

wake of the recent directives on health and safety this research is accorded high priority in both British and several other European Government circles but, as yet, there is a demonstrably unco-ordinated research response from the fire community; the availability of research and education for Fire Engineers and related consultants varying from member state to member state.

What is necessary is for all concerned with this important area to realise, firstly, that fire related activities within the European Community fall under the remit of several of the Directorates General. More particularly, health and safety legislation, of which Fire Safety Law forms a part, is the responsibility of Directorate General V. In contrast, the majority of the scientific work carried out in the area of Fire Engineering falls within the area supervised by Directorate General XII.

If we are to consider both the policies and directorates which are intended to implement or extend common EC Legislation and Codes of Practice, then it is necessary to be aware not only of the central issues raised by the EC itself, but also to have knowledge of the way in which Fire Safety is provided for and legislated for across the European States. It is also interesting to note that Governmental Legislation is often supplemented by, or takes account of the existence of, the regulatory effects created by insurance companies in the provision of their financial services and the insurance of fire risk. Movements toward harmonised legislation (subject to all the arguments of subsidiarity, etc.) will only be achieved when we have an adequate knowledge of the European legal and cultural attitudes to Fire Safety and we are able to educate future legislators, lawyers and practitioners to review and adapt fire and safety legislation and associate policies in the light of knowledge of other member state systems and of the level and sophistication of Fire Engineering knowledge.

Of particular concern to any researcher in this area, is the extent to which policies relating to harmonisation or improvement of legal, quasi legal and professional practice standards throughout the European Union are implemented. These can be reviewed only on the basis of a thorough knowledge of the cultural, economic and legal aspects of commercial and construction processes across Europe. Attitudes vary considerably; approaches to standardisation and implementation of regulations differ from state to state; The influence and responsiveness of insurers (and their loss prevention strategies) also vary. Any move towards harmonisation, even within the principal of subsidiarity, pre-supposes an understanding of its impacts on those cultures and its interaction with national character in order to identify those areas where changes are possible. Presently, however, there does not seem to be any commonality of approach to, for example, the enforcement process; nor is there, yet, any comparative work available on variations in the Fire Safety Provisions across Europe.

The European Basis
The basis of Fire Safety Legislation in the European theatre is Health and Safety Provision. The primary cause of European Activity is a desire to ensure better standards of safety in the workplace. The originating proposal arises out of the Health and Safety provisions of the Single European Act.

In England, a review[1] of its fire legislation was triggered, not just by a concern about the complexity of the English legislative framework but also by a reaction to the Home Office proposals in 1992 for new Fire Precautions (Places of Work) Regulations intended to implement the requirements of EC Directives, all of which arise out of the health and safety provisions of the Single European Act.[2]   The reaction, from commerce and industry, was one of horror at the potential cost of implementation. The Single European Act introduced a new Article, retrospectively, into the original Treaty of Rome[3].

That Article, 118a, reads:

(1)   *Member states shall pay particular attention to encouraging improvements, especially in the working environment, as regards the health and safety of workers, and shall set as their objective the harmonisation of conditions in this area while maintaining the improvements made.*

(2)   *In order to help achieve the objective laid down in the first paragraph, the Council, acting by a qualified majority on a proposal from the Commission ..... shall adopt, by means of directives, minimum requirements for gradual implementation, having regard to the conditions and technical rules obtaining in each of the Member States.*

Such directives shall avoid imposing administrative, financial and legal constraints in a way which would hold back the creation and development of small and medium-sized undertakings.

The intention here is clear; and the point most relevant to our present discussion lies in the last sentence of this extract. However, the ideal of common standards, making it more straightforward to construct or operate in other Member States, is some way off. Even if the regulations are harmonised, it is likely that compliance and enforcement practice will be subject to the vagaries of local cultural influences.

The first of the expected directives[4] was issued in 1989 - to be implemented by the end of 1992. This so-called "framework directive" sets out health and safety principles that are to be applied to all industries. These are that:

1. risks should be avoided
2. if they cannot be avoided they are to be evaluated and combated at source.
3. the dangerous should be replaced by the less dangerous
4. measures that protect groups of workers should have priority over measures that protect only individuals.

Here, then, is an EC directive which makes explicit for all work related activities the risk assessment and risk management approach already evident in some of the UK regulations like, for example, COSHH[5]. This first directive was quickly followed by a second[6] - the "workplace directive". However, this second directive, whilst also referring to a risk based approach, explicitly requires arrangements for first aid, fire precautions and emergency regulations. Hence the Home Office proposal and hence, subsequently, the review and consultation process recently undertaken.

The English proposals, if implemented in their present form, would achieve two objectives. First, they would facilitate the implementation of EC Directives and, second, they would enable a consolidation of fire safety legislation and the establishment of national standards. Implicit in this statement must be, of course, the Article 118a caveat: "so as not to hold back the creation and development of small and medium sized undertakings." This implied term poses great problems. Risk identification, quantification and management techniques are available - but when economic considerations are allowed to outweigh safety considerations we are in danger of going backwards rather than forwards. Note, for example, the recent disaster in China - of which more later.

Action has already been taken in the UK to commence the process of implementation of these directives. It has taken the form of the issuing of a sequence of three consultative documents all intended to stimulate debate and to, ultimately, bring into force a new regime of Fire Safety Regulation. This process is still continuing.

However, one of the major areas of concern within the UK is the main thrust of recommendations within one of these reports that, in general, Fire Safety should lie within the ambit of Health and Safety at work. This area of general health and safety in the UK falls not within the ambit of the Fire Services nor even the building control or planning functions of the local authority but remains with an independent body - the Health and Safety Executive. It is argued, therefore, that the expertise found within the British Fire Service would not be available to Fire Safety Inspectors if this area of activity were to be transferred to, and become, the sole responsibility of the Health and Safety Executive. No doubt similar debates are occurring in other European states.

An Opportunity

The advent of the European Directives provides a Europe wide opportunity for constructive action and a new approach to Fire Safety Legislation. What is needed, in all nations, is a comprehensive approach to Fire Safety provision. What is presently obvious is that the level of provision varies at present from member state to member state and the extent of compliance (or non-compliance) also varies significantly. Quite apart from the variations in approach between those countries such as England where there is a common law basis for all implementation of legislation and other European states in which there exists a codified system of law derived from Roman Law or Cannon Law, it should be noted that the extent and nature of member states' Fire Safety Legislation varies and is a reflection of not just the cultural attitude but also the level or prosperity and economic activity.

It is often argued that the underlying reason for the existence of Fire Safety Legislation (for both suppression and prevention) is to enhance the protection of human life. This is an interesting humanistic argument but one that does not hold if we look at the facts. In the United Kingdom there are, on average, about 35 deaths each year as a result of fires in industrial or commercial premises. By contrast there are over 600 deaths resulting from fires in domestic premises. If Fire Safety is about the protection of human life then why is the main thrust of both domestic and European legislation concerned with fire protection in industrial and commercial property and why is its cost of provision always a factor to be taken into consideration?

The Purpose of Legislation

In addressing the need for new legislation, the starting point should be to look at the purpose of legislation. All societies make laws for their self regulation - government of the people by the people. It is also an axiom of legal theory that enforcement of law requires the acceptance of society in general. A bad law, no matter how strict the attempt at enforcement, will be quickly repealed. Society, all sections of society, must see and appreciate the need for a piece of legislation for it to be acceptable. It must be to the benefit of all.

From what point, therefore, should the need for new fire safety laws be addressed? The point is often made (and I have made it myself on many previous occasions) that the historical background to current fire legislations relating to occupied premises, across Europe, shows the majority of those pieces of legislation to have been introduced as a response to major fire disasters and not as a response to any long term strategy or plan. "Emergency legislation" introduced in this way always seems to create additional problems as a result of overlaps with other areas or gaps left due to the indecent haste of its introduction. This is the present position in England and the time is right, therefore, for a strategic review.

The starting point for this review, however, should be to examine the underlying purpose of fire safety legislation. English fire safety courses often draw attention to the Great Fire of London (1666) as being the catalyst for fire safety legislation - legislation intended only (or primarily?) to save human life. There had been major fires in the metropolis on many occasions prior to 1666 but these had passed without such an outcry. For example, in London in the 13th century, one fire alone cost 3000 lives. What was different about the Great Fire? It only lasted for three days and only three people died! However, thirteen thousand buildings were destroyed and over sixty percent of the country's wealth was destroyed along with them! This resulted in the introduction of a requirement for the construction of full height dividing walls between properties (known as "party walls") - in itself an excellent idea - but it was for the protection of *property* against fire, not for the protection of people. It was not until the mid nineteenth century, with its social change and concern with public heath, that fire death was really considered - in Regulations designed to control the nature of buildings and to promote the health and efficiency of the workforce.

Today, we are all taught that the express purpose of fire safety legislation should be to protect *persons* from fire risk. However, if the protection of human life is paramount, we would legislate (and design) for zero risk. Yet, in recent European legislative proposals and directives, we are being asked to consider a risk-based, goal-oriented approach to fire safety.

Before considering the need for new legislation as a result of this change of emphasis, consider the possible extreme effect of de-regulation. It may be summed up in the example of the factory fire in Shenzhen Province in China, last year. Hundreds of people died in that fire because, to maintain productivity, all of the doors and windows had been secured to prevent the ill disciplined peasant workforce from wandering out during working hours!

The European Effect on the U.K. Scene

Although comprehensive data is not to hand regarding the extent of variations in European legislation, it is interesting to note that the recent studies in the UK point towards the fact that:

1.
   a. domestic and domestic/European legislations are overlapped and lack clarity; and
   b. these overlaps lead to confusion and put unnecessary burdens on business; and
2. there is a need to
   a. rationalise, simplify and modernise domestic fire safety regulations and procedures; and
   b. fill gaps in the coverage of fire safety provisions throughout Europe.

Currently, to address these findings in the U.K it is proposed that:
   a. general fire precautions should be in the same framework as other health and safety legislation *but* enforcement should be by fire authorities;
   b. in general, owners and occupiers should provide *and maintain* adequate fire precautions;

c. fire certificates should only be retained for higher life risk premises;
d. compliance should be by means of a self authorised risk assessment and emergency plan (certification, where required, to be based on the risk assessment);
e. national standards should be introduced to ensure uniformity of approach; and
f. a national advisory committee should be formed.

The intention of these proposals is to:

- reduce the compliance burden without adding to the administrative burden
- enhance levels of fire safety
- reinforce existing duties

and in moving in this direction it is suggested that no additional resourcing should be necessary for enforcement. The country's Health & Safety Executive (HSE), however, has other views.

It is also stated that implementation of these recommendations should result in:

a. a matching of precautions to risk
b. the creation of a general duty of care, similar to that found in the existing English general health and safety legislation[7]
c. avoidance of duplication of risk assessments for different purposes
d. reduction in required consultations between the Local Authorities' Building Control Officers (BCOs) and the Fire Authorities (FAs)

A streamlined building control process (plus new national standards) should facilitate the development and use of innovative techniques and approaches. However, it also raises many questions.

Suggestions that the necessary change can be best brought about by the re-use, or continued use, of some of the existing systems ignores the opportunity for comprehensive legislative change and the opportunity to modernise some, if not all, of the approaches to implementation of fire safety in England.

The Situation in the Rest of Europe

The main area of Fire Regulation in the UK lies, presently, within the Building Regulations and within the regulatory provisions of the Fire Precautions Act although, as has been previously mentioned, there are regulatory provisions in over 60 Acts of Parliament and some 80 or more sets of Regulations. In contrast, the Fire Safety Regulations of Denmark are to be found predominantly in the Building Regulations[8] issued by the Ministry of Housing. These date from 1982, although a new set is currently in preparation (publication iminent). These are quite clear and fairly stringent and it is interesting to note that the number of fire related deaths in Denmark is significantly lower than in many other European countries. Germany has a reputation for having a strict written code of fire related regulations[9] covering both the construction and occupation of buildings and, whilst no detailed investigative work has been done on the German Legislation, it has been noted that there is a strong insurance based control overlying the statutory framework.

France and Spain also have fairly stringent codes of fire safety. However, these are in many cases very detailed and, in the case of France (where they are contained in a series of separate documents[10]), apply at a national level. Investigations reveal, however, that there are relaxations and variations at local level which lead to variations in the level of implementation although this does not appear to be official. The national code of fire regulations[11] in Spain, NBE CPI-91, is supplemented by local provision at both regional and large town level and, therefore, there exists a variation in the system from place to place such as existed in the UK prior to the introduction of nationally based building regulations in 1965. The Spanish Fire Safety Regulations were upgraded locally[12,13] for the "island" site in Seville for the construction of the buildings for EXPO '92 in an attempt to have a strict fire safety regime on a site that was subject to world scrutiny and this exercise involved the re-writing of materials specifications and standards to cover some 85 different building techniques (from different countries) and some 110 cultural approaches to fire safety. Of considerable interest in this activity was the way in which materials test standards were harmonised.

It points to a problem which needs to be thought through. The harmonising of standards at European level is all very well but the requirements specified by national legislations also need to be aligned. Otherwise, because local regulations are written to satisfy local conditions, there could be a situation in which, for example, an M2 material might satisfy the European Test Requirements but, because of the non-alignment of local legislations, might not satisfy the requirements in one or two European states whilst more than satisfying the requirements in others. The fault would not be the adequacy of the material but the misalignment of the legislations. This problem is currently being experienced in Portugal which is in the process of reconstructing its code for fire safety. New fire codes exist for commercial buildings[14], industrial buildings[15] and public service buildings[16] and work is progressing on codes for administration buildings and school buildings. In Portugal, the influence of the insurance companies is fairly weak and the underlying fire engineering science is largely lacking in its underwriting professions. There is, therefore, a tendency towards strictly drafted laws which appear, then, to be more honoured in the breaking than the keeping.

The current situation in Spain is similar. The general code of fire protection[ibid.] for new constructions was introduced in 1991 and is similar in scope to the U.K. Building Regulations. There is, however, no current legislation dealing with fire standards in occupied buildings - other than a recently introduced set of regulations requiring introduction and maintenance of firefighting equipment[17]. Neither of the Iberian Peninsula countries appears yet to have taken any step towards implementation of the European Directives.

Conclusion

At European level, therefore, the matter of implementation of directives needs to be pursued with vigour but it is pointed out that there also needs to be co-operation between the several European Governments on matters of both general and process fire precautions, legislative and regulatory implementation. For those in fire related academe in both research and teaching and for those in fire engineering consultancy it is very important to realise that the engineering solutions and the materials standards can not exist in isolation; their acceptance, their implementation and their continued development depend just as much upon the existence of an adequate and compatible legislative and regulatory framework and on the economics of the environment in which they are placed.

But perhaps this paper should conclude with an observation on the matter raised initially, i.e., the cost to European industry. If the proposed changes to fire safety legislation and enforcement procedures are implemented in the wake of E.C. Directives, what are the real financial and legal implications for small and medium enterprises?

It should be noted that, in a parallel issue, the British construction industry is already predicting an increase of some 20% in its requirements for health and safety personnel as a result of the proposed implementation of the Construction (Design and Management) Regulations (CONDAM)[18].

Changes in fire safety requirements will make it necessary for there to be a substantial increase in the provision of education in fire safety, design economics and risk assessment/management in all relevant professional and vocational courses. This will, itself, give rise to a need for a radical review of educational provision in these areas throughout the European Union and, no doubt, yet another increase in the costs to industry and commerce.

References

[1] Crichley, J., Scott, S., Swift, N. & Terry, H., **Fire Safety Legislation and Enforcement: Report of the Interdepartmental Review Team**, The Department of Trade and Industry, London, 1994.

[2] European Parliament, **The Single European Act, 1986**, The European Commission Publications Office, Brussels, 1986, OJ 1987 L169/1.

[3] European Parliament, **Treaty Establishing the European Economic Community (EEC) (Rome, 1957)**, The European Commission Publications Office, Brussels, 1957, Treaty Series No.1 1973 Part II, Cmnd 5179-II.

[4] European Commission, **Council Directive (EEC) 89/391 on the Introduction of Measures to Encourage Improvements in the Safety and Health of Workers at Work**, The European Commission Publications Office, Brussels, 1989, OJ L183 29/6/89 p1.

[5] British Parliament, **Control of Substances Hazardous to Health Regulations, 1988**, Her Majesty's Stationery Office, London, 1988, Statutory Instrument No. 1657/1988.

[6] European Commission, **Council Directive (EEC) 89/654 on the Minimum Safety and Health Requirements for the Workplace**, The European Commission Publications Office, Brussels, 1989, OJ L393 30/12/89 p1.

[7] British Parliament, **Health and Safety at Work, etc., Act, 1974**, Her Majesty's Stationery Office, London, 1974, Statute Chapter 37, 1974.

[8] Danish Government, **Building Regulations, 1982**, Ministry of Housing, Copenhagen, 1982

[9] Deutches Institüt für Normung e. V., **Deutche Normen DIN 4102, Teil 4. Brandverhalten von Baustoffen und Bauteilen,** Berlin, 1981

[10] *for example:*
Government of France, **Décret N°. 73-1007, du 31 octobre 1973, relatif à la protection contre risques d'incendie et de panique dans les établissements recevant du public**, Journal Officiel de la République Française, Paris, 1987
Government of France, **Arrêté du 31 janvier 1986, relatif à la protection des bâtiments d'habitation contre incendie. Titre VI "Parcs de stationnement"**, Journal Officiel de la République Française, Paris, 1986
Government of France, **Arrêté du 31 janvier 1986, relatif à la protection des bâtiments d'habitation contre incendie. Titre III "Dégagements"**, Journal Officiel de la République Française, Paris, 1986

Government of France, **Circulaire du 7 juin 1974, relatif au désenfumage dans les immeubles de grande hauteur**, Journal Officiel de la République Française, Paris, 1974

[11] Spanish Parliament, **Norma Básica de la Edificación. Condiciones de Protección contra Incendios, 1991**, Madrid, 1991, Real Decreto 279/1991

[12] Sociedad Estatal para la Exposicion Universal de Sevilla 92, S.A., **General Regulations / Special Regulations (SR8)**, 1990

[13] Sociedad Estatal para la Exposicion Universal de Sevilla 92, S.A., **Instrucción de Desarrollos, N°. 2.06, Para la Protección contra Incendio en los Edificos**, 1991

[14] Portugese Council of Ministers, **Regulamento de Segurança contra Incêndio em Edifícios de Habitação**, Decreto-Lei N°. 64/90 de 21 Fevereiro, 1990

[15] Portugese Council of Ministers, **Normas de Segurança contra Riscos de Incêndio aplicar em Estabelecimentos Comerciais**, Decreto-Lei N°. 61/90 de 15 Fevereiro, 1990

[16] Portugese Council of Ministers, **Medidas Cautelares Mínimas contra Riscos de Incêndio a aplicar aos Locais e seus Acessos Integragos em Edifícios ondé estejam Instalados Serviços Públicos da Administração Central, Regional e Local, Instituições de Interesse Público e Entidades Tuteladas Pelo Estado, Resolucao do conselho de Ministros, N°. 31/89**, 1989

[17] Spanish Parliament, **Reglamento de Instalaciones de Protección contra Incendios, 1994**, Madrid, 1994, Real Decreto 1942/1993

[18] Anon. (consultation document), **Proposals for Construction (Design and Management) Regulations,** The Health and Safety Commission, London, 1992.

EC PERSPECTIVES – NEW HORIZONS?

32 OPERABILITY ANALYSIS AS A TOOL FOR FIRE RISK EVALUATION

N. PICCININI
Politecnico di Torino, Dipartimento di Scienza dei Materiali
e Ingegneria Chimica, Torino, Italy

Abstract

This paper illustrates the execution of an Operability Analysis (OA) directed to the elaboration of logic trees. It is shown that both Fault Trees and Event Trees descend almost automatically from a well-structured AO, though it is also pointed out that this transition is facilitated by the intermediate construction of an Incidental Sequences Diagram (ISD), a logic tree that provides a clear indication of all the links between primary events and their unwanted consequence.
Keywords: Operability analysis, Fault Tree, Event Tree.

1. Introduction

In their book, T.J. Shields and G.W.H. Silcock adfirm: "The development of useful, reliable fire safety evaluation techniques is proving elusive and is still very much in the embrionic state" [1].
The author thinks instead that the set of methodologies currently used in the process industries for a Probabilistic Risk Analysis (PRA) are well consolidated and that their application provides a real knowledge about fire safety evaluation [2].

2. Development of an OA

OA is a qualitative method whereby critical examination of deviations from the normal operating conditions is used to identify different forms of plant malfunctioning, the risks thus created, and the management problems involved.
This well-known method, however, is on many occasions still utilised in accordance with the procedures codified when it first appeared or the few modifications that have since been introduced [3,4].
Briefly, an OA executed in this way represents the stage in which a project is checked by a group of experts who examine it critically in the light of their own engineering experience. The forms summarise the results of their discussions in the shape of actions to be taken later so as to obtain a plant with a greater degree of safety.
In situations when quantification of a certain number of

Fire Engineering and Emergency Planning. Edited by R. Barham.
Published in 1996 by E & FN Spon. ISBN 0 419 20180 7.

potential accidents (i.e. estimates of their probable frequency of occurrence) is also required, logical models should be constructed from the information included in the operability forms. This information, however, must be interpreted by an analyst (not always an easy task) and translated into fault logic diagram terms.

The methods described in this paper have been worked out so as to automate the logical model construction stage and thus reduce the costs associated with the safety analysis and quantitative checking of the project. There was thus a need to define a well-structured procedure for execution of the OA.

Moreover, the way in which the form is filled in (explained in para 2.1 below) takes account of the logical links inevitably present in a correct OA. Indeed, it is the presence of these links that enables both an FT and an ET to be extracted from an OA [5, 6].

The proposed OA has been illustrated by way of example with reference to a "frying-pan". The plant is illustrated in Fig. 1, with its possible critical points or "Nodes" [4]. Deviations from normal operating conditions are sought in these nodes.

2.1 Procedure

To make the analyst's task easier and provide a certain degree of formality, an OA is elaborated on forms such as that shown in the table of Fig. 2. This is divided into eight columns corresponding to a precise logical pattern permitting subsequent graphic reconstruction of sequences of events identified in the OA.

The execution of an OA may be heavily dependent on the proper division of a plant into sub-systems. For this reason, boxes A, B, C and W in the flow diagram (Fig. 3) are dedicated to a check for the above aspects.

Another preliminary operation is identification of the boundary nodes and those internal to each sub-system - |Box D|, i.e. the points where deviations of a process variable (temperature, pressure, etc.) may develop or propagate. These points must be suitably numbered. This opeartion must naturally go hand in hand with identification of the process variables regarded as significant for the analysis -|Box E|.

The OA itself actually begins at box F. The details of the process from this box onwards will now be described.

- |Box F| - A plant is usually analysed according to its flow lines. It is thus desirable to take a boundary node of the first sub-system as the starting point.

- |Box G| - The first process variable chosen to start the analysis is selected. Its significant deviations from the stable operating conditions are examined -|Box H|- and then the regulating (or shut-down) systems able to intervene - |Box X|.

- |Box I| - Identification of a "Deviation" (column 1) is followed by the search for all its "Possibile causes" (column 2).

In Fig. 2 these causes are separated from each other by a dashed line if they are linked by an OR gate, and joined by the ampersand (symbol &) if they are linked by an AND gate.

The "Consequences" expected from the "Deviation" if the protection devices fail or are not present are placed in column 3.

The analysis is now continued by tracing back to the origin of the cause identified. This is done by moving from one node to the next

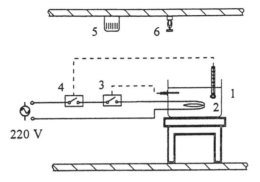

Figure 1 - Plant diagram:
1, Frying-pan; 2, Oil; 3, Thermostat; 4, High temperature switch; 5, Smoke detector; 6, Sprinkler

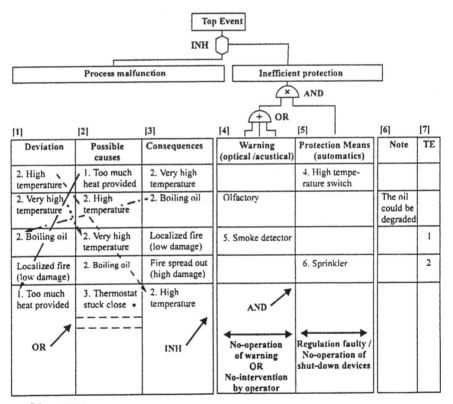

• Primary event

Figure 2 - Logic links in the construction of an OA and the final part of an FT

along the flow lines of the process, using simple logic rules applied
to columns 1, 2 and 3.
- |**Box J**| - If the causes identified in box I are regarded as primary
at the depth of analysis obtainable at this level, they are given an
appropriate distinguishing mark, such as an asterix - |**Box L**|, if not,
they require further analysis.
- |**Box K**| - Non-primary causes are further examined by regarding
them as deviations. The content of the item in column 2 is thus
included in column 1 as well [————▶] and since this "Deviation"
must have a "Consequence", column 1 must be shifted to column 3
[-------▶]. In this way a new cause is identified.
- |**Loop K-J**| - Systematically one works up from a cause as a deviation
to the causes regarded as primary with reference to the consequence
identified in box I.

At this point, having traced the origin of the causes - |**Box
L**|, one investigates the gravity of the consequences - |**Box M**|. If one
consequence is critical, i.e. appears as a TE, it must be indicated in
column (7) with a progressive number. In any event, to be identified
as a TE a consequence must be the outcome of deviations for which
protection systems have failed to come into action, or were not
provided (Fig. 2).
- |**Loop N-M**| - In the same way, if the first consequence is not a TE
and hence conclusive for the analysis, it must be further analysed
until either the consequence regarded as final is reached, or all the
consequences regarded as possible and likely for the entire plant have
been examined.
- |**Box N**| - The intermediate consequence is now converted into a
deviation by shifting column 3 to column 1 [-·-·-·-▶] and its
deviation is then transformed into a cause by shifting column 1 to
column 2 [·········▶]: a new consequence is identified.

One can therefore get to all the possible TE's by proceeding in
this way along the flow lines of the plant. It is worth noting that
this developement of the analysis automatically leads to the boundary
nodes, which means that the adjoining sub-system must be analysed.
It should also be noted that combination of the two procedures Loop
K-J and Loop N-M both ensures that the analysis is congruent and
permits connection between branches in the subsequent logic tree
development.
- |**Box O**| - Completes the inquiry for all the process variables
identified in box E.
- |**Box P**| - The analysis ends for all the nodes comprised in box D.
- |**Box X**| - Each time a deviation is noted in box I or in the course
of Loop K-J and N-M, columns (4) and (5) must show:
  - the optical and acoustical devices installed to give warning
    of the deviation,
  - the automatic protective or shut-down means provided for each
    deviation.
- |**Box Y**| - Moreover, a completeness check to ensure that all the
nodes identified and their process variables have been duly analysed
can be made by scanning the head of each form. An OA congruence check
must also be carried out by following the column 1, 2 and 3 shifts for
each deviation, cause and consequence.

An initial result with regard to the safety of the plant is

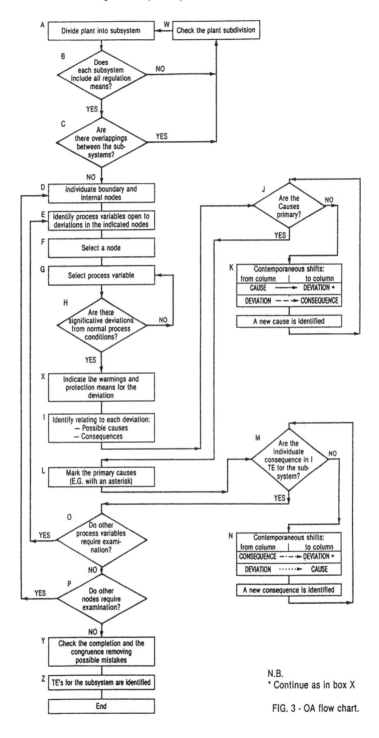

FIG. 3 - OA flow chart.

N.B.
* Continue as in box X

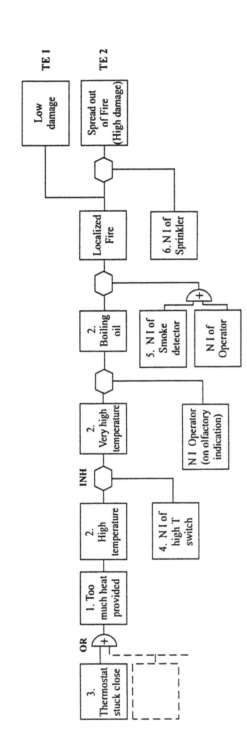

N I = No-intervention

Figure 4 - Incidental sequences diagram

immediately evident when the forms filled in are examined at the end of the OA. The TE's for each sub-system are identified, in fact, and columns (4) and (5) show whether or not alert signals or protection systems are envisaged for the deviations giving rise to such TE's.

3 Construction of logic trees starting from an OA

The first logic tree that can be extracted from an OA, and from which all the other trees (FT, ET, etc.) can be derived, is the Incidental Sequences Diagram (ISD). This is a graphic representation providing an easy link between an OA and an FT or ET [5,6]. Its construction should be the result of simple transcription of the contents of the forms, using well-known logic gates already implicit in a correct OA (Fig. 2).

As previously indicated, the ISD obtainable from the type of OA described is in general a series of elementary ISD's each providing a logical description of the occurrence of the TE (Fig. 3).

It should also be noted that construction of one ISD for each TE serves as a further check of the completeness and congruence of the OA itself.

The transcription procedure will be easier if certain points concerning the logical meaning ascribable to each column of the form are borne in mind. As can be seen in Fig. 2, columns 1, 2 and 3 refer to process faults, whereas columns 4 and 5 are concerned with failures of the systems employed to prevent such faults.

An ISD analysis stops at the depth reached in the OA. It is only in an FT that the primary events are fully developed so that they can also be analysed quantitatively. To do this, it is enough to develop the statements comprised in the boxes, which refer to column 2 of the OA form, into primary events associated with failure rates.

A similar procedure can be applied to draw an ET (Fig. 5) [7].

4 Conclusions

As far as the OA is concerned, a very innovative procedure has been presented to allow filling up of the form to the point of systematic and definite completion of the analysis by means of a recursive mechanism.

In addition to enhancing the strong points of the OA method, i.e. its systematic nature and completeness, the new procedure enables the actual attainment of these objectives to be checked in a simple manner, while direct extraction of the ISD from the OA provides a check on the congruence of the analysis that would otherwise be impossibile and a simple way to draw FT or ET.

The method is illustrated with a simple example and proposed as powerful tool for fire risk evaluation.

4 References

1. Shields, T.J. and Silcock, G.W.H. (1987) "Building and Fire", Longman Sc. Tech., Harlow.
2. Lees, E.P. (1981), "Loss Prevention in the Process Industries", 2 vols, Butterworths, London.
3. Lawley, H.G. (1974), "Operability studies and hazard analysis", Chem.Eng.Progr., $\underline{70}$ (4), 45-56.
4. "Guidelines for Hazard Evaluation Procedures", AICHE, New York, 1985.
5. Ciarambino, I., Scarrone, M., Piccinini, N. (1991), "Quantitative Operability Analysis: a study of a Pressure Regulating Installation on a city mains", proc. Int.Conf. Probabilistic Safety Assessment and Management, Feb. 4-7, 1991, Beverly Hills, G.Apostolakis ed., Elsevier, New York, 625-630.
6. Piccinini, N., Scarrone, M. and Ciarambino, I., "Operability analysis as a tool for an easy construction of logic trees", European Meeting on chemical Industry and Environment, Girona 2-4 June 1993, J. Casal ed., UPC, 187-200.
7. Piccinini, N., Scarrone, M. and Ciarambino, I. "Probabilistic analysis of transient events by an event tree directly extracted from operability analysis", J. Loss Prev. Process Ind., 1994, $\underline{1}$ (7) 23.

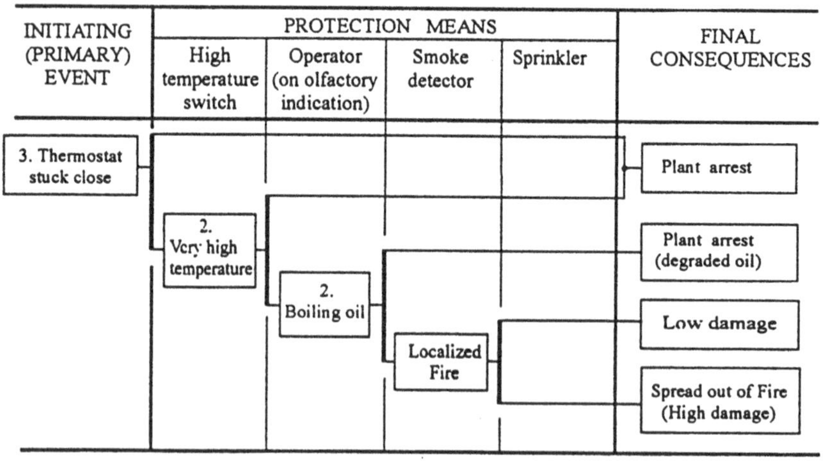

INITIATING (PRIMARY) EVENT	PROTECTION MEANS				FINAL CONSEQUENCES
	High temperature switch	Operator (on olfactory indication)	Smoke detector	Sprinkler	
3. Thermostat stuck close					Plant arrest
	2. Very high temperature				Plant arrest (degraded oil)
		2. Boiling oil			Low damage
			Localized Fire		Spread out of Fire (High damage)

Figure 5 - Event tree

33 DEVELOPING A GEOGRAPHICAL INFORMATION SYSTEMS-BASED DECISION SUPPORT SYSTEM FOR EMERGENCY PLANNING IN RESPONSE TO HAZARDOUS GAS RELEASES

S. CARVER
School of Geography, University of Leeds, UK
A. MYERS
Tyne and Wear Emergency Planning Unit,
Fire and Civil Defence Authority, Newcastle upon Tyne, UK

ABSTRACT

This paper outlines a prototype emergency decision support system for use in the event of a release of hazardous gas. The system is being developed as a collaborative project by the University of Leeds and Tyne and Wear Emergency Planning Unit using data from a case study of the Sterling Organics' CIMAH site at Dudley in North Tyneside. The design of the system is based around the pcARC/INFO GIS package and the GASTAR dense gas dispersion model developed by Cambridge Environmental Research Consultants Ltd. The system provides for data aquisition, visualisation, predictive modelling of gas dispersion, analysis and training. A demonstration facility is also included for illustrative and introductory purposes.

1. INTRODUCTION

The work described here arises from discussions between members of the Tyne and Wear Emergency Planning Unit and the University of Newcastle upon Tyne (now at the University of Leeds) regarding public perception of major hazards and the potential offered by Geographical Information Systems (GIS) in the emergency planning process. A subsequent Home Office research grant allowed for both a doorstep survey of public perceptions to be conducted in the Newcastle area (Carver and Myers, 1993) and for an investigation of the feasibility of linking advanced gas dispersion models to proprietary GIS software for decision support in emergency planning (Carver et al. 1992). The latter has given rise to a more extensive research and development programme involving the University of Leeds, Tyne and Wear Emergency Planning Unit, the University of Cambridge and Cambridge Environmental Research Consultants Ltd. This work is ongoing and is the subject of this paper.

1.1 GIS, Emergency Planning and decision support
GIS has long been considered a potentially useful tool in both the development of emergency plans and in supporting emergency decisions in the field (see for example, McMaster and Johnson 1986, Dunn 1989, Fedra and Reitsma 1990). Indeed, in their review of the role of GIS in managing natural and technological hazards, Gatrell and Vicent (1991) go as far as to suggest that "few areas of the application of GIS technology can be as socially significant or environmentally relevant, as the management of emergencies and disasters due to natural and technological hazards" (p.148). Examples cited by Gatrell and Vicent (1991) include building databases of hazards and emergency management resources, optimal routing for the

Fire Engineering and Emergency Planning. Edited by R. Barham.
Published in 1996 by E & FN Spon. ISBN 0 419 20180 7.

safe transport of hazardous materials, monitoring of the health implications of disasters, developing evacuation plans, etc. as well as modelling the dispersal of toxic gas plumes. Clearly, this list is not exhaustive and may include many more areas where GIS can provide an input. One area of recent interest in the GIS research community is the development of GIS-based decision support systems (see for example, Carver 1991 and Clarke 1990) whilst a few examples look specifically at GIS-based decision support systems for emergency planning (e.g. Fedra and Reitsma, 1990). This paper focuses on the development of such a system by linking a GIS with a sophisticated dense gas dispersion model.

1.2 Linking GIS and dense gas dispersion models
A number of gas dispersion models have been written for use in research and emergency management. These range from very simple models such as half-angle and sector models through to complex finite element analyses capable of making highly accurate and very detailed predictions of plume development. However, the usefulness of a model to the emergency planner is closely linked to its ability to place its results in a real world geographical context for realistic decision support and make them accessible to further analysis. A few dispersion models do make some attempt at linking with geographical information, but usually only via purely visual means of displaying outputs over a background map. As they stand, however, most gas dispersion models are limited in their usefulness to the emergency planner. The visualisation of model predictions is an important ability in real-time decision support which should not be underestimated, but on its own it does fall a long way short of the full potential of linking GIS with dispersion models in providing much wider decision support not only in real-time, but also for pre- and post-event analyses. Fully fledged GIS-based gas dispersion models should provide not only for visualisation of model outputs in a geographical context but should also allow for the integration of environmental information to both improve model predictions and assess likely impacts (e.g. the effects of terrain variables plume development and time of day, week, etc. on predicting the population at risk), the design of emergency plans (e.g. evacuation priorities, resource allocation, etc.), the definition of risk maps and zones of consequence, real-time changes affecting model predictions (i.e. changes in wind direction, etc.) and realistic error analyses.

Of course, there are many proprietary GIS packages on the market, but gas dispersion modelling remains such a low key and esoteric application as to make the incorporation of such models within GIS software as standard tools a poor business proposition. It is therefore, largely upto the research and user communities to provide their own customised and hybrid systems for this purpose.

2. SYSTEM REQUIREMENTS
The requirements outlined here are for a system that integrates both the functionality of a GIS and a dense gas dispersion model into a single, easy to use, decision support system running on a PC. Consideration of system requirements are split here into those concerning system hardware, software and data.

2.1 Hardware
A key requirement of the system under design is that it should run on notebook and desktop PC. This will ensure both portability (for use in either the field or office) and low-cost of hardware, whilst ensuring the largest possible market for the finished product. Using a high-end system (e.g. 486 or Pentium) will provide sufficient computing power for models to be run interactively in the field in order to make real-time predictions.

2.2 Software
The question of system software requirements is somewhat more complex. It is clear that the prototype system will necessarily involve a modification of existing software, both in terms of the GIS and the dispersion model used. In the final version it may be more effective if the software is bespoke, adapting parts of off-the-shelf packages wherever possible. New code would only be written from scratch where functionality cannot be effectively adapted from existing systems. This is, however, beyond the scope of this paper and the following design focuses firmly on the prototype system.

The development environment utilised is the pcARC/INFO GIS package running under DOS using the pcARC/INFO Simple Macro Language (SML) and FORTRAN as the primarly system development languages. GASTAR, developed by Cambridge Environmental Research Consultants (CERC) Ltd., is the gas dispersion model used by the system. Output from GASTAR can be quite easily incorporated within pcARC/INFO through a loose coupling approach. This may be accomplished at a number of levels from a basic displaying of model outputs within the GIS to a dynamic link between GIS and model using the GIS database to improve model predictions by including spatial variability in factors such as terrain, surface roughness and albedo. Details of how this can be done with specific reference to pcARC/INFO and GASTAR are contained in a separate paper (Carver, 1994).

2.3 Data
The data requirements of the system are equally complex. Two basic types of data are defined: those which can be predetermined and stored in the GIS database as a series of map layers such as terrain and population distribution (i.e. fixed data); and those which vary between incidents, and even within the time span of the incidents themselves, such as meteorological conditions and conditions of release (i.e. dynamic data). The latter can only be input in real-time although sample datasets, for example, those containing meteorological records for the area, can be included.

The data used by the system cover a wide range of environmental, demographic and physical phenomena. Included within these categories are data pertaining to the conditions of the gas release (e.g. wind direction and speed, type of release, etc.), factors affecting model predictions (e.g. terrain, locations of sinks and barriers, surface roughness, atmospheric stability, etc.) and those factors determining the likely impact of the gas release (e.g. population distribution, traffic flows and local infrastructure). A full list of data included in the prototype system is included in table 1.

3. SYSTEM DESIGN AND FUNCTIONALITY

The prototype system is modular in its overall design and based on the ARC/INFO vector data model. In this model data is stored as individual thematic layers, such that there exists one layer for each dataset in the system (i.e. a population layer, a terrain layer, a landuse layer, etc.). This facilitates the easy management of data (including input, update and manipulation) and analysis between different data layers. In a modular system individual sub-sections of the system perform specific tasks such as data aquisition, visualisation and analysis. Apart from providing a logical structure to the system, this approach makes system programming and subsequent alteration/update much easier. The modular architecture of the system is outlined in table 2.

In addition to those modules available within the main system, demonstration and training modules are also provided. The training module provides access to a range of example environmental and release datasets facilitating the generation of realistic incident scenerios which can be used in place of actual incidents when training staff in the use of the system. The demonstration module provides a non-interactive overview of the system's functionality via a simple slide show.

3.1 Data aquisition

The data aquisition module provides basic data input functionality in two areas important for the running of the system: environmental (fixed) data input; and release (dynamic) data input. For a system such as described here to function properly a certain amount of environmental data needs to be included within the package. In this context environmental data is taken as referring to the range of relevant factors both affecting and being affected by the dispersion of the body of gas released. As such this includes data relating to those physical factors affecting gas dispersion (i.e. terrain, surface roughness, meteorological conditions, etc.) and those data relating to the likely impacts of a gas release (i.e. population figures and characteristics, transport networks, local infrastructure, etc.).

The system provides for the input of fixed environmental data from a number of sources. These include raw digital co-ordinates (e.g. ARC/INFO Ungenerate format data), ARC/INFO Export format data, DXF data (AUTOCAD), etc. so as to make data input from a wide variety of other systems as easy as possible. Naturally, it is not feasible to include all known formats in any one system. However, the inclusion of a simple raw co-ordinate based format should enable users with even the most basic of programming skills to convert their data into a format compatable with their current system. In the case of the final product it is envisaged that software for converting between various input formats could be provided on an individual user basis.

The provision of an environmental data input module allows the user to rapidly create the necessary environmental database required by the system, to update existing data as and when changes occur, and to extend the database to cover new areas of interest. Access to the environmental data input functions is via a custom menu/text based interface giving the user access to different directories in which data may be stored/imported as well as providing complete access to the range of pcARC/INFO data transfer formats, editing and display functions at a system level.

The system provides facilities for the quick and easy input of release data, both direct from the keyboard and from the system database. Release data is taken here to mean that data relating to the conditions and type of gas release which need to be input directly from observations made at the time of release. Data on conditions of release include meteorological data (i.e. wind speed and direction, stability, temperature, etc.) as well as the conditions relating to the release itself (i.e. presence or absence of fire, type of release, etc.).

Direct input from the keyboard is obviously necessary in real-time modelling situations as and when an incident occurs. In other situations (i.e. risk mapping, designing zones of consequence and training) it is often more appropriate and flexible for data to be stored as a file in the system database. Two options are therefore provided within the release data input module; one allowing direct input via the keyboard in response to prompts from the system, the other from files stored in the system database.

3.2 Dispersion modelling

The system described here uses the GASTAR model to predict the pattern of dispersion followed by a mass of dense gas in the event of an accidental release. It is noted that although the current system is designed to incorporate the GASTAR model, the modular design of the system means that future versions could easily switch to using alternative or additional models if necessary. Future versions could therefore also be adapted by including models of other potential hazards such as radiation, explosion, fire, etc.

3.2.1 Model description

The GASTAR model consists of a suite of programs for simulating the dispersion of dense gases released into the atmosphere. The name is an acronym derived from **GAS**eous

Transport from Accidental Releases. The model represents the latest in "state-of-the-art dispersion physics and advanced computational methods" (CERC, 1993, p.?) and is aimed at providing an efficient and powerful tool for risk assessment and emergency management. The model was originally developed for the Health and Safety Executive but is also sold for use in industry, consultancy and research. The uses listed in the GASTAR user manual include land use planning/zoning, emergency response planning and active emergency management.

The model simulates the dispersion of dense gases for a wide range of meteorological conditions, material properties, source models and dispersion models. The output from the model is also available in various forms to facilitate ease of interpretation, validation and application of the results. To date, however, the model has only been used as a stand alone system and no previous attempts have been made to link it directly to a GIS, although some of the model's outputs are inherently spatial and easily incorporated within a vector GIS data model.

3.2.2 Integration of the model into the main system
The GASTAR model produces a number of outputs in the form of simple ASCII text files. These include a file giving the distance of the gas cloud from the point of release along the wind vector together with several variables relating to cloud dimensions, gas concentration, etc. This information is used here to create an ARC/INFO map data layer describing the model's predictions which is subsequently used for visualisation and analysis purposes (Carver, 1994).

Whilst the most simple incorporation of the GASTAR model into the current system is simply to run the model from within the main system shell and use the model predictions as they are output, a number of modifications and enhancements are envisaged. The first of these involves by-passing the data input sections of the GASTAR and using the input forms and menus of the main system. This would create the impression to the user of a seamless join between GASTAR and the main system. This can be achieved simply and with little alteration of the original GASTAR source code. Suggestions for further enhancements are based around alterations of the GASTAR code to enable the model to make predictions using the environmental data available in the main system rather than assuming that the conditions at the point of release apply across the whole of the release area. Experiments are in progress to determine whether an approximation of this may be achieved without altering too much of the original GASTAR source code.

3.3 Visualisation
Simple visualisation of data and model outputs is regarded as an important part of the whole system. The ability to map model outputs onto environmental data, for example, provides the user with an indispensible tool for estimating likely impacts of a gas release. Even without more advanced analyses such as population data retrieval within the predicted extent of the gas cloud, useful first impressions of impact can be gained just by looking at model predictions in the context of ground information.

3.3.1 Release data
Data relating to the initial conditions of release can be shown on the screen in map form along with base map data to place it in a geographical context. Release data which can be illustrated in this fashion include position of release (e.g. static storage tank/pipeline or tanker) together with wind speed and direction (represented as symbols on the base map). Other release data, including type of gas, type of release, presence/absence of fire, etc. is illustrated in a text window at the side of the map display (see figure 1).

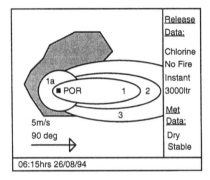

Figure 1. Schematic screen display of release data showing evacuation priorities

3.3.2 Model output

It is vitally important that model outputs can be viewed in the context of base map and environmental data. The first operation carried out on the model output should therefore be a simple visual overlay of the predicted gas cloud on to an annotated base map showing key geographical features such as roads and built-up areas. This allows the user to orientate themselves in regard to the model predictions and make first assessments of the likely impact of the release.

When the emergency services first arrive at the scene of a gas release, a single model run using current release data followed by viewing the predicted plume/cloud extent in the context of the base map data may be an extremely useful first analysis. Because of the immediate nature of the threat, there is a need to enable this initial run to be performed as quickly as possible while retaining the integrity of a full calculation. This enables quick decisions to be made by experienced operators allowing rapid action regarding evacuating those at greatest risk, diverting traffic, setting up of command and control points, etc. It is suggested that the initial model run should include not just a simple prediction of the area likely to be impacted by the gas plume but also areas adjacent to the plume and areas immediately surrounding the point of release. This allows for prioritising the evacuation of people in these areas (as shown by the numbers in figure 1) and for both errors in the model itself and changes in release conditions during the incident (i.e. sudden changes of wind direction). Later detailed analysis could provide information regarding numbers of people at risk, etc. once immediate concerns have been dealt with.

3.3.3 Environmental data

Visualisation of detailed environmental data is potentially a very useful part of system's functionality. The ability to examine model outputs in the context of environmental factors which are both likely to affect the model's predictions and be affected by it is a key aspect of the visualisation process. For example, viewing model predictions in the context of terrain information containing details regarding locations of barriers and sinks may reveal oversights in the model predictions (i.e. areas where a gas plume/cloud is likely to be diverted or retained). Another example may be viewing model predictions in the context of impacted features such as areas of housing and the road network (see figure 2 below).

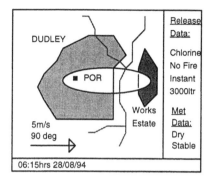

Figure 2. Suggested visualisation of model outputs in the context of impacted features such as areas of housing (Works Estate) and the road network (shown in grey)

3.4 Analysis
Beyond simple visualisation of model outputs in the context of ground information more advanced analyses may be performed using the system. These include integration of model predictions with environmental data to retrieve information, generation of risk maps and emergency plans and error estimates. This covers both instances of pre-event planning and post-event clean-up/evaluation as well as real-time decision support. Special consideration is given to the system's role in these areas are given separately.

3.4.1 Integration of model outputs with environmental data
After visualisation, model outputs can be analysed in the context of relevant environmental data. Information within the boundary of the predicted gas plume/cloud relating to relevant data can be retrieved and displayed for use in decision support. This is achieved simply by overlaying the map layer containing the model predictions ontop of the map containing the relevant environmental data and retrieving the information from within the plume/cloud boundary. This is illustrated in figure 3.

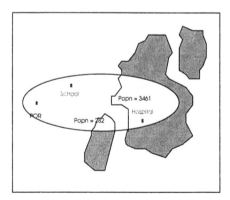

Figure 3. Retrieval of population data and other relevant features from within mapped gas plume/cloud boundary

Of particular relevance to the proposed system is population distribution. Figures for total population within the predicted gas plume/cloud based on 1991 Census data can be retrieved and displayed, along with associated information regarding demographic structure which may be deemed of relevance. For example, the age distribution and proportion of unemployed within an area is useful in estimating the likely numbers of people in residence at the time of day/week when a release incident occurs. Supplementing maps of population distribution by residence are the locations of schools, hospitals and other areas where concentrations of people are likely to be found. Indications as to likely diurnal and longer term fluctuations are also included. The spatial resolution of population data is often a problem with the finest level of resolution provided by Census data being the Enumeration District (ED). EDs contain, on average, 500 people in urban areas and 150 in rural areas, but when mapped may be largely open spaces such as fields and parks. The distribution of population within EDs is often assumed to be uniform giving rise to underestimates of population density in the populated areas. Here a technique known as dasymetric mapping is used to obtain finer resolution data on population distribution by combining map and satellite data on built up areas with ED boundaries (Langford et al. 1990).

Other environmental data which may be commonly integrated with model predictions in this fashion include roads and associated traffic figures, railway lines, airports, etc. Typical interogations of the data and model outputs can include such questions as:

1. how many people live within the predicted plume/cloud boundary?
2. which roads will be affected?
3. how much traffic will have to be diverted?
4. which schools, hospitals, residential care homes, etc. lie within the plume/cloud's predicted path?
5. where are the nearest evacuation centres and are they available?

3.4.2 Pre-event analyses
It is intended that the system be used for pre-event analysis as well as decision support in the field. In this context the system can be used in the delimitation of zones of consequence and for risk mapping around static hazards (i.e. CIMAH sites). When used in conjunction with longitudinal meteorological records for the area of interest, the system database and models provide all the necessary tools and much of the information required to define zones of consequence and draw risk maps for the hazard in question.

Zones of consequence can be defined according to the type of hazard (toxicity of the gas, volatility, etc.), its sphere of influence and the proximity of population (both residential and transitory). Inside the maximum likely limit of hazard, as determined by the gas dispersion model and the characteristics of the gas in question, then different zones of consequence can be defined according to population distribution and characteristics. In general, the greater the population density, the greater the impact or consequence of a gas release. This is illustrated schematically in figure 4. Detailed zones can therefore by defined according to population density and distance from the hazard (since gas concentrations will decrease with distance from their point of release as a result of dispersion in and mixing with the surrounding air) this can of benefit when pre-designating evac centres, etc.. The relationship between distance from point of release and gas concentrations can be determined using the dispersion model and transfered to the zone map.

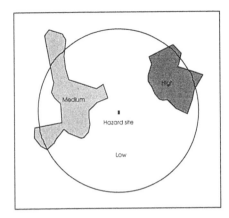

Figure 4. Zone of consequence map based on 'sphere of influence' of hazard and density of populated areas

Risk maps for hazards of interest can be defined utilising meteorological records and the zones of consequence defined above. It is suggested here that a suitable way of determining reliable risk maps would be to use the system described here to make multiple model runs according to the range of meteorological conditions, principally wind direction, wind speed, air temperature and stability, found at the site. The results of the multiple model runs could then be compiled into a single map layer. Those areas most frequently affected by gas plume/cloud predictions are used together with the zones of consequence map to determine areas of varying risk. This is illustrated in figure 5.

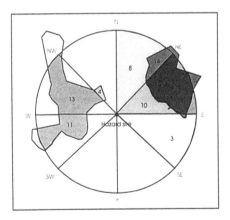

Figure 5. Schematic risk map based on figure 4 and wind direction

3.4.3 Real-time analyses
Analyses carried out in real-time as an incident is actually occuring obviously need to be both quick and reliable. For this reason it is suggested that they should be limited to visualisation, model predictions and information retrieval in support of emergency decision making. More complex analyses, such as those described above in the context of defining zones of consequence and risk mapping, should be restricted to pre-event analyses and post-event

clean-up scenarios.

3.4.4 Post-event analyses
The system described here may be used over a range of post-event analyses. Information contained within the system databases can be used together with data relating to the incident to assist in post-event evaluation of the performance of both the model and the emergency services. In this manner problem areas and possible improvements to both the system and emergency plans may be identified and implemented prior to another incident.

4. DECISION SUPPORT

The decision support role of the proposed system can be divided into the three main areas of its functionality in terms of analysis, i.e. pre-event, real-time and post-event.

4.1 Pre-event decision support
Pre-event decision support is mainly concerned with making plans for evacuation, traffic control, defining planning policy and identifying the need for public information campaigns and is based on the assumption that wherever there is a hazardous gas storage tank or where gas is transported, there is the chance that a release of gas could occur. Decisions made regarding these issues need to be made by individuals involved with the emergency services, planning authorities and the chemical industry. Information provided by pre-event analyses using the system database and model can be used to support these decisions. it should be noted that the system is not being designed for use as a decision making system, but purely for use in decision support.

Pre-event decisions which may be supported by the system may include how to:

1. assign evacuation priorities to areas in the vicinity of CIMAH sites
2. define evacuation corridors along existing road networks
3. locate safe evacuation reception centres
4. locate traffic control points at critical points in the road network
5. locate command and control units to oversee and co-ordinate emergency operations
6. modify local structure plans regarding development in risk areas
7. target areas for public information campaigns

4.2 Real-time decision support
Real-time decision support is concerned solely with providing the emergency services at the scene with up-to-date information on which to base decisions regarding evacuation, traffic control and location of command and control units. Emergency planners would find this most useful in the pre-event planning and post-event clean-up and evaluation phases of an incident. As outlined above, operations should realistically be confined to simple visualisation, modelling and information retrieval. However, it should be obvious that only a limited amount of forward planning can be carried out with the system prior to the event. Plans made earlier on the basis of pre-event analyses, etc. will, therefore, need to be modified in real-time using predictions of gas plume/cloud development and other relevant information (i.e. time of day/week, meteorological conditions, etc.) should an incident occur.

Real-time decisions which may be supported by the system include:

1. which areas are at greatest risk and need priority evacuation
2. how to prioritise further evacuations
3. which routes to use for evacuation
4. which evacuation reception centres to use
5. where to locate traffic control points
6. where to locate command and control units

4.3 Post-event decision support

Post-event decision support benefits largely from hindsight; the incident has already occurred and has been dealt with. Utilising the existing environmental database of the system together with data regarding the incident itself (i.e. release data, plume development, casualties, property damage, etc.) a post-event evaluation of the problems faced by both the model and the emergency services can be carried out. The results of such an evaluation can then be used as positive feedback in redesigning emergency plans.

Post-event decisions which may be supported by the system include:

1. is the system and dispersion model used adequate?
2. are existing emergency plans effective?
3. are the resources available to the emergency services adequate?
4. can the system and emergency plans be improved, and if so, how?
5. are existing risk maps and zone of consequence an accurate representation of reality?

4.4 Training

Training is an essential part of designing and using a decision support system such as that described here. The system is intended to be used for training as much as for its use as a tool for analysis and pre-event, real-time or post-event decision support. The training element is essential in that it both enables the user to familiarise themselves with the system and with the decisions that are likely to be required in the event of a real incident. The training module of the system provides the user with access to the full range of the system's functionality and databases together with a scenerio generator which can be used to create realistic release incidents on which the user can practise.

5. DISCUSSION AND CONCLUSIONS

This paper has introduced and outlined in detail a prototype GIS-based decision support system for emergency planning in response to hazardous gas releases. The system is still in the development stage and it has yet to be fully tested. Work to date has however shown that a link between proprietary GIS packages and gas dispersion models can be made and used to create an effective and workable decision support tool. This also illustrates that other similar hazards models could easily be incorporated into such a system, for example radiation relases to be used with LARRMACC (Local Authority Radiation Radioactivity Monitoring Advice and Collation Centre).

Many of the advantages and pretexts of adopting a GIS-based approach applying not only to emergency planning in response to dense gas dispersion modelling but also to a whole range of other emergency planning situations from explosions to flooding. In this context the current system, and other examples cited in the growing literature on GIS and decision support for emergency planning, is illustrative of the wider advantages which can be gained through an integrated GIS and modelling approach.

In specific reference to gas dispersion the research and development process outlined here has identified a number of basic problem areas which need to be addressed in further developing this and other systems. These can be summarised as follows:

1. Many gas dispersion models fail to adequately take spatial and temporal variation of relevant parameters and variables into account by assuming that conditions at the point of release apply over the whole of the impact area. Integrating models with GIS and attendant databases within real-time systems is perhaps the best way in which these problems can be addressed.
2. The 'usability' of gas dispersion models is seriously curtailed if they do not allow for the direct integration of model outputs with the ground information with which emergency planners and emergency services are used to working on a day-to-day basis.
3. Where dispersion models have been linked to some form of geographical database, analyses are largely limited to simple visualisation and information retrieval. The full potential of GIS-based analyses is being missed and this needs to be addressed by moving research on dispersion modelling and GIS closer together through co-operative interdisciplinary programmes with close links to the end-user community.

Further research and development work in this field is clearly necessary before a practical system can be produced. In the context of the system described in this paper, development work is planned to continue on the prototype system with the aim of securing further funding to allow work to proceed on the development of a full working system which would then be available for use by emergency planners throughout the UK and abroad.

References

Carver, S.J. (1994) Linking dense gas dispersion models to the Arc/Info GIS. Working Paper No. ??, School of Geography, University of Leeds.

Carver, S.J. (1991) Integrating multicriteria evaluation with Geographical Information Systems. In: *International Journal of Geographical Information Systems.* 5(3), p.321-339.

Carver, S.J & Myers, A. (1993) The role of public perception in the response planning for major incidents: public perception and memory retention questionnaire survey. *Final report to Home Office.* November 1993.

Carver, S.J., Myers, A. & Newson, M. (1992) The role of public perception in the response planning for major incidents: a proposal for a GIS-based strategy. *NorthEast Regional Research Laboratory Report No.92/1.* University of Newcastle upon Tyne.

Clarke, M. (1990) Geographical Information Systems and model-based analysis: towards effective decision support systems. In: H.Scholten and J.C.H.Stillwell (eds) *Geographical Information Systems for Urban and Regional Planning.* Kluwer Academic Press, Netherlands. p.165-175.

Dunn, C. (1989) GIS in emergency planning: the Cumbria Geographical Information System. In: *Mapping Awareness.* 3(6). p.15-19.

Fedra, K. and Reitsma, R.F. (1990) Decision support and Geographical Information Systems. In: H.Scholten and J.C.H.Stillwell (eds) *Geographical Information Systems for Urban and Regional Planning.* Kluwer Academic Press, Netherlands. p.177-188.

Gatrell, A.C. and Vicent, P. (1991) Managing natural and technological hazards. In: I.Masser and M.Blakemore (eds) *Handling Geographical Information: methodology and potential applications.* Longman, London. p.148-180.

Langford, M., Unwin, D.J. and Maguire, D.J. (1990) Generating improved population density maps in an integrated GIS. In: *Proceedings of the First European Conference on Geographgical Information Systems.*

McMaster, R.B. and Johnson, J.H. (1986) Assessing community vulnerability to hazardous materials with a Geographic Information System. In: N.Chrisman (ed) *Proceedings of AutoCarto 8.* p.471-480.

Acknowledgements
The authors would like to acknowledge the help of the Home Office in providing the initial funding for this project, Dr Rex Britter, Cambridge University and CERC Ltd. for providing the GASTAR model and Tyne and Wear Fire and Civil Defence Authority EPU, the University of Leeds and the University of Newcastle-upon-Tyne for providing general research support.

Table 1. Prototype system database

The data included in the prototype system can be described at a number of levels. The first level describes the thematic contents of an individual data layer and the second level its individual attributes and complexities.

Level 1: Thematic layers

Name	Description
CIMAH	Location of CIMAH sites
TERRAIN	Digital terrain model
LANDUSE	Land use data
SURFACE	Surface roughness index
ALBEDO	Surface albedo index
BARRIER	Barrier locations and type
SINKS	Sink locations
STREET	Street level data (selected areas only)
ROADS	Road network
RAIL	Rail network
BOUNDARY	Boundary data (including study area)
POPFIXED	1991 Census of Population data mapped by Enumeration District
POPTRANS	Estimates of population adjusted according to time of day/week
POPCEN	Concentrations of people (i.e. hospitals, schools, etc.)
POPDASI	Dasimetric map of fixed and transitory population density
EVAC	Location of evacuation reception centres

Level 2: Data characteristics and complexities

Name	Type	Source	Scale	Attributes
CIMAH	point	TWEPU	1:25,000	Type of hazard, etc.
TERRAIN	polygon	IoH	50m grid	Altitude, Slope, Aspect
LANDUSE	polygon	LANDSAT	25m grid	Land cover type
SURFACE	polygon	derived	25m grid	Surface roughness index
ALBEDO	polygon	derived	25m grid	Surface albedo index
BARRIER	line	various	various	Type, Height
SINKS	polygon	various	various	Type, Estimated volume
ROADS	line	OS map	1:25,000	Type, Traffic count
RAIL	line	OS map	1:25,000	Type, Passenger figures
BOUNDARY	polygon	OS map	1:25,000	Type
POPFIXED	polygon	OPCS	1:25,000	Total, + various others
POPTRANS	polygon	various	1:25,000	Total transient population
POPCEN	point	various	1:25,000	Total
POPDASI	polygon	various	various	Total
EVAC	point	OS map	1:25,000	Capacity

Table 2. System architecture

Module	Functionality
Data aquisition	Fixed data input
	Dynamic data input
Dispersion modelling	Running of gas dispersion model
	Provision of numerical model outputs
Visualisation	Visualisation of fixed and dynamic data
	Visualisation of model predictions
	Integration of visual displays
Analysis	Integration of model output and environmental data
	Information retrieval
	Generation of evacuation plans
	Generation of risk-maps and zones of consequence
	Generation of impact scenerios
	Error analyses

34 APPLYING ARTIFICIAL INTELLIGENCE TO THE SCIENTIFIC ANALYSIS OF TIMBER IN FIRE

C.A. GREEN, R. FOSTER and P. SMITH
School of Computing and Information Systems,
University of Sunderland, UK
G.W. BUTLER and M.T. NIELSEN
Tyne and Wear Metropolitan Fire Brigade,
Newcastle upon Tyne, UK

Abstract

It is widely recognised that timber plays an important role in both preventing and determining the development of a fire. Conventionally, forensic analysis of timber in fire investigations is carried out by means of destructive test methods such as wire brushing and full-scale furnace testing. Analysis of the results of such testing relies heavily upon the expertise of the human investigator. These methods are currently used during both fire investigations and fire door inspection and testing.

The removal of the charcoal from the softer species of timber may result in slightly higher values for the depth of char, due to the lack of definition between charred and uncharred timber, causing inaccuracies in the data collected.

This paper discusses new methods of acquiring data which are representative of the effects of fire on timber through non-destructive methods, thus leaving the physical evidence intact. New methods for analysing such data will be identified, particularly through the use of Artificial Intelligence (A.I.).

Keywords: Artificial Intelligence, Depth of Char, Pyrolysis, Scientific Analysis.

1. Introduction

Regardless of the introduction of man-made substitutes such as plastics, timber still remains one of the most extensively used materials in building construction and furniture production. Therefore, it is inevitable that timber in some form will be found present at the majority of fires, and as such, will contribute to the development of the fire as a source of fuel. Conversely in the form of a fire door it will form a barrier, thus

Fire Engineering and Emergency Planning. Edited by R. Barham.
Published in 1996 by E & FN Spon. ISBN 0 419 20180 7.

restricting the development of the fire from one compartment to another to allow sufficient time for the occupants to escape to safety.

Evaluating the effect of fire on timber can provide valuable clues as to the nature of the fire. Conventionally, forensic analysis of timber in fire investigations is carried out by means of destructive test methods such as wire brushing and full-scale furnace testing. Analysis of the results of such testing relies heavily upon the expertise of the human investigator. These methods are currently used during both fire investigations and fire door inspection and testing.

Research being carried out by the University of Sunderland and the Tyne and Wear Metropolitan Fire Brigade in the United Kingdom, is looking at the possibility of using Non-Destructive Testing (NDT) methods on timber during fire investigations, and the suitability of such methods for fire door inspection and testing. Artificial Intelligence techniques such as Neural Networks are being considered as methods for analysing data from the NDT. This paper highlights areas currently being considered for the application of this technology and describes the potential benefits which could be gained from carrying out this type of work.

2. Current Methods for Evaluating the Effects of Fire on Timber

The effect of fire on timber has for many years been used as an indicator of the type of fire development, by both the fire services and the forensic science services. One of the main combustion properties of wood is charcoal, which is formed by pyrolysing the non-volatile constituents of the wood. The depth to which the pyrolysis action of the fire has converted the wood to its volatile fractions and charcoal is known as the 'char depth'. This action on the wood is progressive and occurs at a predictable rate which leads to zones of char and pyrolysis being formed, these become clearly visible when a cross-section of burned timber is studied and are used to calculate the char rate of timber. What can be seen from examination of a cross section of timber involved in a fire is a line between the charred and the uncharred timber. This line of demarcation differs depending upon whether it has been a fast or a slow burn. The area known as the "Pyrolysis Zone" will be very narrow in a sudden and fast moving fire, giving a sharp line between charcoal and timber [1].

A lot of emphasis has been placed on the charring rate of timber to calculate the time duration of fire exposure to a structure. The figure generally used is approximately 0.6mm per minute or 1/40 of an inch per minute. This figure is a result of the mean rate of char depth development measurement in a standard test furnace. Fire service manuals, such as the *Manual of Firemanship Book 12,* document the British Standard as being between 30mm to 50mm per hour depending on the specie of the timber [2]. However, work carried out by the Timber Research and Development Association (TRADA) identifies a wide range of charring rates dependent upon species, ranging from a high of 0.8mm per minute for Abura to a low of 0.425 per minute for Teak [3].

The depth and pattern of the char are also used by investigators for determining the development of the fire, which may provide significant clues as to the possible cause. Wire brushing is done to remove the charcoal which has formed to uncover the charred timber. The depth and surface pattern may be indicative of the type of fire (i.e.

slow or fast burn). It is therefore important that this information is recorded. The depth is measured manually by inserting a small probe into the char, whilst the size and shape of the cracks which appear in the char are noted [4].

A number of inaccuracies can occur whilst collecting information of this kind during investigations, which are further compounded by:

- The inaccuracy in the method of measuring depth of char, rather than the depth of the remaining timber (the charcoal layer having shrunk when compared with the original wood).
- The use of wire brushing as a method for removing charcoal to measure the depth of char, which may result in slightly higher estimates for softer species.
- The density and moisture content of timber and the intensity of the flames which affect the char rate of timber.

These factors increase significantly the difficulty of the task faced by the fire investigator in making proper assessments of the data [4].

3. Fire Safety- Fire Doors

Fire doors form a barrier restricting the development of a fire from one compartment to another, allowing sufficient time for occupants to escape to safety.

Fire door testing in the U.K. is carried out to the approved British Standard (BS 476 Part 8) which requires a door is tested by destructive methods using a furnace [5]. This test will certify a particular door for a set period of fire resistance i.e. Half-hour, Hour or Two Hour doors. Providing doors are made to the same standard specification they are accepted as qualifying for that purpose and no further test is required.

Any imperfections in the timber and its inherent reduction in quality can go unnoticed. Subsequently this may lead to a possible reduction in the standard of fire resistance. Quality control of this product is essential if we are to maintain the high standard of fire resistance provided by this product.

4. Non-Destructive Evaluation of Timber

Non-Destructive Evaluation (NDE) of materials is defined as the science of identifying the physical and mechanical properties of a piece of material without altering its end-use capabilities [6]. Evaluation of this kind requires suitable testing methods. A number of Non-Destructive Testing (NDT) methods are being put to use in a wide range of areas. Methods such as ultrasound, X-ray and microwaves are being used for testing of materials such as metals, concrete, ceramics, fabrics and wood.

NDT techniques for wood differ from those homogeneous, isotropic materials such as metals, glass, plastics and ceramics. In non-wood materials, whose mechanical properties are known and tightly controlled by manufacturing processes, NDT techniques are used to detect the presence of discontinuities, voids or inclusions. Because wood is a biological material, these irregularities occur naturally, and may occur because of agencies of degradation in the environment. Consequently, a lot of NDT techniques have been used on wood to measure how natural and environmentally-induced irregularities interact in a wood member to determine its mechanical properties [6].

Whilst many of these techniques are successful in a wide range of applications, ultrasound has attracted most interest and consequently become the most widely used for non-destructive inspection, due to its ease and safety of handling [7]. Ultrasonic testing is widely used in industry for quality control and equipment integrity tests; using ultrasound it is possible to detect flaws, showing their size and location, and also to determine differences in material structures and physical properties [8]. All of these features support the decision to use ultrasound for the NDE of timber which has been subjected to fire.

5. Ultrasound For Analysis of Timber in Fire

Ultrasonic scanning is being considered as a tool to assist in the determination of the effects fire has on timber. Two areas are being considered during this study, in fire investigations looking at charring of timber, and for quality inspection and testing of fire doors.

5.1 Fire Investigations
As has been stated above, current methods for collecting and interpreting data from charred timber present a number of problems to investigators trying to establish an accurate picture of the development of fire.

Ultrasonic scanning of the timber during fire investigations is seen as a way in which the investigator may examine the remaining timber without destroying the evidence, building up a clearer representation of the fire damage. Possible applications of this technique could be:
- Measuring the depth of char for particular species without the need to remove the charcoal and thus remove important characteristics of the charred timber;
- To identify the char pattern;
- Cross-sectional examination which highlights the pyrolysis zone, ultrasound would allow for this type of examination without destroying further the timber structure. Results of such an examination may show that the depth of char does not affect the integrity of the structure removing the need to replace that particular structure.

5.2 Fire Safety Inspection and Testing
The aim of using ultrasound in this area is to identify the internal characteristics of fire doors in order that a 'prototype' for a suitable fire door might be established. Through scanning the door prior to furnace testing, it is hoped that relevant data relating to the door's composition will be captured, which would under normal circumstances not be available. This data, and data collected during furnace testing of fire doors could then be used to establish whether or not any particular features (i.e. flaws, timber defects etc.), are responsible for the failure, or success of a particular fire door. By establishing such criteria, improved fire door standards could be set and used to assist in further fire door inspections.

Many benefits are proposed relating to:
- Identifying features of door in 'old and listed buildings' for their integrity;

• Establishing criteria on which to base quality control for companies manufacturing fire doors;
• Providing companies with guidelines using a fire door 'prototype';
• Reduce the amount of furnace testing.

6. Artificial Intelligence for Timber Analysis

Work is currently looking towards the use of Artificial Intelligence techniques as a suitable method of interpreting and representing data obtained by ultrasonic testing of timber which has been subjected to fire.

Artificial Intelligence is the field of advanced computing which attempts to empower computer systems with the ability to automate the process of reasoning and decision-making [9]. Whilst a number of these techniques are being employed in a wide range of disciplines, this research project will be considering one, namely Neural Networks, for the interpretation and representation of data obtained through ultrasonic testing.

Neural Networks may be described as *"massively parallel interconnected networks of simple (usually adaptive) elements and their hierarchical organisations which are intended to interact with objects of the real world in the same way as biological nervous systems do"* [10]. This may be simplified such that we recognise neural networks as systems which consist of a large number of simple processing elements (known as neurons), highly interconnected, and which respond to an input pattern of some description by altering the states of their interconnections. Unlike conventional computer systems, neural networks do not need to be programmed, instead they learn by experience. This act of learning is effected by altering the strength of the connections between the neurons across the network in response to the input pattern. The knowledge of this pattern is effectively 'stored' as a pattern of energy in the connections between neurons.

Neural networks can successfully deal with noisy, incomplete, or vague data on the strength of previous experience [11]. This makes the technique robust for many real-world applications, including ultrasound testing [12,13], where the majority of data encountered contains noise and contradictions. Work has been carried out in applying neural networks to the analysis of ultrasound data in timber related projects, showing the potential of the technique [14,15,16].

7. Conclusions

The collection of accurate and representative data in the areas of fire investigation and fire door testing, and the subsequent analysis of such data by human investigators, are areas which are both problematic and complex. These difficulties have provided the motivation for examining the potential benefits of applying ultrasonic NDT and A.I. data analysis techniques in these areas.

This paper has discussed the potential for the application of AI techniques in this area. There is clearly much scope for work in this area, and many benefits to be gained form applying such advanced technology.

8. References

1. DeHaan, J.D., *'Kirks Fire Investigation'*, 1991

2. *'Manual of Firemanship'*, Book 12 - Practical Firemanship, 1983.

3. Hall, G.S., 'The Charring Rate of Certain Hardwoods', Timber Research And Development Association, Buckinghamshire, U.K., Research Report WT/RR/10.

4. Cook, R.A. et al, *'Principles of Fire Investigation'*, The Institution of Fire Engineers, 1985

5. British Standard 476 Part 8

6. Ross, J.R., *'NonDestructive Testing of Wood'*, USDA Forest Service, Forest Product Laboratory, Madison, Wisconson. 1992

7. Yoshiro, T. et al, *'Nondestructive Inspection of Hidden Knots in a Japanese Cedar Log Using Ultrasonic Computed Tomography'*, Ultrasonic Technology 1987

8. Mix, P.E., 'Introduction to NonDestructive Testing - A Training Guide', J. Wiley & Sons Publications, 1987, pp 104-157.

9. MacIntyre, J.D., *'Condition Monitoring and Artificial Intelligence'*, University of Sunderland, U.K. September, 1993

10. Kohonen, T., *'An Introduction to Neural Computing'*, Neural Networks, Vol 1, pp 3-16, 1988, Pergamon Journals.

11. Hinton, G., *'How Neural Networks Learn from Experience'*, Scientific American, Sept 92, p105-109.

12. Taylor-Burge, K. et al, *'The Real-Time Analysis of Acoustic Weld Emissions using Neural Networks'*, Proc. of 6th International Conference on Joining of Materials (JOM-6), Helsingor, Denmark, April 1993.

13. Damarla, T. et al, *'Application of Neural Networks for Classification of Ultrasonic Signals'*, Proc. of Artificial Neural Networks in Engineering (ANNIE 92), St. Louis, Missouri., 1992.

14. Occena, L.G., Chiu, C., *'Neural Models of Defect-Driven Hardwood Log Sawing'*, Dept. of Engineering, University of Missouri-Columbia, 1994.

15. Zabel, A., *'Integrating of Neural Networks and Fuzzy Logic Based Classification for a Common Quality Assessment System of Different Surfaces'*, Proceedings of 2nd International Conference - Computer Integrated Manufacturing, 6th - 10th Sept, 1993, Singapore.

16. Simula, O., Visa, A., *'Self-Organizing Feature Maps in Texture Classification and Segmentation',* Proceedings of ICANN'92, Brighton, U.K. 1992

35 APPLICATION OF EXPERT SYSTEMS AND MACHINE LEARNING IN FIRE INVESTIGATION

P.J. IRVING and S.L. KENDAL
School of Computing and Information Systems,
University of Sunderland, UK
G.W. BUTLER
Fire Safety, Tyne and Wear Metropolitan Fire Brigade,
Newcastle upon Tyne, UK

Abstract

It is generally recognised that the careful investigation of a fire incident is of great importance since this can reveal much about the cause and nature of the fire, its development and effects. Although there are a number of very experienced and expert human fire investigators, the vast majority of fires are investigated by the fire brigade officers who have fought the fire. Training courses are available to extend the scope and improve the quality of fire investigations by fire service personnel but this can be a major financial burden on fire brigades and does not provide the complete answer, since experience also has a major part to play.

In other fields investigative or diagnostic problems have been tackled by use of expert computer systems, which can emulate a human expert's problem solving techniques rather than the conventional computer system's use of mathematical modelling.

This paper describes the collaborative work that is taking place between The University of Sunderland and Tyne and Wear Fire Brigade in the application of expert systems to fire investigations. The project will provide fire brigade investigators with a portable system which will, after rudimentary training, guide a fire officer through complex fire investigations. Further development is aimed at producing an expert system which "learns" new facts and rules from investigations. By the use of networking the system should be capable of sharing any new knowledge with investigators in other geographical locations.

Fire Engineering and Emergency Planning. Edited by R. Barham.
Published in 1996 by E & FN Spon. ISBN 0 419 20180 7.

1 Introduction

The systematic investigation of unplanned and uncontrolled outbreaks of fire is of fundamental importance for the protection of lives and property from fire since it is only through an accurate determination of the cause and responsibility that future fire incidents can be avoided [1]. Proper fire investigation is also essential if accurate statistics on fire occurrence, development and spread are to be collected and mathematical fire models validated. An investigation may also be necessary to establish whether a crime has been committed and to assist in the gathering of evidence to lay before the courts.

Within the United Kingdom there is no one body responsible for investigating fires, the main organisations involved are the fire service, the police and the insurers. Each have there own reasons for investigating fires and there is much overlap in the information collected.

The fire service is concerned with the collection of data to enable statistics to be collated nationally. They are also concerned with the behaviour of fire in buildings and the behaviour of materials, fire safety systems and people in real fires, since this will inform fire safety advice, standards and techniques. Fires are investigated by the police when there is a suspicion of arson or where the fire has involved a death or serious injury. The problem of arson in buildings is of particular concern, in the geographical area covered by Tyne and Wear it is estimated that arson accounted for 72 percent of all fires in 1993 [2]. In the whole of the United Kingdom, arson was estimated in 1992 to be the cause of one fifth of all fires nationally, it was also estimated that arson represents between 40% and 50% of all insured losses and is in the region of £500m per annum in direct cost, with the real cost much higher [3].

Clearly insurers are interested in investigating fires which have presented them with a financial liability, particularly if arson is suspected.

Since the fire service is the only body which investigates every fire, it is highly likely that when fires are investigated in a thorough and systematic manner that significant factors may be revealed. Guidance to fire services and to police forces on the liaison arrangements for the investigation of suspicious fires have been issued by the Home Office, which is the Government department responsible. This guidance also stresses the importance of sound training in fire investigation methodologies and techniques [4].

Although a small number of United Kingdom fire brigades have officers designated to fire investigation duties, the majority rely on the officer who has fought the fire to then carry out the investigation. This can result in investigations of variable quality since not all officers will have received specialist training and the large numbers of officers involved, dilutes the fire investigation experience of all.

If fire brigades are to achieve the ideal of every fire receiving a consistent, thorough and careful investigation. Based on sound scientific principles, there is the need for a major training investment and perhaps an extension of the practice of using small specialist teams. Most United Kingdom fire brigades would find this prohibitively expensive and so there is a need to develop an alternative approach to fire investigation.

To be attractive to fire brigades any fire investigation system should:

- ensure that investigations were carried out thoroughly,
- incorporate the knowledge of the best fire investigators,
- learn from and preserve, all relevant experience,
- be flexible enough to allow for new investigation topics and criteria to be easily incorporated,
- be capable of being operated by inexperienced personnel,
- be inexpensive to set up and operate

The collaborative research project being undertaken between the University of Sunderland and Tyne & Wear Metropolitan Fire Brigade in the UK is an attempt to address these issues through the use of artificial intelligence.

2 Artificial Intelligence Technology

The field of artificial intelligence, is an area of computing which has been around since the mid 1960's. Stanford University developing their landmark system DENDRAL in 1965 and *"successfully demonstrated that it was possible for a computer program to rival the performance of domain experts in a specialised field"* [5].

However, it was not until the 1980's that artificial intelligence technology took off. The advent of PC's and their development into powerful machines coupled with the development of artificial intelligence languages in the academic and commercial sectors have brought forward this technology. The area of artificial intelligence is broad and includes vision systems (helping robots "see" objects), learning systems and expert systems. It is the latter two areas which are of specific interest to this project.

3 Expert System Technology

The first phase of the project is the development of an expert system for fire investigation. Jackson [6] distinguishes expert systems from more conventional applications programs in that:

1. They simulate human reasoning about a problem domain, rather than simulating the domain itself. Distinguishing expert systems from payroll and fire simulation systems which require mathematical modelling.
2. They perform reasoning over representations of human knowledge, in addition to mathematical functions. Demonstrating that the system contains a representation of human knowledge (specific to the domain).
3. They solve problems by heuristic or approximate methods, distinct from algorithmic functions.

The third point demonstrates two important features of expert systems to this project: The fact that they can emulate a human's ability not to make a decision if there is insufficient information and that like a human they are capable of working with imperfect knowledge. Both very important points in their application to fire investigation since in a fire evidence is often destroyed.

4 Problems with Fire Investigation

Experience of the expert fire investigators is very much a personal attribute based upon the incidents investigated and personal research. It is also a volatile resource which can be lost at any time due to, for instance, retirement. Similarly the experience is only available to that individual when, if it were encoded into an expert system it would be permanent and available to all.

A hybrid of the experiences of a collection of expert investigators backed up by scientific fact would be an even more valuable resource, although it must be recognised that there may be problems in collecting the knowledge due to conflicting experiences.

It is such a combination of experience from the best fire investigators and scientific fact which is planned for the project.

5 Expert Systems aiding Fire Investigation by the Fire Service

This application of expert systems is quite unique. Current systems can diagnose faults, etc. based on existing symptoms. Fire investigations however, have to be carried out with imperfect symptoms given that vast amounts of relevant information are destroyed in the fire or simply unknown. Here we are working not with the ability to fault find on an appliance to determine if it is overheating, but rather what is left of an appliance to determine if it overheating was the cause of the fire.

Clearly expert fire investigators are well placed to ascertain such causes, however it may be the officers at the incident who investigate the fire. In addition, evidence and clues essential to the fire investigation process may deteriorate or be removed before the expert investigator is able to carry out the investigation.

Although it is highly desirable to equip each fire engine with a fire investigator this clearly poses a huge ongoing training and education programme. An expert system carried on each fire engine or on specialist fire investigation units could supply not only the required knowledge but also acquired expertise. Acquired expertise could also easily be modified and distributed to other copies of the system.

A prototype system has been developed. This Fire Investigation Research Expert system (FIRE) demonstrated that expert systems could contribute significantly in this area.

This was the first step in a development programme the intention of which is to assist fire investigators. The investigation of really difficult fires may take the human expert but, if the expert system can investigate the lesser ones then it frees the time of the human expert. A second much more powerful system is now being actively developed which will deployed by the brigade.

In the introduction 6 points were highlighted as being necessary for fire brigades in any fire investigation system. Below each is discussed with reference to the project under development:

1. *Ensuring investigations are carried out thoroughly*
   The expert system is being constructed using the knowledge of experts and scientific fact in both ideal and real fire conditions. Being reviewed by experts before incorporation into the system has allowed an "ideal" path to be determined. Once incorporated into the system this path will then be followed systematically every time the system is used.
2. *Incorporating the knowledge of the best fire investigators*
   It is intended to collect a hybrid of expertise married with scientific fact. The system will therefore incorporate the consensus knowledge of a panel of experts and scientific facts. The panel providing expertise in excess of anyone persons'. Once incorporated into the system the knowledge will never be forgotten or deteriorate over time.
3. *Learn and preserve all relevant experience*
   Should any of the experts leave or retire their knowledge will be preserved. As new developments are forthcoming they can be incorporated and all copies of that system updated. Learning in its own right is to be addressed later in the project.
4. *Allowing for new investigation topics to be incorporated*
   With new scientific developments it is important that new investigative topics can be incorporated into the system.
5. *Operation by inexperienced personnel*
   The system will hold the knowledge and expertise allowing it to be used by inexperienced personnel. Therefore if the users can answer questions and follow instructions then they can use the system. If they can use the system then they can investigate a fire.
6. *Be inexpensive to set up and operate*
   It is intended that the system developed will run on an ordinary Personal Computer (PC), be inexpensive and thus universally available.

It is intended that the system will have a role in training, since it will be able to explain why it asks each question and justify its conclusions the human investigator will learn. Anyone wishing to enhance their knowledge in fire investigation can use the system as one source of knowledge.

The system being developed is intended to be able to complete a fire investigation report suitable for national statistics and also pass on electronically, these statistics reducing paper work and administration costs.

6 Uncertainty In Fire Investigation

Because of the destruction of evidence in the fire often a single cause cannot be determined. Instead there may be several possible causes or a "feel" for the cause. Rather than giving a single conclusion or series of conclusions the system is being built to determine possible multiple causes and to give the probability of these being the cause in an incident. The system therefore will rank probable causes. Essentially there are two main ways of ascertaining the ranking. Certainty Factors which represent a human judgement of likelihood and Bayesian theory which works from gathered statistics.

Bayesian theory seems the more appropriate for the system under development as in this field certainties are difficult to ascertain but statistics from previous cases are easy to collect.

7 Learning Systems

Expert systems are capable of being updated, however this must be achieved through revising the system with new knowledge. A system which could actually learn and pass on that knowledge is the ultimate aim of the project.

One of the most successful machine learning mechanisms to date has come about by simulating the biological, though not cognitive, structures of the brain. The discovery of the neuron, around the turn of the century, allowed researchers, in the 1960's, to develop simplified models of the neuron, to connect these to form neural networks [7] and to simulate the behaviour of these networks using computers. Research was held up until the 1980's when methods for training these networks were developed.

The brain because of its massive neural network is extremely good at pattern recognition tasks, an area where computers have been traditionally weak. The implementation of neural networks has provided a mechanism where, like humans, machines can have very good pattern recognition abilities. Over the past few years applications of this technology have been introduced to the market place that demonstrate a learning ability at least comparable to that of humans, arguably better.

However, neural network technology does not at the moment provide a mechanism capable of general intelligence. Such systems are extremely good at pattern recognition bringing advances in visual and speech recognition. Whilst such systems can spot patterns in, for instance insurance claims, it is expected that pattern recognition ability will not provide a complete learning mechanism suitable for the problems under consideration. Fire investigation is an abstract planning function that typifies the so called high order functions of the brain. The human brain is not yet fully understood but it is clear that it is not homogenous. Substructures within the brain exist and these have not been simulated on machines. Perhaps it is for this reason that neural networks are not yet capable of simulating the higher order brain functions of planning and controlling goal oriented behaviour.

One machine learning mechanism who's authors claim that it is capable of doing this is 'SOAR' [8]. Work on Soar began in the early 1980s and was a direct attempt to model the cognitive processes of the brain. While its development was not yet complete in 1991 significant claims were made about its problem solving abilities and of its learning mechanism [9]. It was claimed to be so successful that it represents an architecture capable of general intelligence, i.e. it would perform a full range of tasks, using any problem solving methods suitable for those tasks and employing a general learning mechanism capable of enhancing the future performance of the mechanism when solving similar tasks.

Therefore given a new problem Soar should construct a representation of that problem and obtain information about the legal operators that can be applied. This is akin to humans building a mental model of a task and determining possible steps that may lead toward a solution. After this initial model has been built the next step would be to select the best operator to apply, apply this to generate a new problem state, and repeat this process until the goal is achieved. The intelligence within the system is shown in its ability to choose the operator that will most quickly lead to the goal.

Fire investigation is a thorough planned process. A process that should change in light of new experiences and knowledge. It is our intention to develop a high order learning mechanism capable of enhancing and guiding the process of fire investigation.

8   Conclusions

Fire brigades are facing major problems to which training and education pose only a partial solution, since loss of expertise is inevitable. Expert systems on the other hand are a more cost effective solution which preserve knowledge.

The research expert system, FIRE, proved conclusively that expert systems can be effective in this domain; even though due to the destructive nature of fire demands it differs from current expert systems.

Also due to destruction of evidence, the systems will be able to work with imperfect knowledge and capable of recognising when there is insufficient information to make a decision.

Based upon imperfect knowledge it is important the system will investigate multiple causes and give the probability of these being the cause. The probability will be based upon statistics to provide accuracy.

Knowledge of experts and scientific fact are being built into the system to ensure it has a large and accurate knowledge base upon which to base its decisions. The knowledge base will be frequently updated ensuring maximum effectiveness.

A system is planned which will not require such formal updating but, which will learn itself. This system will be based around a learning mechanism akin to the so called high order functions of the brain. Again expertise from this system will be appended to the expertise of other copies of the system to produce an extensive knowledge base on fire investigation.

Both the expert and the learning system will provide the fire brigade with an aid which:

- carries out investigation thoroughly;
- incorporates the knowledge of the best fire investigators;
- learn and preserves experience;
- flexible to incorporate new knowledge;
- capable of being operated by inexperienced personnel; and
- be inexpensive.

It is felt that the collaborative research project being undertaken between the University of Sunderland and Tyne & Wear Metropolitan Fire Brigade in the UK offers a positive approach to the problems being faced in fire investigation.

9 References

1   The investigation into causes of fire (NFPA 921) The National Fire Protection Association, 1992
2   Tyne & Wear Fire Brigade Annual Report 1993
3   Home Office circular No. 106/1992 The Investigation of Fire of Doubtful Origin, The Home Office, 1992.
4   IBID
5   Jackson, Peter. Introduction to Expert Systems 2nd Edition. Addison Wesley 1990. Page 35.
6   Jackson, Peter. Introduction to Expert Systems 2nd Edition. Addison Wesley 1990. Pages 4 & 5.
7   Hertz J, Krogh A, Palmer R (1991). Introduction to the theory of Neural Computing'. Addison Wesley.
8   Laird J E, Newell A, Rosenbloom P S (1987). SOAR: An architecture for general intelligence. Artificial Intelligence Vol. 33 pp164.
9   Rosenbloom P S, Laird J E, Newell A, McCarl R(1991). A preliminary analysis of the SOAR architecture as the basis for general intelligence. Artificial Intelligence (Netherlands). Elsevier Science pp 289-325.

36 MAKING FIRE MODELLING SOFTWARE MORE ACCESSIBLE TO END USERS

A.N. NDUMU, D.T. NDUMU, J. ROBERTS
and A.K. PLATTEN
Department of Built Environment,
University of Central Lancashire, UK

Abstract
This paper discusses the need and techniques for making complex fire modelling software more accessible to end users. Following a characterisation of typical users against types of fire modelling software, a hybrid software architecture posited to meet the requirements of the full spectrum of users is presented. The architecture comprises a knowledge-based system for qualitative reasoning and the production of explanations, an artificial neural network layer for qualitative reasoning and approximate quantitative analysis and a finite difference solver for full quantitative analysis. The neural network layer is based on a novel application of artificial neural networks; and we present some preliminary results on its functioning.
Keywords: Artificial neural networks, end-user, fire modelling, finite difference solver, knowledge-based systems.

1. Introduction

In recent years, highly sophisticated software tools for deterministic fire modelling have been developed. These tools can be broadly classified as (i) zone models [1,2] and (ii) field models [3]. The latter have the advantage that they provide detailed quantitative information of the fire scenario. Both types of software require users with expertise; making it difficult for novice users to obtain reliable results. Therefore, there is a need for a tool that is flexible and more accessible to the novice user. We propose a system that exploits the advantages of field models and meets the needs of the full spectrum of users. This is a hybrid software architecture, implemented using a blackboard framework, that comprises a knowledge-based system (KBS), an artificial neural network (ANN) layer and a finite difference (FD) solver.

Fire Engineering and Emergency Planning. Edited by R. Barham.
Published in 1996 by E & FN Spon. ISBN 0 419 20180 7.

In section 2, we the describe the characteristics of end users and identify their needs. Section 3 discusses the various components of the hybrid architecture while section 4 discusses the functioning of the ANN layer and reports on preliminary results.

2.   Characterisation of end users

Users have different needs when modelling a fire scenario. At one extreme, some users require only qualitative information which generally take the form of fire risk assessment. This could involve the use of relatively simple equations to determine a given parameter [4,5]. Also, qualitative knowledge can be obtained by using decision trees to represent the knowledge embodied in building codes and/or from a panel of experts. This mode of use is particularly suited for implementation on knowledge-based systems and some work has been done along this line [6,7,8]. At the other extreme, some users require detailed quantitative information in order to get a comprehensive description of the fire scenario. Field models have been developed to meet this need and because they are based on techniques in computational fluid dynamics, users need a considerable level of expertise to fully exploit the potential of the software tool. In between these two extremes are users of zone models who need qualitative and some quantitative information at the same time. The quantitative information is not as detailed as that of field models but it provides enough qualitative information so that reliable results can be obtained within certain constraints [9].

Most users however, need to have access to different fire modelling tools in case their requirements change, thus creating the need for a hybrid software tool that meets the requirements of all user types. In our proposed system, the KBS component is able to provide, by reasoning, qualitative information on different aspects in fire modelling. This could range from the provision of advice on the best optimisation strategy for the numerical solver, to the determination of the fire risk in a compartment. Furthermore, users can refer to the KBS's explanation facility to determine its reasoning in arriving at any given advice. Users also need the ability edit and update the knowledge-base so that the system's performance reflects the state-of-the-art in fire modelling techniques.

The ANN component provides the facility for users to obtain results quickly and fairly accurately without recourse to the numerically intensive finite difference solver. This could be for preliminary analysis and provides enough qualitative and semi-quantitative information allowing the user to gain a reliable understanding of the fire scenario.

3.   The hybrid architecture

The proposed system is based on the blackboard problem-solving architecture [10]. The blackboard architecture consists of a global database called the blackboard which receives and provides data for several modules called knowledge sources. Each knowledge source consults a partition of the blackboard for relevant data, and if present, the data is processed and the results are posted on the blackboard (not necessarily on the same partition). The consultation process is co-ordinated by a specialised knowledge source called the scheduler. By constantly monitoring the blackboard, the scheduler is able to draw up a list of knowledge sources to be activated according to some predetermined criteria. In this way, the knowledge sources act independently and can only affect others through the blackboard. One advantage of the blackboard architecture over other software architectures is that it simplifies the implementation of a large program by providing an excellent mechanism in which modules with very different characteristics can be co-ordinated. This allows the use of different programming methodologies to implement the various modules. In our proposed system (Figure 1), the knowledge sources are represented by a User Interface (UI), a KBS, an ANN component and a FD solver.

The UI is based on a virtual reality graphics [11] which greatly enhances the user's visualisation of the problem. It enables the user to specify the problem domain in terms of simpler pre-defined objects like fire sources, vent openings, walls or cells for the FD solver which are stored in the object database. The objects and their default values (stored in the parameter database) can only be retrieved through the blackboard.

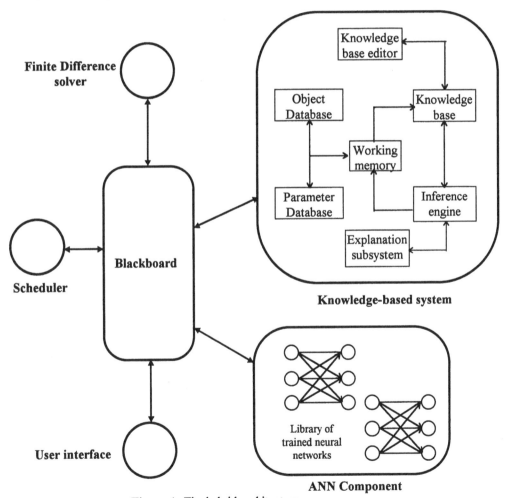

Figure 1: The hybrid architecture.

The KBS drives the system and its actions influence all other modules. It is capable of holding knowledge relevant to fire modelling encoded in the form of *if ...then* rules which are heuristic in nature. Using these rules, the KBS is able to reason about the course of the modelling process. In this way, the KBS incorporates intelligence into the overall system. Some work has been done in incorporating intelligence in computational fluid dynamics software [12], but because they rely exclusively on a FD solver, the user does not have the option of obtaining qualitative results quickly. The provision of an ANN layer improves on this, whereby the user can obtain qualitative and approximate quantitative information rapidly, and at a significantly lower computational cost. The functioning of this component, which forms the core of our current research, is described in the following section.

4.   The ANN component

Artificial neural networks consist of many simple processing units connected to each other. Each unit receives signals from other units and after processing, sends an output signal to the other units. The neural network paradigm used in most science and engineering problem-solving is the multi-layer feedforward network [13]. It consist of units arranged in a layered structure with each unit receiving signals from the layer below and sending signals to the layer above (Figure 2). The input and output layers receive signals from and send signals to the external environment respectively. The number of unit in these layers depend on the specific problem at hand while that of the hidden layer is chosen to optimise the network performance.

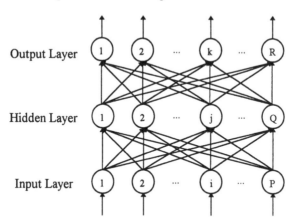

Figure 2: A feedforward neural network architecture

Each unit (except units in the input layer) carries out processing according to Equation 1:

$$a_j = \frac{1}{1+e^{-net_j}} \tag{1a}$$

where

$$net_j = \sum_{i=1}^{i=P} w_{ji} a_i \tag{1b}$$

w_{ji} is the weight of the connection from the ith unit on the previous layer to the jth unit, a_i is the output of the ith unit on the previous layer and net_j is the net input to the jth unit.

It has been shown that the multi-layer feedforward neural network can be trained to perform any mapping from an input space $\mathbf{i} \in \mathbf{R}^n$ to an output space $\mathbf{o} \in \mathbf{R}^m$ [14]. Network training is carried out using a set of characteristic training pairs $(\{\mathbf{i}\},\{\mathbf{o}\})$ describing the mapping. Given an input $\mathbf{i}$, and denoting the actual network output for this input by $\mathbf{y}$; training is a process of minimising the error $(\mathbf{o} - \mathbf{y})^2$, where $\mathbf{o}$ is the target output for the input $\mathbf{i}$. Thus, network training attempts to minimise the function:

$$E = \frac{1}{2} \sum_p \sum_k \left(o_{kp} - y_{kp} \right)^2 \tag{2}$$

where p is the number of training pairs and k is the number of output nodes. Many network training algorithms have been proposed for multi-layer feedforward networks [13,15,16].

4.1   Standard model

One approach to incorporating an ANN layer into the hybrid architecture is to create a library of networks, each trained to solve a typical parameterised problem. Each network is trained for a specific physical boundary condition e.g. Neumann conditions and geometric boundaries are expressed in parameter form. Thus, faced with a modelling problem, the KBS component selects an appropriate network from the library, and executes it with suitable parameters to provide a solution to the problem. We will illustrate the results of one such network with the following problem:
Given the differential equation

$$\frac{\partial \phi}{\partial t} + \frac{\partial}{\partial x_i} \left(u_i \phi - D \frac{\partial \phi}{\partial x_i} \right) = 0; \quad i = 1,2 \tag{3}$$

where D is the diffusion coefficient and ϕ is the required scalar parameter (e.g. temperature) subject to Neumann conditions $\partial \phi / \partial x_i = 0$ at the boundaries on the domain shown in Figure 3. The following fields are also defined:

Velocity field, u_i (doesn't change with time):

$$u_1 = -\cos(\pi\alpha_1)\sin(2\pi\alpha_2)$$

$$u_2 = \sin(2\pi\alpha_1)\cos(\pi\alpha_2)$$

where

$$_i = \left(\frac{x_i - x_i^c}{x_i^d}\right)$$

Scalar field; ϕ at t = 0 is the Gaussian:

$$\phi(x_1, x_2) = \exp\left(-\left(\frac{\left(x_1 - x_1^c\right)^2 + \left(x_2 - x_2^c\right)^2}{r^2}\right)\right)$$

where r is any real number that determines the shape of the Gaussian and

$$x_i^d = x_i^{max} - x_i^{min}$$
$$x_i^c = \frac{1}{2}\left(x_i^{max} + x_i^{min}\right)$$

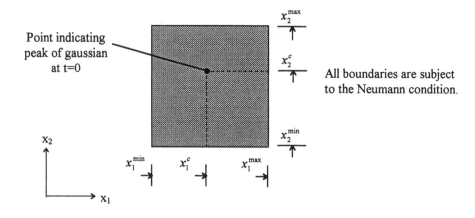

Figure 3: The problem domain

From Equation 3 it can be deduced that for steady state solutions $\phi = f(D, x_1, x_2)$. A neural network with 3 nodes (for the inputs D, x_1, x_2) on the input layer, 10 nodes on the middle layer and one node (for the output ϕ) on the output layer was trained for the above problem. Training data for the network was obtained from solutions generated by a FD solver that solves Equation 3. Twenty percent of the data at grid points where used for training and the network error $E = 0.1\%$ was achieved (as given by Equation 2). After training, the network was able to compute the values for the whole domain with minimal computational costs. Figures 4a and b shows a comparison of the solutions obtained by the FD solver and the trained network.

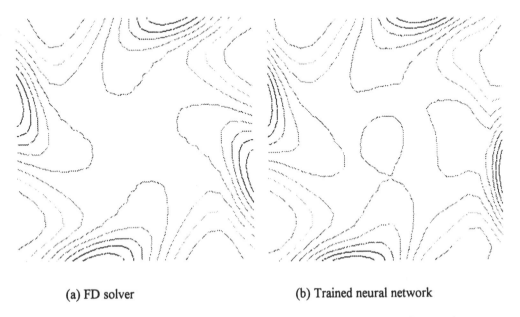

(a) FD solver   (b) Trained neural network

Figure 4: Comparison of solutions obtained by FD solver and trained neural network

4.2   Improved model
The approach outlined above has one major drawback. In some problems, the physical process being learnt is governed by a very large number of parameters or can be subject to a wide variety of boundary conditions. This leads to large networks where training can be difficult. Research is currently underway to rectify this problem whereby the network is trained at a deeper level. Rather than using a large network to learn the global mapping region, a series of smaller networks are trained to learn the underlying process. Each of the small networks represents a sub-component of the global process and interacts with its nearest neighbours in a recurrent fashion.

5.   Conclusions

We have proposed a hybrid system based on the blackboard architecture to cater for all user types. We also show that by including an ANN layer, qualitative and approximate quantitative solutions can be obtained rapidly. Problems encountered with the ANN component where outlined and an improved model is suggested. Research is currently underway with the improved model and results will be reported in due course.

6.   References

1. Jones, W.W. and Forney, G.P. (1993) Improvement in Predicting Smoke movement in Compartmented Structures. *Fire Safety Journal*, Vol. 21, No. 4. pp 269-297.
2. Charters, D.A., Gray, W.A. and McIntosh, A.C.(1994) A Computer Model to Assess Fire Hazards in Tunnels (FASIT). *Fire Technology*, Vol. 30, No. 1. pp 134-154.
3. Cox, G. and Kumar, S. (1987) Field Modelling of Fires in Forced Ventilated Enclosures. *Combustion Science and Technology*, Vol. 52, pp 7-23.
4. Gupta, A.K. (1994) CALFIRE: An Interactive Model for Fire Calculations. *Fire Technology*, Vol. 30, No. 2. pp 304-325.
5. Babrauskas, V. (1993) Toxic Hazards From Fires: A Simple Assessment Method. *Fire Safety Journal*, Vol. 20, No. 1. pp 1-14.
6. Roux, H.J. and Berlin, G.N. (1979) Towards a Knowledge-Based Fire Safety System, in *Design of Buildings for Fire Safety*, (ed. Hamarthy, T.Z. and Smith, E.E), ASTM, Philadelphia, Pa, pp. 3-13.
7. Amy, M., (1991) Even Experts need an Expert System, *Fire Engineers Journal*, Vol. 51, No. 161. pp 29-30.
8. Dodd, J.F. and Donegan, H.A. (1994) Some Considerations in the Combination and Use of Expert Opinions in Fire Safety Evaluation. *Fire Safety Journal*, Vol. 22, No. 4. pp 315-327.
9. Cox, G.(1994) The Challenge of Fire Modelling. *Fire Safety Journal*, Vol. 23, No. 2. pp 123-132.
10. Jackson, P. (1990) *Introduction to Expert Systems,* Addison-Wesley.
11. Superscape VRT, Version 3.50 (1994) *Reference and user manuals* Superscape Ltd.
12. Petridis, M. and Knight, B. (1993). A Blackboard Approach for the Integration of an Intelligent KBS into Engineering Software. *Third International Conference on the Application of AI in Civil and Structural Engineering.*
13. Rumelhart, D. E., Hinton G. E., and Williams, R. J. (1986). Learning Internal Representation by Error Propagation in *Parallel distributed processing,* Vol. 1, MIT press, Cambridge, MA, pp 318-362.
14. Hornik, K., Stinchcombe, M. and White, H. (1989). Multilayer Feedforward Networks are Universal Approximators. *Neural Networks*, Vol. 2. pp 359-366.
15. Van der Smagt, P. P. (1994). Minimisation Methods for Training Feedforward Neural Networks. *Neural Networks*, Vol. 7, No. 1. pp 1-11.
16. Kasparian, V., *et al.* (1994). Davidon Least Square-Based Learning Algorithm for Feedforward Neural Networks. *Neural Networks*, Vol. 7, No. 4. pp 661-670.

PART THREE

EDUCATIONAL AND OPERATIONAL CONSIDERATIONS

37 INTRODUCTION

B.T.A. COLLINS
Her Majesty's Chief Inspector of Fire Services,
The Fire Service Inspectorate of The Home Office, UK

I am grateful to have this opportunity to speak to you this morning and I feel honoured that I have been invited to do so as we commence the third day of the symposium under the heading of "Bridging the Gap".

Better understanding by the Fire Fighters and Researchers is essential and the fire service in England and Wales is very conscious of its debt to the Institution of Fire Engineers for the work that it does to promote the study and development of the science of fire engineering. For the past two years the fire service has yet another debt to acknowledge because in 1993 the first chair in Fire Engineering was established at the University of Central Lancashire in England.

You have already had the benefit of hearing the opening paper in session one by Georgy Makhviladze, the eminent incumbent of the Chair.

As a representative of the Fire Service I welcome the initiative which has been taken by the University of Central Lancashire and the Institution of Fire Engineers in organising this Euro-conference. I note that this event is billed as the first annual meeting, which presupposes that there will be others. It is a good measure of the confidence and enthusiasm of the organisers as well as an indicator of the importance of the subject and for the need for international co-operation, that this optimistic view can be taken. I welcome the initiative, acknowledge its importance and support the aim for this type of conference to continue in an endeavour to "bridge the gap".

This recognition on behalf of the Fire Service of the work of the Institution and the University of Central Lancashire through its Chair in fire engineering leads to setting the scene for today's presentations.

In the U.K. the cost of fire in financial terms alone is colossal. According to the Association of British Insurers and despite all our efforts, at least £606 million pounds worth of property was destroyed by fire in the United Kingdom in 1994.

Over two thirds of the property loss was in commercial premises. This is just insured loss. The figures take no account of underinsured losses or where people have relied on the "It won't happen to me" philosophy. Although this represents a reduction on losses recorded in previous years there is a continuing need to understand the phenomena of fire and to pursue even greater efforts for more effective fire protection, fire precautions and the planning for fire fighting should this ultimately become necessary. All of this has to be achieved in the most effective way.

The first address on today's programme will deal with education and training. You will have the benefit of hearing a paper by Professor John Roberts. I do not wish to pre-empt anything that might be in that presentation; I suspect, however that there is likely to be some reference to an initiative which the Fire Service Inspectorate in England and Wales support.

I refer to the National Core Curriculum in Fire Safety Studies. Some of you may

Fire Engineering and Emergency Planning. Edited by R. Barham.
Published in 1996 by E & FN Spon. ISBN 0 419 20180 7.

have already heard about the Core Curriculum. It has been developed in response to a recommendation by a Government sponsored report that all professions involved in fire safety, building regulation, design and construction should be educated through a national network of professional development courses, in such a way that they would have a common understanding of the principles of Fire Safety.

Produced by a group representing the major professions as the basis for the courses in fire safety which would be taught in universities and colleges of higher education and which would form the national framework.

The curriculum was designed to be used very flexibly to meet the needs of all the various professions who have to be aware of the principles of fire safety. Among these are architects and other design professionals, building control officers, Fire Safety Officers, developers and Fire safety managers. Specifically within the fire service, there is a growing recognition among fire officers of the value of higher education generally. Fire Safety Engineering degree courses have now been established with the lead being taken by South Bank University in London and the University of Central Lancashire and this should assist the integration of the Research undertaken in the academic arena and the practical considerations which arise at operational incidents.

As society becomes more complex with the rapid development of the technologies on which we have become dependent, so the demands on fire services increase - not only to fight the fires which will inevitably occur, no matter how well-regulated the society in which we live, but also to meet the challenge of the increasing emphasis on preventing fires occurring. I believe the Core Curriculum in Fire Safety Studies is a major development in the field of fire safety and prevention. Earlier I drew attention to its being designed to be used flexibly.

The potential for flexible use of the Curriculum does not apply only to the various professions in the United Kingdom who need to be aware of fire safety. Because the Core Curriculum is concerned with the general principles of fire safety, it can be adopted as a basis for the teaching of fire safety studies in any country and developments have already taken place in that direction. I am sure that John Robert's paper will refer to them and it is appropriate at this Euro-conference to draw attention to the potential value to our colleagues in Europe.

Running in parallel with the academic initiatives on fire safety education is a scheme, started in the mid 80s by the U.K. Employment Department by which industries have been encouraged through lead bodies to formulate standards of occupational competence for their sectors.

Within the fire sector, two lead bodies have undertaken this task. The Fire Industry body has developed standards covering all the more commercial aspects of the sector, including fire safety inspections, insurance surveying, consultancy, fire investigation, systems and extinguisher maintenance and installation.

At the same time the emergency fire service lead body has developed standards of competence for fire-fighters, incident commanders, control room staff and fire safety officers.

Following the development of the standards of competence the next step has been to process them into a suite of qualifications that fit into the National framework.

This work was completed in the summer of 1994 and for the six months up to December the qualifications were subject to extensive field trials. The results of these trials are currently being evaluated and subject to any necessary refinements the qualifications will be submitted to the U.K.'s National Council for Vocational Qualifications for accreditation this summer.

The opportunity to use these competence based qualifications alongside the academic initiatives mentioned earlier in this paper, will ensure that those responsible for fire fighting and safety provision will have, not only the technical knowledge so important to the true professional, but also the ability to deliver the service to a consistently competent standard in the field.

Another paper to be presented by Roger Klein addresses the subject of fire risk assessment. Fire risk assessment is in many ways the rising star of fire safety management. Historically, in the U.K. fire safety standards in new or existing buildings have been assessed according to "the circumstances of the case". That has meant that the assessor has been obliged to set a prescriptive standard around the fire potential found in the building at the time of assessment. Little thought has been given to the need to adjust the assessment to reflect the changes in Risk throughout thr life of a building and Dennis Davis, C.F.O. of Cheshire Fire Brigade (U.K.) will tell us about developments to keep the public up to date with the changes that might affect them in session 9.

Buildings in use present a dynamic situation where risks are constantly changing. Ongoing assessments of Risk are therefore essential if levels of safety are to be maintained. Perhaps it might be helpful to take a little time to set the scene for Roger's paper by explaining why in the United Kingdom, the Home Office needed to develop a guide to fire risk assessment in occupied buildings and to provide a little of the background.

The starting point was two European Council Directives - the Framework and Workplace Directives - which seek to encourage improvements in the safety and health of workers at word These became law in late 1989 and are based on a risk assessment approach and a move to a self compliance philosophy. This meant that in order to make our own fire safety legislation we had to devise a guide to fire risk assessment which could be used by the lay person to assist him with his responsibilities to comply with the requirements of the directives.

Both Directives place the primary responsibility for ensuring health and safety - including fire safety - in the workplace on employers. This requires employers to assess fire safety hazards and to take measures to reduce them.

The main provisions of the Directives, dealing with special fire precautions relating to manufacturing processes, and other safety matters have already been implemented in the U.K. and these Regulations took effect on 1 January 1993.

Because fire safety regulations have not yet been made, the United Kingdom's approach toe fire risk assessment was only published last year - it firstly formed part of Home Office consultation with other Government departments on the Fire Precautions (Places of Work) Regulations and then was part of the public consultation exercise which began last September and finished at Christmas.

However, this approach which is essentially a practical one, took some considerable time to develop. This is because although U.K. fire authorities are used to assessing the risk from fire as part of the enforcement role in certification of specific premises and in the advisory role in relation to many other premises, the lay person who is not experienced in fire safety matters has in the past been largely dependant upon advice from Fire Brigades and other fire professionals.

In order to validate this approach it was necessary to involve other people outside the Home Office who were experienced in risk assessment, particularly risks from fire.

A recent European Commission paper dealing with Fire Risk Assessment used an approach which is similar to the U.K.'s new proposals.

The European formula is considered to be the most feasible way to proceed, particularly as it would be made available to all Member States, and hopefully a cohesive policy regarding Fire Risk Assessment within the scope of the E.C Directives will emerge.

Many different methods of fire risk assessment have allowed fire safety engineers to pitch the fire safety provisions in a building closer to the actual fire safety needs, and provide greater flexibility in the design of buildings. The E.C. DIRECTIVE has for the first time brought the concept of FIRE RISK ASSESSMENT to the level of the USER of a building.

Fire risk assessment has been further developed in the field of Fire Safety Engineering on two fronts:- internationally by ISO and CEN through various TC/92/SC4 committees, as you heard in session 5, and in the United Kingdom by the British Standards Institution through FSM/24 committees. The results of this work should add to the wealth of knowledge on this complex subject in the fire safety world, and perhaps bring a much needed agreed framework to the discipline of Fire Safety Engineering.

The practice of fire risk assessment has many facets and addresses fire safety needs in different ways including the needs of self compliance. A recent report published by the Audit Commission in the U.K. recognises that Risk Assessment may in the future form the basis for determination of the weight of response by Fire Brigades to actual operation incidents. This safety case approach is novel for the Fire Service but is of course a well established practice in Industry.

Therefore, I look forward to Professor Klein's paper presented from the operational fire fighting viewpoint with great interest as I am sure we have much to learn from it.

FIRE TECHNOLOGY EDUCATION AND TRAINING

38 EDUCATING FIRE FIGHTERS FOR FIRE SAFE DESIGN

D. EVANS and J. ROBERTS
Department of Built Environment, Faculty of Design
and Technology, University of Central Lancashire, UK

ABSTRACT

This paper reviews the technological and legislative changes that have generated the need for new wider, and higher level educational provision for fire fighters across the European Union. The new courses developed in response to these needs are reviewed as is their sustainability and likely recruitment. Available academic progression for serving fire-fighters is described as are future developments abd the role of the relevant professional bodies.
Key-words: education, fire, training.

Fire Engineering and Emergency Planning. Edited by R. Barham.
Published in 1996 by E & FN Spon. ISBN 0 419 20180 7.

INTRODUCTION

This paper discusses recent, important developments in Fire Engineering education in the UK. Technological advances in such important areas as fire protection materials, active fire protection systems and computer aided design allow a new, more dynamic, approach to designing for fire safety.

While the experiential and structured training programmes for fire fighters in the UK is second to none, the gap between the level of engineering education required to appreciate and critically appraise novel fire safe building design and that actually provided had become so great as to be of serious public concern.

TECHNOLOGICAL BACKGROUND

Research into passive fire resistance and behaviour of materials has made great progress in recent years particularly in, say, the behaviour of steel framed buildings under fire conditions. There is now a better and more detailed understanding of human behaviour in fire emergencies. There are now more powerful methods of predicting the behaviour of non-uniform structures in fires. These are all significant developments in how the design and use of materials in construction of buildings has changed and novel design can significantly improve fire safety.

But there have been even more rapid developments in active fire protection. What has made novel building designs so readily possible is a much better understanding and appreciation of the protection to be afforded by active fire protection and extuinguishing systems such as sprinklers. It is now possible to better predict the behaviour of flame and smoke within enclosed spaces of complex building forms such as atria using computational fluid dynamics and model the likely patterns of smoke and flame spread.

LEGISLATIVE BACKGROUND

These technological and engineering developments are extremely powerful and have been driving certain aspects of government regulation. New UK Building Regulations were introduced in 1991 which were qualitatively different from earlier Approved Documents and included the very important paragraph: "Fire Safety Engineering. A fire safety engineering approach that takes into account the total fire safety package is now a viable alternative. It may be the only way to achieve a satisfactory standard of fire safety in some large and complex buildings."

With this statement the Regulations moved away from solely prescriptive measures and put in their place Fire Safety Engineering. Those who previously had the authority to say, "No" without challenge can no longer do so. A case must be argued on its merits. This is a revolutionary change for all those in the UK concerned with the control of the design of buildings for fire safety.

Correspondingly, those responsible for assessing and criticising fire safety systems on which human life depends are obliged to discuss the arguments presented on their merits and reply at an appropriate level.

For Fire Brigade's personnel there are two other relevant Acts. The Fire Services Act 1947, which obliges fire authorities to give free advice on request regarding fire prevention measures and means of escape from buildings. That is the responsibility of the Fire Safety Officer [FSO] who is invariably a Brigade Member.

The second important piece of legislation for the Brigades, is the Fire Precautions Act 1971:Chapter 40 which defines the responsibilities of the FSO. The Fire Safety Officer has statutory control of fire precautions in designated premises through a system of fire certificates which concern;

    a) use of the premises

    b) means of escape and their effectiveness in use

    c) appropriateness of fire fighting equipment and fire alarms.

    and in the case of factories

    e) storage and use of explosive or highly flammable materials

The 1971 Act requires the FSO to consult with the Building Control Officer before requiring alterations to any building in connection with certification.

The Building Regulations have already been changed, and the Fire Precautions Act and the Fire Services Act are to be reviewed in the near future. It is only possible to guess at the outcome of those reviews based upon experience in other sectors of the economy and overall government policy. But, if the 1971 act is reviewed then the responsibilities of brigade staff will be significantly altered at a time when the emphasis is on the need for greater technological awareness.

GOVERNMENTAL RESPONSE

The Department of Environment was aware of many of these problems and had commissioned Bickerdike Allen report (1990). The terms of reference specifically included;

      examining means of overcoming the delays and problems that arose in the assessment of new architectural developments and innovative design which cannot comply with prescriptive legislation or codes of practice,

      an examination of the technical and practical skills required to allow authoritative advice to be given on all aspects of fire prevention in buildings,

      consideration of the training ... requirements necessary to secure enforcement of advice once given.

The Bickerdike Allan report made twelve recommendations including; "The educational development of building designers, BCOs and FSOs should be encouraged by the early establishment of a national network of professional development courses in colleges and the then polytechnics ... " The report clearly recommended the establishment of a national network of educational provision.

As a result the UK Home Office initiated a group to prepare and plan what was required from that provision. It is interesting to look at the group that was drawn together,

> Institution of Fire Engineers,
> Fire Service College,
> Institute of Building Control,
> Royal Institution of British Architects,
> Incorporated Association of Architects and Surveyors, (now the Institution of Building Engineers)
> Home Office, and
> Department of Environment

The interesting thing was that no institution of higher education nor Chartered professional engineering body was included. These important omissions can be considered to have robbed the group of a certain wider appreciation of their task. For example, the National Core Curriculum does not contain sufficient "hard" engineering for it, by itself, to form a stand-alone qualifiation giving some form of Engineering Council recognition.

The Working Party produced a document: The National Core Curriculum in Fire Safety by Design which described the breadth of knowledge to be aspired to by all those involved in fire and building regulation and design and contained a recommendation that a course of study at an acceptable national minimum standard be provided nationally through a network of colleges.

EDUCATIONAL STANDARDS AND PROVISION

The background to the educational development was rapid technological advance, changes and potential changes to the legislation which for over twenty years has formed the framework for the fire community in the UK. The University and IFE were faced with providing programmes of study which met the requirements of a new and challenging situation, took on board the aspirations of personnel who wanted to obtain a higher qualification, and the requirements of the professional bodies and employers.

The National Core Curriculum allows institutions to choose their own standards of presentation. The University of Central Lancashire in partnership with the Institution of Fire Engineers determined to develop courses based on the National Core Curriculum which would lead to accepted national qualifications.

When considering education for firefighters and the fire community it was necessary to determine the nature of the courses to be offered and how to provide both academic quality and access for a range of individuals who would be coming to formal higher education for the first time. With the full backing of the University and in partnership with the IFE it was possible to offer new courses, new modes of study, new ways of learning, and new links with the local community and the professions.

After some discussion it is now recognised that minimum standard to which the National Core Curriculum should be taught is that of a BTEC Higher National Certificate (HNC). This is the first step in any structured programme leading to

Chartered Engineer status with the UK Engineering Council and can, itself, give Incorporated (Technician) Engineer status.

The University of Central Lancashire now offers an HNC in Fire Safety Engineering which gives full credit for prior learning, allows workplace assessment and successful completion of which will give exemption from the written examinations of the IFE at corporate member level.

Sponsored by the Institution of Fire Engineers a network of colleges is being slowly put in place. The network now includes colleges of higher education in:

Aberdeen, Eastleigh, Exeter, Glasgow, Leeds, Preston, South London, the Fire Service College and the Washington Hall International Training Centre.

Many of these institutions offered courses of study leading to the IFE's professional examinations, but due to financial constraints not all will be offering the IFE sponsored HNC from September 1994.

The network is based on courses developed and validated by the University with the participation and support of the IFE. Colleges round the country will be franchising the University of Central Lancashire HNC.

An important component in determining the levels of the courses to be offered was consideration practices in other European countries: Denmark, Germany, Netherlands, France, Spain, Italy and Ireland which, together, form the majority of the present EC have the requirement that for promotion to the higher levels of Brigade Command, one must possess either a Degree or Chartered Engineer status. Generally this is achieved through a two or three tier entry system. The UK is unique in not requiring such qualifications of it's senior brigade officers and operating a single tier entry. For comparability with Europe any programme of study must offer the opportunity to attain an Honours degree and contain the National Core Curriculum at an appropriate level.

The IFE are well aware of the European Community Directive of January 1991, requiring member states to recognise professional qualifications and titles authorised by all other countries in the community. It is important because unless UK brigade staff are aware of, and take, the educational opportunities that are open to them, a situation could arise where cross-country promotion at senior level could become a one-way process to the detriment of U.K. brigade personnel.

COURSE CONTENT

For those who have successfully completed any engineering courses to obtain Engineering Council recognition the programmes of study must be accredited by an approved professional engineering body.

The Institute of Energy (IoE) was approached and accepted that Fire Engineering came within its area of responsibility. To be accredited the programmes of study had to contain the necessary broad engineering base as well as the fire engineering specialism. The IoE provides a "model" degree for reference purposes to those wishing to obtain accreditation for a new course.

The model degree is in two stages; Part 1: Technological foundation and an introduction to business and commercial practices. Part 2: General Engineering, the Energy Specialism (in this case Fire Engineering) and supporting studies in law. economics, communication and management. The level, taught time allocation and course content are all specified.

An important and necessary part of all Engineering Council accredited programmes is design work which integrates the individual subjects across the course and gives an applied and challenging element of student self activity. This is an integral part of the Fire courses at the University of Central Lancashire.

COURSE STRUCTURE

The Chief and Assistant Chief Fire Officers Association considers that the one tier entry system is necessary for effective command and leadership. The argument is that "Effective operational command relies on the ability of commanders to demonstrate effective ability to lead and command. This can only be achieved by officers who have a clear understanding of the operational situation that they may encounter. Such an understanding can only be ... obtained through experience and learning whilst in the Service in the rank of firefighter."

That is, unless an officer has gone through the experience of commanding appliances, has been through the testing situation of practical fire fighting, he or she will find it difficult to command the respect of operational personnel and difficult to obtain the practical experience to lead and command effectively. The University has considerable sympathy with this point of view; that those who have the experience of fighting fires will tend to be more effective in commanding personnel to extinguish those fires.

Such an appreciation gave rise to a programme of study which was vocational in nature and significantly different from the only other course available in the UK at that time, the degree in Fire Safety at South Bank University. It was an important development that stemmed from the philosophy of the University, it also fitted well with the requirements of the Institution of Fire Engineers.

But such an approach has placed certain restrictions on the development of the courses; the vast majority of those students joining the new fire degree courses are brigade personnel. Given the support of the IFE it was expected that that there would be an initial a period in which the overwhelming majority of those recruited to the course would be from the fire brigades.

COURSE SUSTAINABILITY

In such a situation it was necessary to analyse the capacity of the brigades to sustain the degree programmes by providing a sufficiently large number of students. In 1992 there were 38,808 full-time fire fighters and 2,397 specialist Fire Prevention Officers. The responsibility of the post should mean that all 2,397 FPOs will aspire to degree status.

Consider a simplified operational tree for brigades. It is not meant to be precise and the figures for the number of persons that a particular officer commands is a typical average.

RANK STRUCTURE
(OPERATIONAL TREE)

CHIEF AND ASSISTANT CHIEF OFFICERS
DIVISIONAL AND SENIOR DIVISIONAL OFFICERS

ASSISTANT DIVISIONAL OFFICER
(responsible for between 40 and 80 persons)

STATION OFFICER
(responsible for up to 28 persons)

SUB-OFFICER (responsible for 5 to 6 persons)

LEADING FIRE FIGHTER
FIRE FIGHTER

To reach Station Officer level the minimum qualification is either Graduate membership of the IFE or having passed the necessary statutory examinations. In the armed forces command of 28 persons would generally be a platoon command, a lieutenant, who would be expected to have attained an educational level equivalent to a University degree. That is by the time an individual reaches Station Officer rank they have a level of responsibility that, elsewhere, is normally associated with a graduate.

There are, thus, some 4,000 fire brigade personnel who should aspire to a suitable degree and be provided with a route to progress in that direction. Similarly it may be argued that there are also some 7,000 brigade personnel who have responsibilities normally associated with a Higher National Certificate.

But, what are the academic qualifications of those entering the Fire Brigades,

QUALIFICATIONS ON RECRUITMENT TO THE SERVICE
INTAKE FOR 1992;

a) Total recruits	769
b) Recruits with 3 GCSEs	393
c) Recruits with 1 or more 'A' levels excluding b) above	88
d) First degree or equivalent excluding c) above	18

Over 50% of recruits have 3 GCSE's or more, 106 - some 14% - have at least one GCE A level, 18 actually have a first degree or equivalent. This analysis showed that there was a substantial potential for degree level qualification within the fire brigade itself without taking into account private consultancies and other public sectors such as Building Control. This University considered that this confirmed its appraisal that those of Station Officer rank, and above, were capable of studying for and obtaining a degree.

AVAILABLE ACADEMIC PROGRESSION

This University has developed a structured programme of academic advancement that by January 1995 will offer The HNC Fire Safety Engineering through an Honours degree Fire Safety Engineering to a Post-graduate diploma/Master's degree in Fire Safety and Risk Management. Each of these courses contains the National Core Curriculum at an appropriate level. However, depending upon the course there will be - to a greater or lesser extent - the additional material required for professional recognition and each has a number of step-out points giving a qualification suitable to the student's attainment.

HIGHER NATIONAL CERTIFICATE
FIRE SAFETY ENGINEERING
(step-out qualifications
ADVANCED CERTIFICATE IN FIRE SAFETY ENGINEERING)

BENG(HONS)
FIRE ENGINEERING
(step-out qualifications
CERTIFICATE OF HIGHER EDUCATION at the end of Year 1
DIPLOMA OF HIGHER EDUCATION IN FIRE ENGINEERING at the end of Year 2
BENG FIRE ENGINEERING at the end of Year 3)

MASTER'S DEGREE
FIRE SAFETY AND RISK MANAGEMENT
(step-out qualification
POST GRADUATE DIPLOMA)

MPhil and PhD research degrees may also be obtained.

WHAT OF THE FUTURE ?

The establishment of the national network of educational provision to offer the Higher National Certificate and National Core Curriculum has not been as rapid as had been originally hoped. The Home Office had expected the network to obtain a significant amount of funding. But cut-backs in expenditure have meant that this has not happened and colleges of higher education and universities are themselves having to pump-prime what is an expensive course. In the present economic climate colleges are reluctant to make such a commitment without guarantees. Thus progress is slower than desirable and additional support is continually sought.

The Government is establishing National Vocational Qualifications (NVQs) to assess workplace competence, and general National Vocational Qualifications (gNVQs) which are based more on examination performance. Pilot programmes are now running within specified brigades and the results will be available early in the new year. These qualifications will impact on all aspects of training and education in the fire community including statutory examinations, training courses at the Fire Service College, membership criteria for the IFE and the academic courses presently on offer.

NVQ's are work-placed competency based, and gNVQ's are more academically biased and tend to take the form of written examinations. NVQ's and gNVQ's are being developed at a rapid pace but there is little correlation or overlap between them. Presently there is no agreed mechanism on how the two are to be related. Will obtaining NVQs give equivalence with gNVQs and vice versa? This is an area where there is very real danger of confusion.

Finally, the European dimension. One thing that the National Core Curriculum has done is to make this Country the first in the European Community to have structured academic provision in fire safety and fire engineering. Countries around Europe are looking to the UK lead in this area and, for example, the IFE degree recruited its first French students September 1993 and recruited three more in September 1994 and 1995.

The IFE and the University of Central Lancashire are providing a one year course for French brigades personnel who have completed a suitable training and education programme of at least two years duration. The first two French students to enrol graduated with good Honours degrees in Fire Safety this summer. Consideration is now being given on how to extend a similar provision to other European countries.

The U.K. has an opportunity that shouldn't be missed. The University of Central Lancashire is contacting relevant institutions across European and is making sure that they are aware of the possibilities that exist within this country to advance the education of their members.

The IFE and other relevant UK bodies must sieze the opportunities now open and ensure that the work done to put the UK in the lead in Europe in terms of European fire education.

39 ADDRESSING THE NEED FOR EUROPEAN INTEGRATED POSTGRADUATE EDUCATION FOR FIRE SAFETY

R. BARHAM
Department of Built Environment,
University of Central Lancashire, UK

Abstract
This paper suggests a novel approach to the postgraduate development of fire protection engineers and of those involved in the assessment and insurance of fire risks. It also describes the nature of a U.K. initiative for the formation of a European network offering courses and co-operative research opportunities at post-graduate level.
Keywords: education, postgraduate, network, fire safety, insurance, economics

1 Introduction

The scale and complexity of recent high-tech construction projects, across Europe, have important implications for those involved with the assessment of fire safety and design risk. In the U.K., the Bickerdike Allen[1] report commissioned by the Department of Trade and Industry found that "**in matters of fire, safety in design is uneven and often inadequate in the light of the many advances in building technology and fire engineering**". Similar conclusions were reached in France in a report by Dupuis[2] made to the Ministre de l'Interieur and other E.U. member states have also carried out similar reviews.

The identified deficiencies are being addressed both in the U.K. and in other European states (primarily in engineering and building control) and several new fire engineering courses have been introduced. However, these do not address the paradox created by a desire to provide buildings at lower end-user cost and the increased cost of satisfying insurers' demands for more safety consciousness in design. A purely regulatory approach encourages speculative property development

Fire Engineering and Emergency Planning. Edited by R. Barham.
Published in 1996 by E & FN Spon. ISBN 0 419 20180 7.

to lowest cost and to the minimum standards allowable by legislation primarily aimed at the preservation of human life. Safety conscious design, however, can also be economics-driven. Reviewing what is technically possible with cost implications, over-specification (and, hence, increased costs) often results from attempts by either designers or insurers, to "ensure" absolute safety and reduce risk. This is not surprising: losses are considerably higher, even from a single incident, where the loss of a building is involved.

By way of an example to reinforce this view, terrorist damage in the U.K. during 1992 created building-related losses originally estimated at £800m. (1.0 Mecu) - good disaster-recovery implementation reduced this figure by over 50%. It should also be noted that recently published statistics reveal that there are now over a quarter of a million reported fires in the E.U. each year. These result in some 2500 fatalities.

Advanced educational provision in fire safety and risk assessment is long overdue; throughout Europe, fire consultants and insurance companies' chief surveyors are actively seeking ways to improve risk assessment strategies. There is a worldwide shortage of specialist staff in this important area. It is vital that an adequate and efficient route be provided into a range of fire safety related vocational careers and universities are encouraged to provide a facility designed to enhance the present and future effectiveness of graduates by further education and advanced training in fire safety based around the assessment of risk of fires and explosion and the related economic effects. However, given that most European fire professionals are already graduates, this requires postgraduate study accessible to both full-time and part-time students. This would be especially attractive, for example, to the insurance industry; it could provide an internationally recognised qualification. It would also have relevance in the field of property protection, where it could provide a specialist qualification for property managers and others working in this specialist vocational area.

Other sources of students exist: the fire protection industry, large industrial companies, government laboratories, architectural and building surveying practices, building services engineering consultancies and local authority building control officers, and there should be an increasing demand from people with varied backgrounds in the property procurement and construction/property management industries, generally. The course would also be of particular interest to serving fire officers wanting to strengthen existing qualifications and, at the same time, provide themselves with a qualification suitable for use in connection with later alternative career routes. Initial enrolments are expected to be members of fire brigadesand the risk surveyor/underwriter employees of insurance companies.

2 Rationale for a course

Intrinsic to the study of fire, its development and prevention are the means by which its consequences may be minimised in human, environmental and financial terms. A course of study must emphasise fire safety within the context of the built environment, particularly its application to buildings. To provide a clear focus for the study of fire safety, it is necessary, also, to emphasise the importance of the assessment of fire risk and the relevance of probabilities of loss.

Society has changed dramatically in the latter part of this century. This process of change has been made manifest in the variety of demands, visions and constraints set. Nowhere has this situation been more true than in the case of health and safety, in which area we have progressed from laissez-faire to a highly structured legal framework in little over a hundred years. The new U.K. Building Regulations (Part B)[3] introduced in 1991, for example, are qualitatively different They include the very important paragraph:

"A fire safety engineering approach that takes into account the total fire safety package is now a viable alternative. It may be the only way to achieve a satisfactory standard of fire safety in some large and complex buildings."

Fire safety solutions must now, therefore, be argued on their merits; however, unique solutions must involve only levels of risk compatible with acceptable levels of security and cost. The proposal and/or assessment of, sometimes novel, solutions within the often contradictory constraints of safety, economy, law and technology pose a challenge and strongly suggest a need for a high level course of study. Integration across diverse fields to propose an acceptable design solution must be a major feature and such a course will also need to develop wide-ranging perceptions of each topic discussed. Internationally, there is currently no such course provision with the required vocational emphasis and the property professions, designers and facilities managers, rely on advice from people having mainly experiential learning.

3 Background

A number of E.C. Directives[4] concerning the minimum safety and health requirements for the workplace, and on technical harmonisation and standardisation, came into force on the 1st January, 1993. These significantly affect the educational requirements of those engineers and surveyors involved with the provision and maintenance of fire safety or with the assessment and management of risks.

There is, therefore, considerable potential for an educational response (in the context of fire safety) to create:-

    i.   an international network of universities to provide a co-operative programme of taught postgraduate courses and facilities for higher degrees by research; and

    ii.   a research and development network, initially across greater Europe, involving both educational and research institutions and the industrial sector.

thus addressing the urgent need for action to facilitate a more rapid and homogeneous interpretation and enforcement of E.C. standards and regulations and assist in the avoiding of possible future differences or gaps.

4 U.K. Provision

The Department of Built Environment has a wide course portfolio and experience, over many years, of providing both full-time and part-time vocational education for the property- and construction-related professions. The building-related professions and fire safety organisations have focussed, traditionally, only upon the engineering basis of fire safety in the built environment but there is, now, a clear need for a wider curriculum for prospective professionals in this field. At postgraduate level, this includes looking at both proposed and existing buildings, including their architecture, materials and construction techniques, in terms of their safety provision and of their viability in terms of acceptable/insurable risk and/or cost. The policy areas in relation to fire safety and risk, including both philosophy and politics, need to be addressed. Finally, there is a management perspective - looking at the balance between the protection, the risk and the cost in both private and societal terms - which needs to be recognised, providing students with the ability to react realistically and flexibly to proposals for engineered solutions to fire safety.

5 Undergraduate course programmes

The University of Central Lancashire has one of the three undergraduate programmes in Fire Engineering in the U.K. (The others being at South Bank University and at Leeds University). The University's BEng(Hons) Fire Engineering course, in the Department of Built Environment, is sponsored by the Institution of Fire Engineers and by Cape Boards Ltdand is specially designed to meet the needs of industry on a national basis. As a result, it is offered as a residential, part-time, block release course with attendance on four occasions in each academic year for a period of two weeks on each visit. The department also offers the degree of BSc/BSc (Hons) in Fire Engineering Management. The first two years of each of these two undergraduate programmes is taught in common - the programme bifurcates at the beginning of the third year and each leads to a separate degree award. These courses provide a natural progression for students who have successfully completed HNC studies or who have an equivalent level of qualification.

6 Supporting research

Significant consultancy and research is necessary to support a suite of courses in fire-related studies. The Department has several current projects in fire related areas: e.g. Prediction and modelling of smoke movement in buildings; Behaviour of concrete when subject to high temperatures; Modelling escape behaviour in fire situations; Comparative Studies of European Fire Legislation; Investigation of two phase flows accompanying fires in enclosures; Modelling of combustion and other effects following the release of fuel into the atmosphere; Study of combustion and behaviour of particulate clouds. Two recent new appointments to the department, as members of staff, will research the behaviour of fires in contained spaces and the relative merits of CFD systems. On-going links remain from three major, privately funded contracts in fire related areas, two of them with British Aerospace on aspects of its aircraft and defence industry activity and the third with Cheshire County Fire Brigade on the effects of gas container explosion. The Department is presently providing short courses for BAe on fire safety.

The industrially sponsored degree offered by the University has led to several other fire-related organisations expressing interest in ways and means of contributing to the success of the University's fire studies programmes. The Association of British Insurers, Warrington Fire Research Centre, Ove Arup, British Aerospace, BNFL, Cheshire, Greater Manchester and Lancashire County Fire Brigades are all supporting the department with offers of student visits, visiting lectures, materials for laboratory tests, advice on course content and structure, etc.

7 Post-graduate taught courses

Success in the provision of undergraduate courses has led to their extension into the post-graduate area. A new MSc./PgDip. in Fire Safety and Risk Management is now available in the department and is provided on the same part-time block release basis.

This course is of particular relevance to those in the insurance sector or to those involved in building control and the implementation of Fire Regulations. A short module of Live Fire Studies is also available and is compulsory for any fire-inexperienced student; all students participate in Research Methods Seminars and must produce a major, research based Dissertation.

In addition to fire studies, it is the intention that, through the medium of Design Economics and Risk Assessment and Management, students should assess the inter-action of areas, appreciate the relevant involvement and contribution of related professions and be capable of formulating a considered response to proposed engineered solutions to fire safety, thereby providing an acceptable level of management of fire risk and loss prevention.

Work is currently proceeding on a MSc./PgDip. in Fire Safety Facilities Management and a Modular Integrated Graduate Development Programme of block release study leading to the award of MSc. has been submitted for approval. The unit availability on the proposed IGDS Scheme is shown in *Table 1*.

8 International activity

The University of Central Lancashire participates in a programme of staff/student exchanges and research/knowledge-transfer networks throughout Europe and is a member of several undergraduate ERASMUS programmes. The broader post-graduate area of Fire Safety and Risk Management forms part of the current programme of European co-operation co-ordinated by the Department of Built Environment of the University of Central Lancashire. A substantial part of that programme area is concerned with facilitating the European goal of harmonisation of Health and Safety standards through increased integration and better exchanges of information, this conference being a demonstration of that. Harmonisation of legislation and codes of practice concerning fire safety, insurance of risks and improvements in health and safety standards will be achieved only by educating future legislators, lawyers and practitioners to view, and adapt, fire and safety legislation, and its associated policing, in the light of the state of engineering knowledge and practice.

The harmonisation of social and cultural "welfare" facilities throughout the international community will be considerably enhanced by exchange of knowledge of economic and legal aspects of fire safety, fire precautions and the various management techniques and policies for coping with emergency planning. Given the cost of fire, in terms of life and property, the impact of the proposed collaborations could be significant.

Safety legislation, especially that relating to built environment, is very specific and specialised. For an understanding of its impacts, and of those areas where changes are possible, it is necessary to have an understanding of the fundamental principles of fire engineering and building construction sciences as well as having a humanitarian view of necessary regulation. The University of Central Lancashire has had several unique experiences in this respect. It has become, at the request of the British Government and the U.K. fire community, the largest provider of education and training in fire engineering, fire legislation and fire safety/risk management, at professional engineer level, in the U.K. Also, its staff have just completed a review of the internationally harmonised fire and safety regulations used in Seville, Spain by the EXPO'92 Authority as an attempt to secure a common standard of safety performance across the 85 varying construction styles and 110 cultural attitudes represented at that exhibition. Much of the material collected during this review will be available for use by students in the form of case studies.

9 European aspects

The links proposed and developed by the University of Central Lancashire, and a number of institutions across the EC and beyond, will improve the exchange of information on the most efficient national methods of implementing and monitoring the European states' applications of relevant legislation and regulations and on the application of fire studies to safety, hygiene and health at work. These links will improve and promote the dissemination of information on occupational fire hazards and their prevention. They will also provide small and medium size enterprises with needed and usable information on the rights and responsibilities of employers, workers and the general public - as well as providing architects and engineers with a sound base from which to suggest engineered design solutions to fire protection problems, and surveyors, inspectors and underwriters with a good understanding of offered engineered solutions.

Negotiations over the last 12-18 months have laid the ground for a Euro-wide network of Universities and other teaching/research establishments to participate in a Masters-level programme of study which will facilitate student exchanges and the provision of study blocks at differing European locations.

10 A research network

The network should also be used to generate transnational research co-operation and knowledge transfer amongst, predominantly, EC member states. However, there is much interest from states outside the EC and, therefore, negotiations are currently taking place with several of the G24 states and states of the former Soviet Union and Eastern Bloc with a view to widening the network and creating even more interest in Health and Safety standards, Fire Engineering and Fire Safety & Risk Management throughout "greater Europe".

Countries participating in the post-graduate education and research programme, or having expressed an interest in participation, are:-

Eire	University College, Dublin	**Switzerland**	E.T.H., Zurich
Portugal	Universidade do Porto	**Austria**	Technische Universitat Wien
Spain	Universidad Polytechnico de Madrid	**France**	Universitie de Rouen
Germany	Technische Universitat Karlsruhe	**Denmark**	Technical University of Denmark

Associated with this group, but currently in a separate knowledge/assistance programme, are colleges/universities in China and Hong Kong and in the former eastern bloc states of Hungary, Russia, The Czech Republic and Slovakia. The membership of this group continues to grow.

11 Summary and conclusion

Throughout Europe, the same concerns as in the Bickerdike-Allen Report in the U.K.have been expressed. Differences in educational provision across Europe need to be addressed in a programme of comparison and analysis to determine their relative advantages and disadvantages and their effectiveness in providing a coherent basis for the implementation of the relevant EC Directives. Particularly in the U.K., fire-related studies properly emphasise the importance of saving and protecting life, but an additional and significant consideration is the economic and environmental consequence of inadequate or inappropriate fire protection or prevention.

An integrated scheme of postgraduate study should be targeted, therefore, at those who now have substantial responsibilities in the field of health and safety at work and, as a result, face particular problems. The rapidity of innovation in both building design and technology and fire engineering combined with the current relatively unstructured nature of education in the same area meant that an important group has had little access to the latest information.

The emergent European network is able to provide, in a variety of ways, for the specialist postgraduate education necessary to develop, in small and medium architectural design and consultancy practices, expertise in design for fire safety in buildings. Once the European network is fully established it should be self-sustaining and the courses should be available on an ongoing basis, from 1995; Thus moving towards the time when postgraduate study of fire-related matters will be possible in a

Two (minimum) Key Modules from:

RISK ASSESSMENT AND MANAGEMENT (CN4106)	STRATEGIC FACILITIES MANAGEMENT (CN4207)
DESIGN ECONOMICS (CN4205)	STRATEGIC MANAGEMENT (MD4020)

Remaining Modules from:

BUILDING MATERIALS, STRUCTURES AND SERVICES (CN2201)	MANAGEMENT THEORY AND PRACTICE (MD3020)
FIRE PROTECTION I (CN2202)	PUBLIC SECTOR MANAGEMENT (MD3005)
FIRE PROTECTION II (CN3201)	FINANCIAL MANAGEMENT (AC3702)
FIRES IN BUILDINGS (CN3203)	ADMINISTRATIVE AND ENVIRONMENTAL LAW (LA37XX)
FIRE SAFETY (CN3920)	EMERGENCY PLANNING (CN3205)

FIRE LEGISLATION (CN2231) (compulsory)

Live Fire Studies (compulsory for fire-inexperienced students)

Table 1. **Integrated Graduate Development Scheme (MSc.) - subjects available (w.e.f. 1995).**

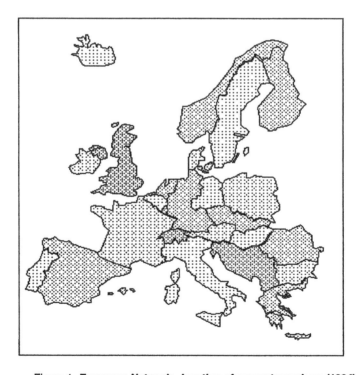

Figure 1. **European Network - location of present members (1994)**

variety of Higher Education Institutions spread across Europe. Then, students will be able to take advantage of the possibility of studying separate parts of their MSc./PgDip. course at differing locations in a range of states during their block attendances.

References

[1] Bickerdike Allen Partners. (1990) *Fire and Building Regulation - A Review.* (London, Her Majesty's Stationery Office)

[2] Dupuis, M. le Prefect. (1990) *Rapport sur la Formation des Sapeurs-Pompiers.* (Paris, Ministre de l'Interieur)

[3] Anon. (Department of the Environment and The Welsh Office) (1991) *The Building Regulations 1991 - Approved Document B - Fire Safety.* (London, Her Majesty's Stationery Office)

[4] per example: (1989) *European Directives 89/391/EEC and 89/654/EEC.* (Brussels, The European Commission)

40 THE DEVELOPMENT OF FIRE ENGINEERING DEGREE COURSES

D.A. EVANS
Department of Built Environment,
University of Central Lancashire, UK

Introduction

I was given the task four and a half years ago to develop a suite of programmes on Fire Engineering in conjunction with the Institution of Fire Engineers and that is my subject today.

Fire Engineering Degree Development

Mr Brian Collins, Head of the Fire Service Inspectorate at the United Kingdom Home Office, has already sketched a broad background to what has already been discussed and what led to the decision to launch the Fire Engineering Degree in the first place. Within the United Kingdom, the South Bank University was the first university in Britain to offer part-time education, or any education, specifically for Fire Engineering, and I am very pleased and proud to say that the University of Central Lancashire followed very quickly indeed. I believe that with the influence of the Institution of Fire Engineers who supported our courses, we now have the only course in the United Kingdom sponsored by a professional body. When people ask me, "How did you do it? I can assure you its has sometimes seemed like being in the centre of a whirlwind and I wonder myself. My answer must be by sheer hard work and a true commitment to the cause of Fire Engineering. The degree course in Fire Engineering was put together with assistance from three local brigades Lancashire, Cheshire and Greater Manchester. It was their help and co-operation that made it all possible. We sat down for what were a series of whirl-wind meetings and very a rapid development took place.

This year sees the first cohort of Honours Graduates. I feel that it has been very rewarding for me personally to be involved with this initiative. It is not often you find very keen students these days to the extent that they will help you drive the course along, which is very useful indeed. We have applied for Chartered Engineer status for the BEng course and we expect an accreditation visit in July, by the Institute of Energy, which is a representative member of the UK Engineering Council. This year will see the University of Central Lancashire together with Leeds University and hopefully the South Bank University obtain between them three accredited Engineering degree courses in Fire and Fire Safety a first for the United Kingdom. The BEng (Hons) Fire Engineering degree course has been accredited by the Association of Building Engineers and graduates will be eligible for professional membership.

On the down side the numbers of employers sponsoring the part time route are not very large, so I must make this plea to Fire Brigades and other employers within the United Kingdom at this moment in time South Bank and ourselves have less than

Fire Engineering and Emergency Planning. Edited by R. Barham.
Published in 1996 by E & FN Spon. ISBN 0 419 20180 7.

twenty students on the First Year of our Honours Degree Programmes. It is a situation which must improve if the future of the course is to be guaranteed.

On a more positive note, what we have achieved is a series of courses which deal with both full time and part-time students from pre-degree right up to doctoral and post-doctoral research work. This was the pattern that we were asked to provide and the origins go back to before any of these courses were in place. It originated from work sponsored by the Institution of Fire Engineers via a firm of consultants who were engaged to find out what was happening in education for the fire service, and what was actually needed.

You can see quite clearly from the chart {Fig 1} we started by offering two part-time degree courses BEng (Hons) Fire Engineering and BSc (Hons) Fire Engineering Management, this was in September 1991. Following the commencement of the degree programme discussion took place with the British Engineering Council. Their recommendations were, that a clear progression route must be available to provide for Technician and Incorporated Engineer grades together with a recognised B.Eng. degree program before chartered engineering status could be considered.
This led to the development of the University Advanced Certificate, HNC in Fire Safety Engineering and HNC in Fire Safety Engineering (IEng).

Non Degree Courses
Having first been given the assignment of developing a degree course in Fire Engineering I set myself the task of producing a series of course notes together with a series of syllabi based on the National Core Curriculum adhering to their same headings. The National Core Curriculum was developed for use by a wide range of audiences and as such I felt it was the ideal starting point at which to develop courses for Incorporated Engineers.

The four basic modules of the National Core Curriculum are now incorporated in the University Advanced Certificate. This course attracts a reasonable amount of interest and as educationalist not only have we to offer courses, we must make sure that they are viable economically because we still have to balance the books at the end of the day. We wanted try and open Fire Education to the largest audience possible. So if a potential student with an interest in fire related topics who has no qualifications to speak of, or with qualification in some other discipline then what better but to offer a University Advanced Certificate in Fire Safety Engineering as an introduction. The next stage along this same line of thought was to link this in with the MIFireE, a professional qualification and perhaps encourage people to become more aware of the Institution of Fire Engineers. Traditionally the department that I represent is in essence a Building Department from a tradition which spans many years and as such I felt confident that as a department we could build Fire Safety Engineering into our existing modules, especially as most fires occur in buildings of one form or another.

After the Advanced Certificate we produced a part-time BTEC Higher National Certificate in Fire Safety Engineering which essentially required an additional four modules, long making it eight modules in all.

Our next task was to find a means of getting external recognition for these courses throughout the United Kingdom and Europe.

It was with this in mind that we decided to add extra modules which contained pure and applied mathematics, engineering science and physics and we produced the BTEC Higher National Certificate Incorporated Engineer (IEng) route which is not only accredited by BTEC but also the Institution of Fire Engineers and the Institute of Energy. This latter body is a member of the Engineering Council of the United Kingdom. With the completion of this route it is possible for any student on this course to step out at the end of the first year with a University Advanced Certificate or carry on and obtain the IEng by completing the two year part time course.

Franchising
We have developed a series of franchising contracts, the HNC IEng route, the BEng (Hons)Fire Engineering and the BSc (Hons)Fire Engineering Management to the City University of Hong Kong. The first year review to vet the course was carried out by our Dean, who chaired the meeting with two other academics to see that everything was in accordance with the University quality assurance systems. The franchise was given a clean bill of health which is not surprising for the people in Hong Kong are extremely dedicated to fire education. I was there a month ago exactly there are 150 applicants for 50 places. The course hasn't been advertised in the public media yet. I must repeat, in the United Kingdom Clive Steele, course leader at South Bank and myself, between us have less than twenty students. I just hope that it is a position that's not going to continue.

Full-Time Fire Degrees
Within our department we have great experience with French nationals who undertake a range of full time diploma courses in France. These students, via various networks, have gained access onto UK degrees in Building Management and Building Services. We were invited and joined a French network that, has been very fortunate for us because it offered students who complete a particular route in fire safety, risk, health and hygiene a route onto the fire degrees. When we assessed the French students who had completed the Fire Safety route we took the decision to accredit them with two years full time study and we put them onto a one year modular degree programme. The BSc (Hons) Fire Safety is a modular degree programme and was developed in response to the economic climate within higher education. This degree commenced in September 1993 and last year two French students each graduated with 2.1 Honours degree. The course was modified to allow entry to students from other countries and other nationalities and from other course routes. Additionally we have developed a series of core modules based on the National Core Curriculum adapted from the original modules of the Advanced Certificate and introduced them onto the full-time Higher National Diploma Course in Building. By requiring additional modules in mathematics, thermofluids and a fire related engineering project we were able to place these students onto the final year of the BSc (Hons) Fire Safety. Currently we have four French nationals, one Hungarian national and three UK nationals taking this particular degree.

Post Graduate Studies
That brings me to the upper echelons of {Fig 1}. Working down from the top, I am very pleased to be able to say that this area of specialised study (i.e. post graduate research) is strictly under the control and guidance of Professor Georgy Makhviladze, who is head of the Centre for Research in Fire and Explosions at the University of Central Lancashire. The Centre is responsible for the post graduates and research into fire studies. Progression from BSc and BEng is, naturally, onto post graduate studies, and we have, therefore, provided a taught route in addition to the research route.

Currently within the department we have a Masters degree in Fire Safety and Risk Management. It is interesting to see how students coming from the BEng and BSc (Hons) Fire Engineering route have asked us to look at a whole range of topics which they can develop. The current engineering students are enquiring from me, post graduate work and the management students are enquiring from me post graduate management studies. I am currently looking at how we can incorporate extra modules for these students. Fire Safety and Risk Management is one possible route.

Future Developments

There are two other developments in the pipeline particularly in the management area, where they will be linked more to nationally and internationally recognised management qualifications such as law, accountancy, personnel management. Those are some of the future development that we have planned, to develop the courses and give ourselves more expertise. Within the department we have, degrees in Building Management, Building Surveying, Quantity Surveying and Service Engineering. Into all these courses we have incorporated different elements of fire. Obviously the Service Engineers want to know about designing for fire. I've been talking to some colleagues in the hall and we were saying about Quantity Surveying, I take the view that if you want to be a Quantity Surveyor you need to know about fire safety as much as anybody else, we all have a role and a duty of care. So within the Building Management, and the other routes, we do have a considerable amount of fire and fire related topics, this enables us to flavour the Department. That broadly, is an outline of what we are currently doing at the University of Central Lancashire, Preston.

Finally in order to give you an idea of the course details, the chart {Fig 2} provides an overview. It starts with the National Core Curriculum, HNC this gives you an exemption from the written papers of Corporate Membership of the Institution of Fire Engineers. So you have either HNC, MIFireE or its equivalent to gain entry onto the degree and that is why we follow a four year part-time degree programme. Interestingly enough, people have advised me that I will have great difficulty teaching physics, chemistry, mathematics to the clientele. I can assure you that is not the picture that I am looking at. What we have found, in fact, is a series of very keen and dedicated students. And, as I said previously, it has been very rewarding for me as an educator to have been involved with these students as a result I have tried to make the courses as dynamic as possible and responded to change and students requests. The first two years are common, at the end of the second year students make a decision whether they want to go Fire Engineering or Fire Engineering Management. Both courses allow for completion and step out at the third year where students will if they are eligible be awarded an unclassified degree. Progression past that stage students are then onto the Honours programme. That again has been very rewarding - the types of projects and topics that we have put together, or students have brought back employer initiated topics, or, "we've got a problem with that ...... is it feasible for us to look at ....?", this has been very useful indeed.

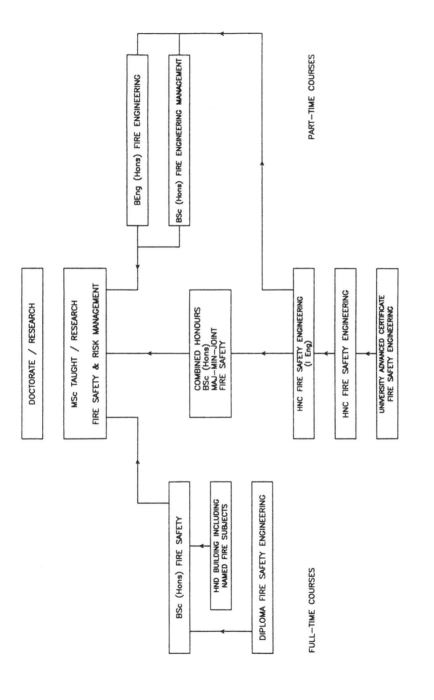

FIRE COURSE PROGRESSION

FIG. 1

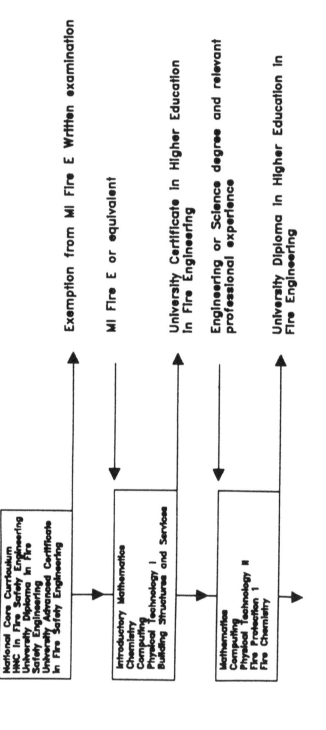

FIG. 2a

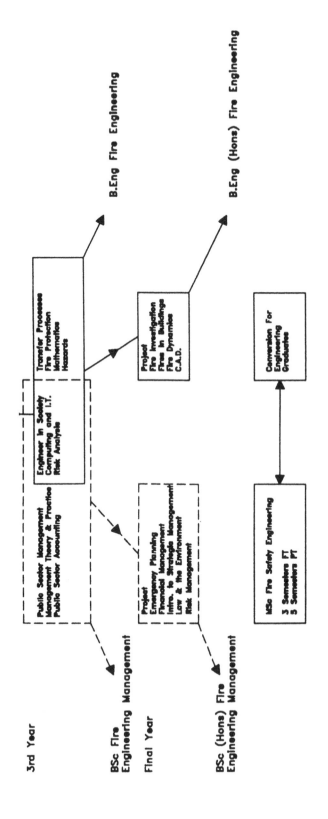

3rd Year

Public Sector Management
Management Theory & Practice
Public Sector Accounting

Engineer in Society
Computing and I.T.
Risk Analysis

Transfer Processes
Fire Protection
Mathematics
Hazards

B.Eng Fire Engineering

BSc Fire
Engineering Management

Final Year

Project
Emergency Planning
Financial Management
Intro. to Strategic Management
Law & the Environment
Risk Management

Project
Fire Investigation
Fires in Buildings
Fire Dynamics
C.A.D.

B.Eng (Hons) Fire Engineering

BSc (Hons) Fire
Engineering Management

Conversion For
Engineering
Graduates

MSc Fire Safety Engineering

3 Semesters FT
5 Semesters PT

OUTLINE COURSE STRUCTURE AND CONTENT

FIG. 2b

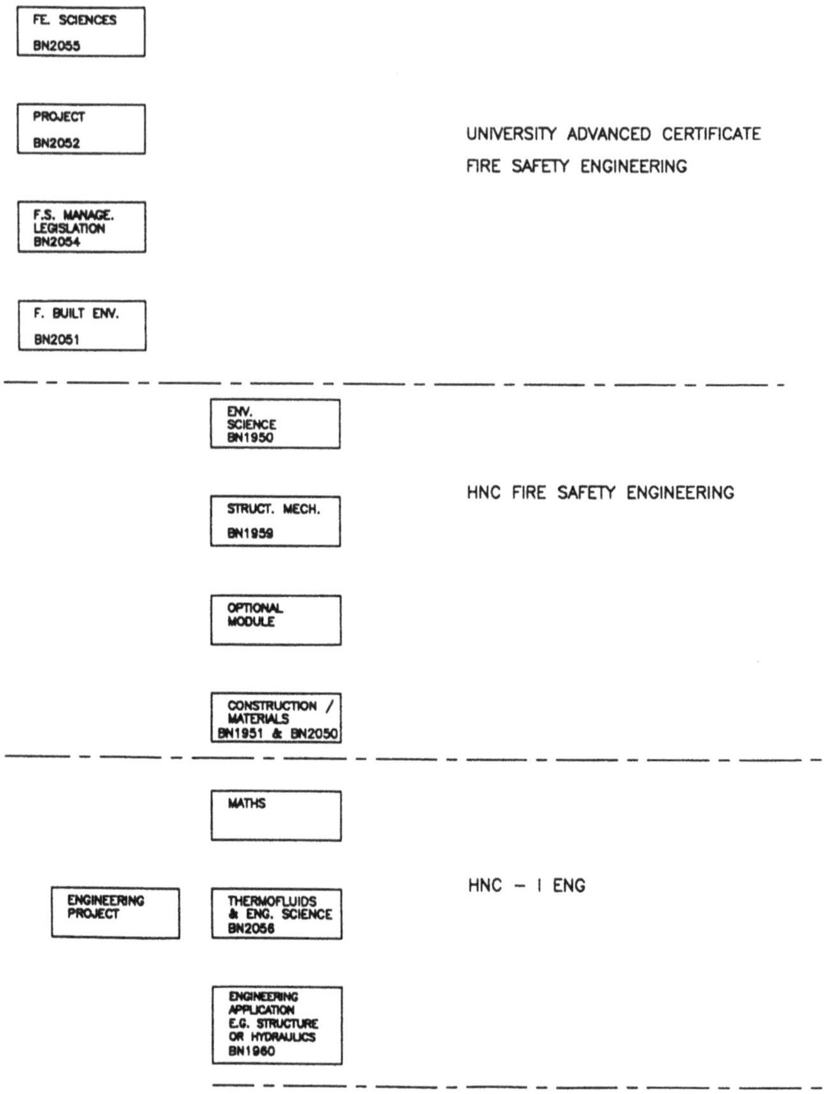

FE. SCIENCES
BN2055

PROJECT
BN2052

F.S. MANAGE.
LEGISLATION
BN2054

F. BUILT ENV.
BN2051

UNIVERSITY ADVANCED CERTIFICATE

FIRE SAFETY ENGINEERING

ENV.
SCIENCE
BN1950

STRUCT. MECH.
BN1959

OPTIONAL
MODULE

CONSTRUCTION /
MATERIALS
BN1951 & BN2050

HNC FIRE SAFETY ENGINEERING

MATHS

ENGINEERING
PROJECT

THERMOFLUIDS
& ENG. SCIENCE
BN2056

ENGINEERING
APPLICATION
E.G. STRUCTURE
OR HYDRAULICS
BN1960

HNC – I ENG

FIG. 3

41 THE IMPLICATIONS OF UK NATIONAL VOCATIONAL QUALIFICATIONS FOR TRAINING AND EDUCATION IN FIRE TECHNOLOGY

B.C. MOTT
Department of Product Design and Manufacture,
Bournemouth University, UK

Abstract: An introduction to the UK National Vocational Qualification (NVQ) developments and a current positional statement of the progress in respect of the UK Construction Industry and for Fire Technology in particular.
Keywords: rules; priorities; standards; lessons; holistic; providers; assessment; resources; current position.

1 Introduction

This paper is written for the benefit of Education and Training providers who will be effected by the introduction of NVQs (to be interpretted to include SCOTVEC - Scottish Vocational Education Council) in the UK construction industry. For a more general background to the subject the construction industry standing conference has produced a paper entitled "Raising standards, raising performance".

Much of the current education and training provision is of high quality, developed over many years. CISC(construction industry standing conference) has not sought to set this aside and assume greenfield conditions, but rather to use the power and value of the NVQ approach to enrich what we already have. Using CISC occupational standards we can make the vital link between the acquisition of knowledge and the achievement of competent performance. I am aware that CISC does not fully cover the Fire Engineering/Technology field completely but CISC are my source for the philosophy of and the development mechanisms for NVQs. Later in this document you will find a position statement for the fire industry.

2 Rules of the game

The fundamental elements of vocational education and training can be related as shown in fig 1.
* Occupational standards reflect competence in the workplace.
* Standards have a power and value for employers independent of NVQs.
* Standards will have a significant effect upon learning provision.
* The link between assessment and learning has been largely unexplored.

The development of occupational standards must reflect the four aspects of the job competence model shown in fig 2. There is a perception that the "reductionist" methodology of standards development will lead to a ragbag of discrete competences whose whole will be less than the sum of its parts. CISC believes that if the **holistic** power of the job competence model is properly harnessed, NVQs will be more demanding and more attractive than many existing vocational qualifications.

3 CISC's work to date

CISC has worked with the construction industry training board and other lead bodies to enable a strategic approach to the five-level framework, figs 3 & 4. NVQ development follows the steps shown in fig 5 and CISC progress and programme is shown in fig 6. When the standards are in place, selections for particular qualifications (or "templates") are made by appropriate interest groups fig 7.

Fire Engineering and Emergency Planning. Edited by R. Barham.
Published in 1996 by E & FN Spon. ISBN 0 419 20180 7.

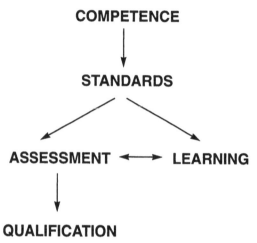

VOCATIONAL EDUCATION AND TRAINING
KEY COMPONENTS

COMPETENCE

↓

STANDARDS

ASSESSMENT ←→ **LEARNING**

↓

QUALIFICATION

Fig. 1.

LEVEL 1 COMPETENCE IN THE PERFORMANCE OF A RANGE OF VARIED WORK ACTIVITIES, MOST OF WHICH MAY BE ROUTINE AND PREDICTABLE.

LEVEL 2 COMPETENCE IN A SIGNIFICANT RANGE OF VARIED WORK ACTIVITIES, PERFORMED IN A VARIETY OF CONTEXTS. SOME OF THE ACTIVITIES ARE COMPLEX OR NON-ROUTINE AND THERE IS SOME INDIVIDUAL RESPONSIBILITY OR AUTONOMY. COLLABORATION WITH OTHERS, PERHAPS THROUGH MEMBERSHIP OF A WORK GROUP OR TEAM, MAY OFTEN BE A REQUIREMENT.

LEVEL 3 COMPETENCE IN A BROAD RANGE OF VARIED WORK ACTIVITIES PERFORMED IN A WIDE VARIETY OF CONTEXTS AND MOST OF WHICH ARE COMPLEX AND NON-ROUTINE. THERE IS A CONSIDERABLE RESPONSIBILITY AND AUTONOMY, AND CONTROL OR GUIDANCE OF OTHERS IS OFTEN REQUIRED.

LEVEL 4 COMPETENCE IN A BROAD RANGE OF COMPLEX, TECHNICAL OR PROFESSIONAL WORK ACTIVITIES PERFORMED IN A WIDE VARIETY OF CONTEXTS AND WITH A SUBSTANTIAL DEGREE OF PERSONAL RESPONSIBILITY AND AUTONOMY. RESPONSIBILITY FOR THE WORK OF OTHERS AND THE ALLOCATION OF RESOURCES IS OFTEN PRESENT.

LEVEL 5 COMPETENCE WHICH INVOLVES THE APPLICATION OF A SIGNIFICANT RANGE OF FUNDAMENTAL PRINCIPLES AND COMPLEX TECHNIQUES ACROSS A WIDE AND OFTEN UNPREDICTABLE VARIETY OF CONTEXTS. VERY SUBSTANTIAL PERSONAL AUTONOMY AND OFTEN SIGNIFICANT RESPONSIBILITY FOR THE WORK OF OTHERS AND FOR THE ALLOCATION OF SUBSTANTIAL RESOURCES FEATURE STRONGLY, AS DO PERSONAL ACCOUNTABILITIES FOR ANALYSIS AND DIAGNOSIS, DESIGN, PLANNING, EXECUTION AND EVALUATION.

NVQ/SVQ FRAMEWORK
LEVEL DESCRIPTIONS

Fig. 3.

COMPONENTS OF COMPETENCE

- **Technical or Task skills:** those specific skills and knowledge which enable the job holder to deliver the key purpose or outcome of the role.

- **Contingency Management:** skills and knowledge needed to manage variance and unpredictability in the job role and the wider environment.

- **Task Management:** skills and knowledge which are 'overarching' and which integrate the various technical and task components into the overall work role.

- **Role or Job Environment:** skills and knowledge which are used to integrate the work role within the context of the wider organisational, economic, market and social environment.

Fig. 2.

A FRAMEWORK FOR CONSTRUCTION

NVQ/SVQ LEVEL	CISC		CITB
	Areas of M & E services and maintenance are represented by HVCA, EIEITO, BPEC, RIB, IETA, IMBM and LGMB, who are all Lead Bodies represented on CISC		
	PROFESSIONAL, MANAGERIAL, TECHNICAL		CRAFT, OPERATIVE
5	'CHARTERED' EQUIVALENT		
4	'INCORPORATED' EQUIVALENT		
3	'TECHNICIAN' EQUIVALENT		CRAFT + SUPERVISION
2			TRADE
1			INTRODUCTORY

Notes
1. Levels will depend on particular occupations, and are illustrative only. The clear relationship between existing vocational qualifications and NVQs/SVQs has yet to be established.
2. A key feature of this framework is the 'bridge' at level 3 between the technician and craft occupations.

Fig. 4

STAGES IN NVQ/SVQ DEVELOPMENT

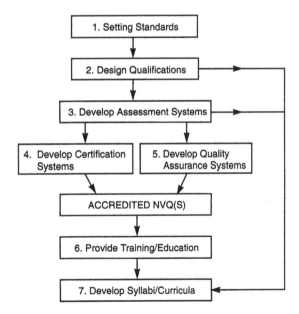

Fig. 5

Published guidance states that "an NVQ is equivalent to the concept of an occupation", and early NVQs at levels 1-3 typically contained 8-10 units. Some of CISC's higher-level templates, however, were as large as 50-60 units, and some way of subdividing this number was necessary, fig 8. This sub-division allowed the early draft of CISC's overall framework, fig 9, which suggests "families" of NVQs. Possible strategies for planning these families is shown in fig 10, and fig 11 suggests what a typical NVQ portfolio for a senior practitioner might look like.

On examination the relationship between NVQs and existing qualifications suggests that much NVQ activity will be in "post qualified" mode. This opens up possibilities for continuing professional development and assessment of "professional competence" figs 13 & 14.

When qualifications have been designed, each standard is supported by a specification of the evidence which the candidate will need to produce. Fig 15 shows how this is derived.

It is now possible to "caricature" the aspects of NVQs as they change from level 1 to 5, fig 16. It can be seen that the public perception of NVQs is close to the level 1 model; the reality at level 5 is a qualification of great richness and power, which will add value to our occupational sector.

4   The priority areas

The work being done does not seek to threaten existing good practice, therfore the most fruitful starting points are those areas where:
*   there are currently no qualifications
*   there are interest groups keen to push the work forward
The first to be covered was the Building Site Management and Supervision (by CIOB/C&G) and Building Maintenance management and supervision (by IMBM/C&G). NVQs at levels 4 and 3 were accredited in 1993 and are now being delivered. An example of one of the CIOB standards is shown in fig 17. Fig 18 shows future trials.

5   Putting standards to work

All those involved in NVQ development are on a grand learning curve. Fig 19 shows a standard developed two years after fig 17 and the refinements are visible. We can now see what evidence the candidate can present, and how this evidence relates in detail to the standards. A single piece of evidence may satisfy many performance criteria and aspects of range across several elements. When a full NVQ has been designed, the **critical** evidence can be determined and the candidate's portfolio kept to a minimum; this will keep assessment costs down.

Standards have many uses in the workplace. Fig 20 identifies 78 uses for standards, only four of which are linked to NVQs. Educators and trainers will therefore be able to put CISC's 1994 standards to work without waiting for NVQs to be accredited.

6   Lessons so far

From CISC's experience to date it is essential to appreciate that:
*   Continuous and wide-ranging consultation has been essential, both to achieve a quality product, and also to create an ownership throughout the industry for CISC's work.
*   Progress will not be made instantly across a wide front, but by targeted pilot operations, whose success will allow a broader advance to follow.
     True added value will be achieved through partnership. This is taking many forms:- partnership between employer bodies and professional institutions;

between employers and academe; between educators and trainers; between lead bodies and awarding bodies; and between those developing NVQs at lower and higher levels. All members of the occupational sector have their part to play; one of CISC's central tasks is to initiate and sustain dialogue and debate, in order to enable its aims and objectives to be met.

7 Implications for the education and training community

The setting of the standards is necessarily **"employment led"**; but educators and trainers, together with existing awarding bodies, have an essential contribution to make in ensuring that resulting qualifications can work in practice. They are and have been keen to be involved in evidence specification and in the designing of assessment systems. CISC has established a national E&T network so that all partners in the development process can be involved in the search for a **least cost, fit-for-purpose** qualification framework. The standards will determine a significant proportion of vocational education provision; but CISC recognises that **academe** has a mission wider than **"competence"**, and does not seek to restrict this. The successful 'formation' of construction professionals depends upon the right **balance** between the **academic** and the **vocational:** CISC needs to form a partnership with higher and further education in order to achieve this.

8 New demands - new responses

The nature of NVQs will cause employers and individuals to place new demands on E&T providers. What might the implications of these demands be?

First, the distinction between learning provision and assessment must be made. F/HE will be in the business of supplying both; and although linked, they are separate activities.

Second, unit-based qualifications will mean that not all students will need full-time courses begining in the Autumn. This will stimulate E&T providers to deliver unit-based learning modules, on demand year round. (Some see this as an opportunity, others as an unrealistic target). Flexibility of provision will be a feature of the successful learning centre.

Third, NVQ units common across construction disciplines will encourage closer collaboration between college faculties in order to provide joint courses.

Fourth, government policy will stimulate courses leading to NVQs, in order to secure outcome-related funding from TECs/LECs.

Finally, the **holistic** nature of NVQs will lead to the need for greater underpinning knowledge of management and finance. If this has to be at the expense of technical input, then the latter may increasingly move into the area of CPD (continuous professional development).

9 What's in it for E&T providers

As we know, underpinning knowledge for NVQs can be achieved by a variety of means. This includes full or part time education, open and distance learning, accreditation of prior experience, etc. It is also becoming clear that qualifications hitherto regarded as "final" (e.g. professional institution membership) will in future be a prelude to a career-long programme of continuing professional development. Since professional status is generally acquired at the age of 25-30, CPD will cover the majority of an individuals working life, and NVQs can play a major part in providing a badly-needed structure for CPD provision and monitoring.

Employers will be reluctant to release staff for long periods for CPD purposes, but the use of open learning techniques would enable candidates to acquire necessary underpinning knowledge towards NVQ units. (In parallel, they would combine this with performance evidence from the workplace). The open learning arrangement would be a tripartite partnership in which candidates gave their time, employers paid for the courses, and the F/HE and training institutions provided the facilities. It would give the open learning industry a great boost, and provide commercial opportunities for the educators and trainers. It would also bring the education, training and employer communities closer together.

10 What's in it for E&T assessment

Assessment is the process of judging evidence presented by candidates as to whether or not they have met the appropriate standards. The analysis of this evidence is shown in fig 15, and we can make the following observations:-
*   Evidence of all kinds is generated both in the formal learning centre and in the workplace.
*   The learning centre will tend to provide the candidate more with knowledge evidence, and the workplace will tend to provide the candidate more with performance evidence.
With regard to assessment, the learning centre staff will be best placed to assess knowledge evidence and employers' trainers and managers will be best placed to assess performance evidence. Employers, will need assessor training in order to provide the necessary competence. Since employers and candidates will be paying for the assessment, the question to be answered is: "How can assessors in the market place provide a service which gives their clients added value and themselves business opportunites?".

11 Summary

To date there is little evidence of NVQ implementation at the higher levels, but the potential benefits of NVQs for the E&T community may be summarised as follows:-
Benifits for F/HE:
*   Occupational standards specify the necessary knowledge evidence, and this will help to plan learning provision.
*   Potential access to a bigger market:
     generalists and specialists
     the part-qualified and the unqualified
*   Need for generic/cross-disciplinary competences:
     more opportunity for joint and collaborative work
     more viable student numbers
*   Opportunities to be 'first in the field' to help meet identified needs, e.g. management and finance, which have previously not been widely available.
*   Potential growth in demand for structured CPD/staff development - NVQ units.
*   Potential growth in open and distance learning, which employers will prefer for their employees' CPD/staff development.
*   Closer partnership and coherence between F/HE, employers and professional institutions - a culteral shift.
*   The modular nature of NVQs enables an academic course to underpin several NVQs, because many units will be shared.
*   Education and training is more likely to be seen as an investment rather than a cost.
*   Opportunities for delivery of unit-based modules on demand in short courses.
Benefits for training providers:
*   Blueprint for specifying the competence needs for firms and individuals.
*   A format for individual development plans and staff appraisal.
*   A complement to the investors in people initiative.
*   Linking training to business objectives.
*   Targeting training where it is needed, thus eliminating training waste.

* The assessment process itself accelerates the candidates learning/motivation.
* The assessment process itself raises the performance of the line manager/assessor.
* A flexible balance between open/distance learning and on the job training.
* Encourages wide skills to suit a multi-faceted employer.
* Possible eligibility for output-related funding from TECs/LECs.

The question for the E&T community would therefore seem to be:-
* How can we contribute to CISC's work to ensure a practical, high quality, cost effective product which will add value to industry?
* How can we exploit the implementation of NVQs to our commercial advantage?
The answers to these questions will mark the achievement of the objectives of CISC's E&T forum, and CISC hopes that delegates will play an active part in providing them.

12 Conclusion

Many of the problems faced by CISC and the E&T sector are the result of ignorance and misunderstanding. For example, NCVQ (national council for vocational qualifications) are perceived as treating knowledge as a side issue. Education (sometimes) treats competence as a side issue. One of the key tasks of the forum is to break down this polarisation, so that the resulting NVQs combine the best of both aspects. CISC acknowledges that much excellence already exists in the E&T sector. The aim is to build on this excellence, and produce a VQ framework that is truly **"world class"**.

13 Fire industry current position

There are three lead bodies which have developed and are developing further NVQs for the industry. These are the Fire industry lead body (FILB), the Security industry lead body (SILB) and the Emergency fire services lead body (EFSLB). To date they have developed and are currently in the process of piloting the following NVQs:
Level 2 qualifications -
    Maintaining portable fire extinguishing appliances
    Maintaining fixed fire protection systems
    Maintaining security and emergency systems
    Installing security and emergency systems
Level 3 qualifications -
    Fire safety inspection and audit
Level 4 qualifications -
    Providing fire safety advice and guidance
    Fire safety systems and design
The FILB have produced an implementation action plan for April 94 to June 95 as follows:
* Complete the development of the level 2 qualifications by the end of June 94.
* Establish assessment centres by the end of June 94.
* Undertake field testing of all five qualifications from July to December 94
* Review the results of the field testing by January 95
* Submit the final qualifications to NCVQ for accreditation by January/March 95
* Launch the qualifications May/June 95

14 Personal comments/reflections

I am aware, like most in HE, of the arguements that have been widely circulating regarding these new NVQ developments. From the construction industries point of view I have been aware of the very close collaboration of both employers, professional institutes and academics working closely with the CISC team, and it is my opinion, that although much of the criticisms, for example, that which emanated from Prof Smithers last Autumn in his investigatory report comparing the UK with Europe etc, may be valid, much of it is not, some of it is a direct result of misunderstanding and ignorance. Furthermore I believe that the UK has got the opportunity to get it right. Once the piloting of the current NVQs have been completed there will be an opportunity to achieve very high quality products which will bring all the benefits envisaged.

However this must include the resouces which are necessary to implement NVQs effectively. This is probably the area where much needs to be done, the resource implications to achieve a good quality result rely heavily on cooperation between academic staff and employers and this means a massive cultural change from what is perceived as currently normal. This is going to be necessary to enable the implementation for the **best quality at the cheapest price!** Why load HE with increasing equipment and other resource costs when the resources and facilities are already in place at the place of employment.

Lastly it is not going to be an immediate implementation, but is likely to be a rather slower than anticipated implementation due to the need to educate educationalists and more impotantly the employers. However there will be a gathering momentum which will become increasingly stronger as time goes by following the initial spate of NVQs. Currently I am aware of several F/HE Colleges where they are taking advantage of the opportunity to be leaders in this field and I am aware of the progress that has been made at levels 1, 2 and 3. We shall be faced soon with GNVQs (general national vocational qualifications) as equivalent to two A level **entry for degree courses**, and no doubt future employers are going to be more interested in graduates who have proved themselves worthy of an NVQ qualification as well as having a degree qualification, such graduates may have a **distinct advantage** over others!

15 References

1. CISC document "Raising standards, raising performance"
2. CISC document "A Hitchhikers guide to NVQs" by Richard Larcombe
3. Fire industry lead body publications
4. SCOTVEC documents on "Qualifications for the fire industry"
5. Useful names and addresses:
   David Smith, FILB Secretariat, 10, Bluebell Drive, Burghfield Common, Reading, Berks. RG7 3EF
   Launa Pettigrew, Secretary, EFSLB, Home Office Fire & Emergency Planning, Room 662, Horseferry House, Dean Ryle St, London, SW1P 2AW
   NCVQ, 222, Euston Rd, London, NW1 2BZ
6. The visual figures that are not published in this paper due to lack of space will be made available at the Symposium or they can be obtained from B. C. Mott, Dept of Product Design and Manufacture, Bournemouth University, Studland House, Christchurch Road, Lansdowne, Bournemouth, Dorset, UK.

BCMott/August 94.

CISC PROGRESS AND PROGRAMME AT JANUARY 1994

	1990	1991	1992	1993	1994	1995
CISC Established	▼					
Mapping Project		▬▬▬				
Standards Development			▬▬▬			
First Working Draft Standards and National E & T Conference			▼			
Consultation on Standards				▬▬▬		
Evidence Specification				▬▬▬		
Second W.D. Standards					▼	
NVQs/SVQs: CIOB/IMBM			▬▬ ▬▬	▬ ▬ ▬	▬▬	
:2nd Tranche				▬▬ ▬▬	▬▬ ▬	
:3rd Tranche				▬▬	▬▬	▬▬
: Future Work					▬▬ ▬▬	▬ ▬▬
E & T Regional Conferences Round 1					▬▬▬	
Round 2						▬▬▬
CISC Transition to OSC						▬▬ ▬▬ ▬▬

Fig 6.

SELECTION OF UNITS ("TEMPLATING")

TO DESCRIBE AN OCCUPATION

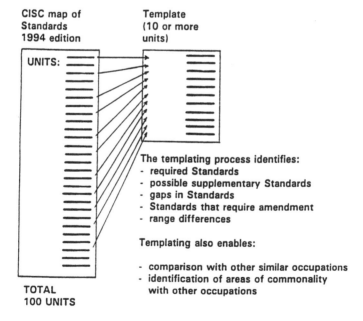

CISC map of
Standards
1994 edition

Template
(10 or more
units)

UNITS:

TOTAL
100 UNITS

The templating process identifies:
- required Standards
- possible supplementary Standards
- gaps in Standards
- Standards that require amendment
- range differences

Templating also enables:

- comparison with other similar occupations
- identification of areas of commonality
  with other occupations

The building block of the NVQ/SVQ system is the UNIT.

A higher-level NVQ/SVQ is likely to comprise 10 or more units.

Fig 7.

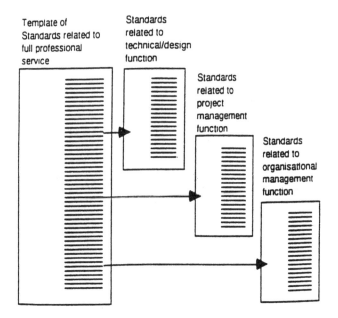

Template of Standards related to full professional service

Standards related to technical/design function

Standards related to project management function

Standards related to organisational management function

Segmentation of professional roles by discrete functions

Fig 8.

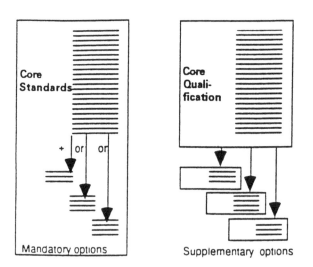

Core Standards

Core Quali-fication

+ or or

Mandatory options

Supplementary options

Strategy for dealing with minor variations between qualifications   Fig 10.

MANAGING	POLICY, PLANNING AND DESIGN	TECHNICAL PROVISION	CONTROL
ORGANISATIONS Within larger businesses Small business 1	**SETTING POLICY** Development Improvement Use 6	**MEASUREMENT AND ANALYSIS** Land Buildings Quantities Mapping Testing Research 10	**REGULATION** Development Building Health and Safety Environmental Health (housing) Statutory Grants Valuation 14
PROJECTS for Clients/Employers 2	**PLANNING** Town & Country Transportation Infrastructure 7	**TECHNICAL SUPPORT** Estimating Buying Surveying Planning 11	**INSPECTION** New Work Maintenance Services Health & safety 15
CONTRACTS & PRODUCTION Construction Installations Maintenance 3	**DESIGN** Buildings Structures Civil Engineering Transportation Services Installations Landscape 8	**RESOURCE PROVISION** Plant Materials Personnel Information 12	
PROPERTY Facilities Buildings Land 4	**MAINTENANCE PLANNING** Property Transportation Services 9	**PROPERTY EVALUATION & AGENCY** Need assessment Investment Appraisal Valuation Procurement & disposal Advice services 13	
SYSTEMS Services Transportation 5			

Fig 9.

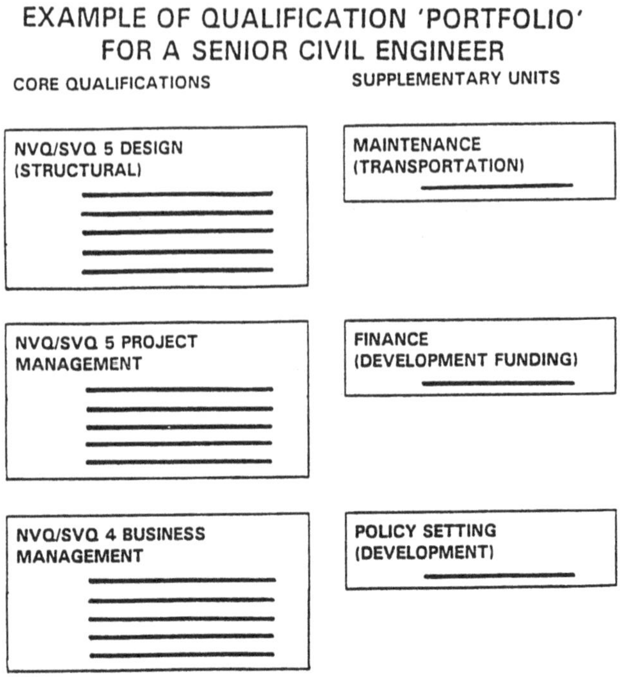

EXAMPLE OF QUALIFICATION 'PORTFOLIO' FOR A SENIOR CIVIL ENGINEER

CORE QUALIFICATIONS   SUPPLEMENTARY UNITS

NVQ/SVQ 5 DESIGN
(STRUCTURAL)

MAINTENANCE
(TRANSPORTATION)

NVQ/SVQ 5 PROJECT
MANAGEMENT

FINANCE
(DEVELOPMENT FUNDING)

NVQ/SVQ 4 BUSINESS
MANAGEMENT

POLICY SETTING
(DEVELOPMENT)

Fig 11.

HOW NVQs/SVQs MAY BE ADOPTED BY PROFESSIONAL INSTITUTIONS

STATUS	EXISTING QUALIFICATION	LEVEL	NVQ/SVQ
"Chartered"		5	
"Incorporated"		4	
"Technician"		3	

Notes

1. The clear relationship between existing qualifications and NVQs/SVQs is yet to be established.

2. ▨ ▧ ▥ equates to a professional Institution's technical requirements. These can be satisfied by equivalent NVQ/SVQ units.

3. ■ represents an Institution's "professional" requirements. These can be satisfied at interview.

4. ☐ represents further units required for full NVQs/SVQs. This will form a post - chartered CPD programme.

5. The principle demonstrated here at Level 5 will also apply at lower levels.

Fig 12.

THE PROFESSIONAL COMPETENCE MODEL

Unit F226

Provide solutions to and advice on complex, indeterminate problems within an ethical framework.

Element F226.1

Exchange information and provide advice on matters of technical concern.

Element F226.2

Identify, re-frame and generate solutions to complex, indeterminate problems.

Element F226.3

Contribute to the protection of individual and community interests.

Unit F227

Contribute to advances in the body of knowledge and practice.

Element F227.1

Contribute to advances in knowledge and theory which underpin occupational practice.

Element F227.2

Contribute to advances in occupational practice.

Element F227.3

Contribute to advances in construction-related technology.

Element F227.4

Enable others to learn and benefit from one's experience.

Fig 13./14.

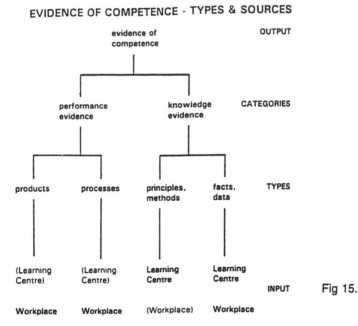

EVIDENCE OF COMPETENCE - TYPES & SOURCES

Fig 15.

	LEVEL 1 ———————————————→	LEVEL 5
STATEMENT OF COMPETENCE	Few units	Many units
	Simple pcs	Complex pcs
	Narrow range	Wide range
	One occupation = one NVQ/SVQ	One occupation = more than one NVQ/SVQ
EVIDENCE OF COMPETENCE	Emphasis on performance	Emphasis on knowledge underpinning performance
PERFORMANCE EVIDENCE	Emphasis on products	Emphasis on processes
KNOWLEDGE EVIDENCE	Emphasis on facts & data, related to discrete standards	Emphasis on theories, principles & methods, related to groups of units
SOURCES OF EVIDENCE	Few and restricted	Many and diverse
ASSESSMENT	Emphasis on workplace observation. Few assessors.	A balance of many techniques and many assessors.
	Judgement of simple activities.	Judgement of complex, combined activities.
	Limited use of inference.	Substantial use of inference.
	Little knowledge assessment.	Much knowledge assessment, carried out separately.

CHARACTERISTICS OF NVQs/SVQs AS THEY CHANGE WITH LEVEL

Note: This is a broad generalisation; the characteristics of individual NVQs/SVQs will vary from this model.

Fig 16.

Performance Criteria

i) Resource requirements are identified using all appropriate sources of information.

ii) Where resource requirements are unclear or ambiguous clarification and/or expert advice is obtained.

iii) Unusual/special/critical resource requirements are identified, recorded and appropriate action taken to confirm their availability.

iv) Where specified resources do not comply with statutory requirements and/or organisation policy relevant personnel are informed and recommendations are made.

v) Where specified type, quantity and/or application of resource is inadequate and/or inappropriate relevant personnel are informed and recommendations made.

vi) Where appropriate, alternative resources are recommended.

vii) Resource requirements and any special considerations are recorded in appropriate format and processed according to administrative requirements.

Range Statement

All resources: Plant, labour, materials, equipment

Sources of Information: Mats - BoQ and specifications
P.ant - pretender plan or plan
Sub C - contractual conditions
Equipment - contractual agreements
Direct labour - company policy and est

Resources not meeting statutory requirements: Plant and materials not complying with:-
Health and Safety legislation (Construction Regulations);
Control of Pollution legislation, Control of Substances Hazardous to Health regulations
Inform line manager, client representative, supplier, as appropriate

Inadequate resource specification: Estimated quantity required inaccurate. Type of materials specified unobtainable or unsuitable for purpose. Notified in writing to line manager, client representative or supplier as appropriate.

Supplementary Evidence Required

Knowledge of methods to:

* calculate resource levels
* resource level programmes of work
* select and analyse resource information
* present resource information

Knowledge of data/information about:

* available resource information
* availability of resources
* health, safety, welfare regulations pertaining to resource use
* security arrangements for resources on site
* contractual conditions (main and sub-contract)
* plans, programmes and methods conditioning resource selection
* statutory, contract and working role conditions of employment
* supplier conditions of contract and previous commitments
* company/contract procedures for acquiring resources

Performance Evidence Required

Resource schedules (All pc's)

Assessment Guidance

Resource schedule for all range resources for a project period of no less than one month. Schedules to be related to information sources in the range and differences between the information and the scheduled materials identified and explained.
Minimum of one case each confirming availability of unusual/special/critical range.
Two recommendations for overcoming resource specifications not meeting statutory requirements.
One recommendation for overcoming resource specifications is:

* Inappropriate for type of material
* Inadequate in quantity
* wrong for application

Notes:

Fig 17.

Item	Level	Item	Level
FIRST TRANCHE – JAN.–JULY		*SECOND TRANCHE – APRIL–SEPT.*	
Contractors' Technical Support	3, 4	Building Conservation	3, 4
Project Management	5	Building Design	4
CE Site Management	3, 4	Facilities Management	4
Highways and Transportation	4, 5	Town Planning	3, 4
Plant Management	3, 4	Construction Materials Technology	3, 4
Organisation Management	4		
Building Services	3, 4, 5	Fig. 18.	

D114.2   Assess the resource
requirements
and costs within an estimate, bid
and tender

(a) a proposed method statement and draft
programme are developed which meet
key client requirements

(b) a complete and accurate statement of the
quantitative and qualitative resource
requirements and their availability is
produced and accurately calculated in a
format which facilitates costing and
planning

(c) an accurate cost is calculated for each
item of resource requirement using
sufficient relevant sources of information

(d) costings are suitably modified to account
for external factors which may effect cost
projections

(e) the overall costing is complete and
accurate and in a form suitable for
adjudication

(f) explanation and clarification of projected
costs are provided to support calculations

(g) realistic schedules of payments are
proposed to meet known cash flow
requirements

Range Indicators

1 Type of tenderer: contractors;
sub/works/trade contractors; suppliers:
consultants

2 Resources: labour (in-house, external);
plant and equipment; materials; finance;
time

3 Resource requirements will be ascertained
from: client brief; site measurements;
scaled drawings; schedules; method
statements; programmes

4 Costing includes: cost based on a
quotation; unit cost built up from basic data;
internal and historical cost data; published
cost data

5 Costing methods: manual; computerised

6 External factors: variations over time;
geographic location; statutory and
contractual requirements; special working
conditions and methods; special resourcing
conditions

Evidence specification

Product Evidence
1 Proposed method statement and draft
programme developed (a)
2 Overall costing complete for adjudication
(e) (all range; which includes:
• resource requirements and availability
identified and calculated (b) (range 2, 3)
• costs calculated for each item of resource
(c) (range 4, 5)
• costs modified (d) (range 6)
• explanation and clarification (f))
3 Proposed schedules of payments (g)

Process Evidence
1 Explanations and clarification of projected
costs (f) (range 2, 3, 4, 6)

Theories, Principles and Methods to
• Develop method statements and
programmes (a)
• Identify resource requirements and
availability (b) (range 2, 3)
• Calculate costs (c, e, f) (range 4, 5)
• Modify costs (d) (range 6)
• Propose schedules of payments (g)

Facts and Data about
• Key client requirements (a)
• Resources (b) (range 2, 3)
• Cost calculations (c) (range 4, 5)
• External factors (d) (range 6)
• Cash flow requirements (g)

Critical Evidence

Fig. 19.

78 Uses of Occupational Standards

RECRUITMENT AND SELECTION

1. preparing recruitment specifications
2. preparing job advertisements/details
3. a format for collecting information from referees
4. identifying the required components of a current role/job
5. identifying the required components of an anticipated/future role/job
6. an interview checklist for selectors
7. advance information to job candidates/interviewees
8. specifying induction and initial training

JOB DESIGN AND EVALUATION

9. development of job/role specifications
10. regular updating of job/role descriptions
11. monitoring the pattern of role/job responsibilities in sections/organisations
12. job design and redesign
13. criteria for job evaluation
14. criteria for the grading of staff
15. criteria for payment and reward systems

ASSURANCE OF PRODUCT AND SERVICE DELIVERY

16. quality specification for work processes/outcomes
17. structuring and 'loading' production systems
18. monitoring of work processes
19. guaranteeing customer service quality/standards by licencing job holders
20. specification for contract tendering
21. monitoring contract delivery/compliance
22. evidence of competence for compliance with international standards (BS5750/ISO9000)

LOCAL/NATIONAL LABOUR MARKET PLANNING

23. analysing and quantifying skills availability within local labour markets
24. monitoring national and local skill supply shortages
25. providing training/learning guarantees

IDENTIFYING INDIVIDUAL/ ORGANISATIONAL TRAINING NEEDS

26. specifying the skill/competence needs of an organisation
27. identifying individual learning needs
28. a format for individual action planning
29. identifying group/organisational learning needs
30. identifying previously acquired competence
31. developing a strategic view of future learning requirements
32. coordination of different HRD processes

STRUCTURING LEARNING PROGRAMMES

33. linking training to business objectives
34. linking training to national economic needs
35. increasing the relevance and credibility of training/learning programmes
36. allowing new learners to see the 'whole picture' in a simple and convenient format
37. enabling learners to match the relevance of off job training programmes
38. broadening the scope and relevance of traditional skills based training
39. identifying learning opportunities in the work environment
40. coordination of on and off job provision

41. development of learning contracts
42. development of specific learning objectives
43. development of knowledge content for learning programmes
44. specifying off job content and learning processes
45. specification of required outcomes and targets for external training providers
46. monitoring external training providers
47. evaluation and selection of learning resources against organisational requirements

DELIVERING AND EVALUATING LEARNING PROGRAMMES

48. a format for structured learning in the work environment
49. identifying progression routes for learners
50. providing clear goals for learners
51. highlighting opportunities for transfer between jobs/occupations
52. evaluating individual/ group training programmes

CAREERS GUIDANCE AND COUNSELLING

53. a basis for information / advice for people entering a first career/job
54. a basis for information and advice for people changing to new careers/job
55. assessing aptitude and potential for careers/ occupational areas
56. identifying common and potentially transferable skills in different careers/occupations
57. analysis of local and national career opportunities in outcome terms

ASSESSING ACHIEVEMENT

58. identification of assessment opportunities

59. specifying the methods and processes of assessment
60. a specification for formative assessment
61. a format for the collection of evidence for National Vocational Qualifications
62. a specification for summative assessment for public certification
63. a specification for internal assessment and appraisal
64. a format for joint review of learner progress
65. a format for individual review of progress/ achievement
66. criteria for the recording of achievement
67. a specification for self-assessment
68. a specification for peer/ group assessment

DEVELOPMENT OF PUBLICLY FUNDED TRAINING PROGRAMMES

69. assessing requirements for national and local training provision
70. assessing funding requirements for national training programmes
71. allocating funding for national training programmes
72. monitoring success of publicly funded programmes
73. providing coherence for national provision of qualifications
74. development of formal assessment systems
75. monitoring and assessing priorities for the development of new qualifications
76. development of National Vocational Qualifications
77. updating of National Vocational Qualifications
78. providing criteria for equivalence between national and international qualifications

© *PRIME R&D 1992*

Fig 20.

42   IMPLEMENTING A DISTANCE LEARNING COURSE IN FIRE SAFETY FOR TECHNICIANS

K. PARSONS
Department of Built Environment,
University of Central Lancashire, UK

Abstract

All too often open learning schemes are mooted and packages developed without the proper feasibility study taking place and without any real reference to good practice in educational planning. This paper seeks to lead the reader through the stages involved in the implementation of this type of course. The importance of planning, implementation, development and learner support are addressed with specific relevance to distance learning. The paper discusses the need to establish a formal course team, involving teaching and support staff, and formulate realistic schedules for the production of learning materials, the pace of study and for contact time. Advice is given on developing the teaching materials with regards to good practice in open learning. Different types of learning package and teaching media are considered along with the implications they hold for production and delivery.

Key words: open learning, distance learning, fire safety education, implementing.

1 Introduction

Open learning is an approach to education that is currently becoming more and more common in the Higher Education Sector. It is important to understand the purpose and nature of open learning and to consider how open a specific scheme is intended to be (see 2). This paper seeks to offer guidance to institutions who may be considering implementing fire safety education by distance learning. It attempts to address some of the issues that need to be considered before the teaching materials are developed. It is based on the background research for a recent feasibility study for just such a proposal.[1] It lays down some guiding principles for the planning, implementation and production stage and discusses learner support.

2 Background

There is no accepted definition of open learning but one was offered by the Council for Education and Technology in 1980: "An open learning system is one which enables individuals to take part in programmes of study of their choice, no matter where they live or whatever their circumstances." [2] This is not necessarily the same thing as distance learning, which usually involves a 'learning package' where the

Fire Engineering and Emergency Planning. Edited by R. Barham.
Published in 1996 by E & FN Spon. ISBN 0 419 20180 7.

learning takes place at some distance from the teaching (and usually at a later time). One way of appreciating the difference between these two concepts is by considering open learning as both a philosophy and a teaching method. [3] The philosophy of open learning is concerned with answering questions like: 'what are the underlying assumptions and purposes of open learning?' and 'what are the best ways to achieve its aims?' The teaching method on the other hand is the 'package'. These packages may be stand alone packages, supported packages (i.e. backed up with some form of tutorial support) or supplementary packages (that are intended to be used alongside ordinary class teaching - as preparatory material for classwork, follow up work or optional/extension studies material). From this perspective open learning can be understood as an umbrella term for a variety of approaches to education, whereas distance learning is a teaching method that open learning schemes often employ.

Open learning schemes can vary in their openness. One way of determining how effective a scheme is in responding to learners' needs is by referring to the idea of an open learning continuum (see table 1). Openness is assessed by examining each criterion, between theoretical extremes: with closed at one end and open at the other.

Table 1	The Open Learning Continuum (summarised)	
Basic Question	Closed - - - - - - - - - - - - - - - -	Open
who?	only open to select groups set entry requirements	open access
why?	chosen for learner	learner's choice
what?	predetermined content	learner chooses content
how?	one style/method used	variety of styles/methods
where?	fixed place of study regular attendance required	learner chooses place of study
when?	fixed course dates set timetable	start/finish any time learner chooses pace
how is the learner doing?	assessment fixed normative only	assessment negotiable regular feedback
who can help the learner?	no outside support only professionals (teachers)	variety of support
where will it lead?	one destination	various destinations
		(Source Lewis 1986) [4]

3 Planning the Scheme
Certain preliminary issues need to be considered and a policy strategy devised, when planning an open learning scheme. The preliminary issues can be labelled as being organisational, logistical or managerial in nature.

Organisational issues include a consideration of whether the institution is likely to be welcoming or resistant to potential students. In terms of logistics it is important to consider the balance between the 'driving forces' and the 'restraining forces' in order to consider whether the scheme is likely to be a success or not. If the balance appears to be weighted in favour of the restraining forces then it would probably be more prudent to abandon the scheme, unless steps can be taken to alter the situation. For example, there needs to strong support from the senior management, for such a course, and a commitment that extends to providing staff with adequate time and administrative support to make the scheme worth planning. It is important to identify the members of the course team and how they will fit into to the normal management structure of the institution. Ideally a course team should have a dedicated member of staff to perform administrative and clerical duties, including typing, supervising the storage facility for teaching material and organising their distribution. The course team may also need its own direct telephone lines: one for enquiries that the Administrator will deal with and a 'help line' either manned or linked to an answering machine (for students to contact their tutors when they have a problem). Another possibility is that the course team could set up an E mail notice board as an alternative 'help line'.

It is important to have a policy strategy for planning an open learning scheme. One such strategy that has been suggested, is that people who are involved in planning open learning schemes should be clear on the answers to certain questions. [5]

Why are the proposals being made?
What agendas are there - overt and hidden? In planning a scheme the course team needs to address the relevance and importance to the agenda, in their institution, of issues like: widening access to fire safety engineering education, increasing student numbers in the institution, introducing innovatory teaching, tapping into existing funding and 'bandwaggoning'.

Who will the programme be aimed at?
Will it be open to any one interested in fire, or will the programme be geared primarily towards the needs of professionals in the Fire Service, Building Control and the field of Building Design and/or Construction?

How much will be done and when - over what time scale?
The course team must be allowed to work to realistic deadlines, they must be given the time to prepare the programme, and receive adequate support in terms of support staff and funding. Short deadlines can be viewed as 'restraining forces', making it difficult for a scheme to succeed.

What will be focus for such work?
In order to answer this question a course team must resolve certain other issues. Will the course team be a small closed group or will the focus be institutional? Will there be a pilot study or will it be more of a "big bang" approach?

How much will it cost?

Open learning is sometimes seen as a cheap alternative to traditional teaching, but in reality this is a rather naive view. It can be more cost efficient, but without working out the figures and calculating the number of students required to reach the 'break even point' any course team is working completely in the dark.

Implementing distance learning programmes involves a lot of up front capital expenditure, in terms of staff time and production expenditure. There are also the running costs to consider once the course is operational (administration, tutorial and technician support costs and the cost of facilities). It must be decided whether students will pay fees or be charged for 'services'. For a distance learning scheme to be financially viable, income from students must cover the fixed cost as well as the variable costs.

Cost will probably be a limiting factor in the design of a course in fire safety. Since it will be a very specialist course, the number of potential students will be limited, compared to other open learning courses (e.g. Foundation Level Technology at the Open University). The course team should bear this in mind when they are costing the scheme and trying to determine the break even point.

4 Implementing the Scheme

Scheduling is crucial and refers to both the production of the teaching materials and to the pace of study for the programme once it is up and running. During production, there needs to be a schedule drawn up for the draft copies of materials (written, audio or visual), for the critical appraisal of drafts, for producing final drafts and for editing and reproduction. When a course team is producing the materials they should already have decided what the pace of study will be (i.e. how long students are intended to spend studying a given module). The contact time for students also needs scheduling.

There are important aspects to consider in selecting the team. Will it be a large team including a number of subject specialists, their open learning advisors and a variety of support staff; or a small team of generalists? It is easier to keep to the schedule and prioritise with a small team, but the quality, in terms of content and presentation, may be better with a large team.

After selecting the team, recognising staff development needs becomes an important consideration. Will there only be a need for staff development before production commences, before the course is operational, or once the course is running? It is more likely that some staff development input will be required at each stage. Some of the topics that it might be worth considering for staff development sessions include: "change management" skills, team management and partnership skills, distance teaching skills and counselling skills.

Proper scheduling during the production phase is important, in terms of standards, however course teams also need to determine the desired content and form (lectures and/or study guides) and the learning media they will produce. During the critical appraisal period, materials need to be examined from a subject specialism point of view and from an open learning one (see 5), in order to ensure high standards. There needs to be agreement on presentation standards: layout and style and the use of tables and diagrams. Support staff should be involved in the planning of open learning schemes; administrators and secretaries, technicians and librarians should all be included. Librarians should be consulted when decisions are being taken about the form the teaching materials are going to take and library support that will be required. Input and support from technicians also needs to be considered at this stage.

An early decision must be taken on how a course is to be marketed. In some institutions the course team will be responsible for publicising the programme, whereas in others there will be a marketing department that will have to be consulted.

5   Developing the Teaching Materials

One of the distinguishing features about open learning materials is that the learning objectives are made explicit at the beginning of each "lesson". Students then know what they are expected to be able to do by the time they have finished it. In this way they can chart their own progress. The lessons should also include "activities" that are designed to engage the student in active learning and make the material interactive. The learning materials should also use language that is accessible to the learner. [6]

In the UK the learning objectives will presumably be written in such a way that the modules are compatible with the National Core Curriculum in Fire Safety Studies by Design. [7] Therefore, in effect, the decisions about what the students will learn may have been taken already.

There are a wide range of teaching media being developed for use in distance learning packages. Selecting the media is rarely based solely on educational criteria, since the time and funding available are such important considerations. However, from an educational perspective, print would be an appropriate means for conveying much of the factual information in fire safety and for teaching mathematical procedures. Experiments would need to be visual so video would be a fitting medium. Interactive computer based learning is being heralded as the media of the future, but it is expensive and time consuming to produce and has resource implications. Cost may be a limiting factor in the design of fire safety course material due to anticipated student numbers (see 4). Resource implications are considered below (see 6).

There are different types of packages available, that can be categorised by how they have been produced. They tend to be known as: off the shelf packages, bespoke packages, wrap round packages or mixed packages

Off the shelf packages are ones that have been produced externally, are deemed appropriate and have been bought in. Off the shelf packages can be adapted and sometimes they are "badged" as well (i.e. the adapting organisation replaces the logo with its own). Bespoke packages are those that are produced specifically for the course in question. These packages can be tailored to meet the desired objectives and can be produced in house or by commissioned external consultants. Wrap round packages consist of a collection of externally produced materials (e.g. articles from periodicals or extracts from text books) with directed study notes that include learning objectives and activities. The study notes also link the various items in the package to make it a cohesive whole. A mixed package is one in which elements of each style of production are employed. Where off the shelf packages are going to be adapted or materials are going to be reproduced to form part of a wrap round package, the law of copyright must be considered very carefully. However, licences can usually be obtained relatively inexpensively in the UK. [8]

There are obviously time and cost implications for each type of production. Off the shelf packages and externally produced bespoke packages tend to be expensive (making updating a costly consideration). In house bespoke packages are cheaper to produce but are time consuming. They are tailor made, the copyright is owned in house and therefore the package is easy to update. Wrap round packages are quicker to produce and can be tailored to suit the objectives but the copyright does not belong to course team, licences have to be paid for and the material can not be easily altered or updated.

6 Supporting Open Learners

It is important that course teams are aware of what sort of needs their learners are likely to have, when planning an open learning scheme. This involves knowing who the students are likely to be, developing a learner profile and then identifying the kind of support they will probably require. The wastage rate on open learning programmes tends to be much higher than with traditional college based courses. [9] The average applicant for an open learning course tends to be a 'non standard entrant' or an 'Adult Returner'. [10] Such student groups have different needs. Course teams need to be aware of this in designing the course and in planning the support arrangements..

The demographics of the student population will largely be determined by how open the course is intended to be (see table 1). Students' motivation in studying a course has implications in terms of support planning. Research has shown that students with intrinsic motivation tend to achieve higher grades. [11] Therefore students should be encouraged to cultivate an intrinsically motivated orientation, even though many may have initially enrolled on a course for reasons of external motivation (e.g. in the hope that it may help them secure promotion at work).

Learning factors influence the way in which students learn and can determine how much support they need. If many of the students are expected to be coming from a technical background they may see learning as being simply about memorising facts. In this case, they will need educating about the nature of learning. Such students may well have developed a preferred learning style based on concrete experience. Research suggests [12] that students who use a variety of learning styles will be more successful and that courses that cater for a variety of learning styles will be more accessible. Being 'Adult Returners' many of the students may need to re learn their study skills and build up their confidence. [13]

Students' subject background should be borne in mind in developing a course in fire safety. Depending on how open the course is intended to be (see table 1), many of the students studying the course may be building on knowledge that they already have and so they will be able to make valuable contributions to the course. In terms of knowledge and skills, all students should be literate and numerate and have an understanding of basic science. If this is unlikely to be the case then the course team will need to address this issue when they are planning the programme.

Resource factors can affect course planning and students' learning. Different students will be studying in different settings. For example learners who are fire officers may be learning "on station", in a quiet corner of the recreation room or a dormitory/study room. They may not have access to video play back facilities but might have access to a computer in the administrative office (for word processing but probably not CD ROM). By comparison, a Building Control Officer or Architectural Technician would be expected to work at home, where they may have access to video playback facilities but not to a computer. However they may have access to CAD (Computer Aided Design) facilities in their lunch hour at work. Resource factors of this nature impinge upon the type of learning media that course teams can employ in developing the learning package (see 5). If resource factors suggest that portability will be a crucial factor in the design of learning materials, then print may be the best media to employ.

Learner support is another vital component. One perspective is logistical support which relates to arrangements like 'help lines' and resource factors. In view of the likely geographical dispersion of students, library access will also need to be addressed, by establishing postal facilities or an inter-loan network with other libraries across the catchment area.

Human support of learners involves more than just the tutorial and counselling support provided by the course team. To be successful in their open learning studies, students need a supporting infrastructure. They need support and encouragement in their studies from their line manager (and possibly some other work based colleague who can act as a mentor), their family and friends, other students (on the phone at least) and their work colleagues. In the same way that the successful implementation of an open learning scheme depends on the balance between driving forces and restraining forces, whether a student is successful in their open learning studies or does not complete the course, depends on the balance between positive factors and negative factors. [14] These latter factors may include motivation, learning factors and resource factors, but are often underpinned by human support (or the absence of it!). Students must be guided in establishing support infrastructures.

The specific form of educational support that will be required will depend on the learner profiles. This support might include elements related to intrinsic motivation, establishing support networks, other learning styles and study skills (including time management and library skills). Some course teams decide to include a 'block release' induction week or workshop early in the programme and incorporate learner support sessions into the timetable, to provide this support.

7   Conclusions and Recommendations

Some of the recommendations that have been made in the body of this paper are summarised below.

- A course team should be set up. Teaching and support staff should be involved in the planning of the scheme.
- The scheme should be costed to ensure that a financially viable scheme is developed.

- The course team should agree realistic schedules for the production phase, the pace of study and contact time.
- Distance learning materials need to be critically appraised from both a subject specialist and open learning perspective.
- Staff development needs should be addressed.
- The course team must decide on the type of learning package and learning media to be employed, after giving due consideration to time/cost and resource implications.
- Logistical support and access to resources needs to be addressed.
- Students must be guided in establishing support infrastructures and provision must be made for learner support. Towards this end an induction week might be included in the programme.

References

[1]   Parsons K. (1994) "HNC Fire Engineering (by Distance Learning)" unpublished feasibility report, Department of Built Environment, University of Central Lancashire, Preston

[2]   Thorpe M. & Grugeon D. eds.(1987) "Open Learning for Adults".Longman, Harlow, Essex

[3]   Rowntree D. (1992) "Exploring Open and Distance Learning". Open University & Kogan Page, London

[4]   Lewis R. (1986) *What is Open Learning?* Open Learning Vol. 1, No. 2, Open University & Longman

[5]   McNay I. (1987) "Organisation and Staff Development". (in Thorpe & Grugeon)

[6]   Rowntree D. (1990) "Teaching through Self Instruction". Kogan Page, London

[7]   National Core Curriculum in Fire Safety Studies by Design (1992). The Institution of Fire Engineers, Leicester

[8]   Brown S. & Gibbs G. (1994) "Course Design for Resource Based Learning in Built Environment". Oxford Centre for Staff Development, Oxford Brookes University

[9]   Woodley A. (1987) "Understanding Adult Student Drop Out". (in Thorpe & Grugeon)

[10] Northedge A (1987) "Returning to Study". (in Thorpe & Grugeon)

[11] Sagar E. & Strang A. (1985) "The Student Experience: A Case Study of Technicians in Open Learning".

[12] Honey P. and Mumford A. (1986) "The Manual of Learning Styles". Available direct from Dr P Honey, 10 Linden Ave., Maidenhead, Berks.

[13] Northedge A. ibid.

[14] Northedge A. ibid

43 TRAINING FOR COMMAND AT FIRES – THE ICCARUS PROJECT

K. WHITEHEAD
Training Department, Greater Manchester County Fire Service, Manchester, UK

Abstract
Command or Management of Fires is an extremely complex skill. The traditional methods of teaching Fire Service Officers these skills can have some shortcomings. It has therefore been necessary to develop new methods. A computerised simulator has been developed and its use as a teaching aid will be discussed. Progress in teaching methods must continue and options for the future are developing.
Keywords: Command, Computer Simulation, Fires, Methods, Training

1 Introduction

Command and Control, by the Officer in Charge of a major Fire, if it is to be effec-tive and efficient, involves a wide range of actions which are highly complex and often extremely critical with regards to time scale and possible consequences.

When a Fire Service Officer makes attendance at an incident he is presented with a vast amount of data relating to it. Such data may be verbal and or audible. It may be written or in diagrammatic form. It is not unknown for intuition and "gut" feelings to be present. He is expected to understand and prioritise the actions based upon his interpretation of the presented data. Once the actions have been implemented, the Officer in Charge still has to monitor, evaluate and redirect his actions as required.

It is therefore necessary for the Fire Service Officer to have simultaneous control of numerous spheres of responsibility.......this the Fire Service refers to as "Operational Command Skills"

2 Operational command skills

There are three traditional methods that British Fire Service Officers learn their skills.

Fire Engineering and Emergency Planning. Edited by R. Barham.
Published in 1996 by E & FN Spon. ISBN 0 419 20180 7.

1. Formalised,structured courses at the internationally recognised Fire Service college at Moreton in Marsh, in Gloucestershire.
2. By large scale training exercises.
3. By experience gained at actual incidents.

Although providing good knowledge and experience the existing methods do have some shortcomings.

Fire Service College . Although highly beneficial, the training is one off, and rarely is the Officer refreshed or updated.

Large scale exercises - only one individual can actually participate as Officer in charge, therefore the training experience is limited on an organisational needs aspect.
  The training can also be considered costly in terms of personnel and equipment used, although it is acknowledge that on such exercises personnel may be gaining experience in other areas such as the specific technical uses of equipment etc.
  But of course, if real incidents occur during the exercise, then the Training resources will be removed (i.e. by attending the real incident) and the exercise is therefore destroyed and so is the learning experience.

Experience gained at actual incidents - The decisions made by the Fire service Officer are often extremely critical, and can have far reaching consequences in both financial and human terms.
  Financial - It is estimated that the average cost of a major fire before it is brought under control is probably about £10,000 per minute. It is also estimated that an average saving of one minute on each major fire would save the country somewhere between £2 and £80 million per year.[1]
  Human - The Fire Service history is littered with reports of tragic loss of life. Often such losses were due to ineffective or incomplete Command and Control of the incident.
  During the Kings cross underground incident many lives were lost, both members of the public, and Fire service personnel. The subsequent enquiry by Desmond Fennel OBE QC recommended that Fire Brigades should review their management, instruction and training in command and control of emergency incidents.[2]

3 New Methods

  GMCFS and other British Fire Services recognise the shortcomings of the traditional methods of an Officers development, and are determined in their quest to provide an efficient and effective service to its customers - i.e. the population and industry of the Country.
  To achieve these objectives GMCFS has embarked upon a new and radical method of training its officers in Operational Command skills.
  A computerised simulator using an application named ICCARUS is used. ICCARUS stands for Intelligent Command and Control: Acquisition Review Using Simulation.
  The Student is expected to take command of the simulated incident. The computer runs the developing incident in a random and complex fashion, changing the situation obstructively as it develops. There is therefore never a model answer and the

students are required to use their skill, intelligence and judgment on each problem as they present themselves.[1]

The whole package is truly interactive, the student and computer reacting to each others actions. The student is rapidly involved in the simulation and it is not uncommon for the student to be seen talking to the computer generated characters appearing within the simulation.

4 Simulator background

The simulator is based on an Apple Macintosh Quadra CPU, having CD ROM,twin monitors and a laserdisc player. Iccarus has been developed by the Employment department, Home Office, The Fire Service College and Programmers from Portsmouth and lately Brunel Universities.

To reproduce the fire scenarios [1], a disused cinema in the West Midlands was acquired. Initially the special effects team from the TV programme "Londons Burning" were used, until ultimately the building was actually set alight and razes to the ground. Such methods provide realistic fire footage for the simulation. The footage seen is wholly dependent upon the students actions whilst running the simulation.

5 Teaching method

Although GMCFS was presented with the I.C.C.A.R.U.S. simulator as a working tool, it was necessary to develop an effective and efficient method of using the simulator in a training environment.

Prior to commencing the simulation it is necessary to perform an audit of the students existing knowledge, perceptions and possible anxieties concerning Operational Command and Control. The training method is certainly radical within the Fire Service, and amongst many students there exists a fear of technology.

In order to overcome these fears and misconceptions the student undertakes a short semi-structured interview, and positive counselling is given as required. This tactic has proved invaluable in this teaching method and results in the student developing a positive attitude towards the training session.

Following the semi-structured interview, the student is given a short lecture to instruct and refresh them on Operational Command methods and skills. At the conclusion of this input the Student is then expected to put the knowledge into practise within the simulation.

The Officer in Charge of an incident is never truly alone. There will always be other Officers available with whom to consult and seek advice from. This philosophy has been employed when using the I.C.C.A.R.U.S. simulation. The instructor actually sits down with the student and controls the simulation based upon the decisions of the student. The instructor also "role plays" as the assistant on the Fireground. This has distinct advantages over a purely stand alone simulation in that valuable time and resources are not wasted in teaching the student the background and operating methods of the simulator and of course Incident assistance is always on hand.

As the student enters into the simulation, the computer generated characters and actors begin to provide him with information. Radio messages are heard. Officers within the simulation will brief him. Plans and other data relating to the incident are made available. The student is allowed and expected to probe and enquire upon these characters. As decisions are made the simulation constantly reacts and new situations develop.

For the student to be effective and efficient it is considered that control of several functions will be necessary . These functions are Logistics, Planning, Communications, Crew Safety/ W elfare, Post Incident Considerations, Organisation and Liaison.

The simulation will easily last for up to four hours, and as the simulation operates in real time, a realistic fire progress is presented, incorporating the vast majority of problems which would be exhibited at a real incident. The student is subjected to a great deal of decision making, whilst being inundated with information from many sources, however as few system clues as possible are provided. Learning is real and is gained by experience.

At the conclusion of the training session it is necessary to debrief the student in order for learning to be confirmed. The application provides a review of the actions and events which have taken place throughout the simulation.

The student is assisted by means of a another semi structured interview to assess their performance. The instructor never openly makes conclusions as to the performance of the student, he simply makes observations of the actions and decisions taken and then questions and probes the student to ascertain the students views. In other words the instructor simply acts as a facilitator and allows the student to judge their own actions. This achieves a more positive learning process for the student.

Following on from the training debrief, the student is requested to complete an action plan which should highlight the individuals strengths and weaknesses, and the actions which the student intends to take in order to build on the strengths whilst acknowledging and addressing their weaknesses.

The completed action plan is held by the student, in order to maintain confidentiality, and should be brought with them to the next training session. The use of the Action plan enables progress of both the student and the training method success rate to be monitored.

Currently, all Officers within the Brigade will receive dedicated training once per annum.

6 The Way Forward

It is considered that GMCFS is probably the only Brigade in the country at present providing such dedicated training for its officers, although there are several other Brigades involved in the ICCARUS project.

By such methods GMCFS hopes to develop its Officers, and ultimately provide an effective and efficient service to the community to which it is responsible, in a cost effective manner.

Future developments for the simulator are already well underway. Within the very near future the system may be upgraded to provide several scenarios which will be viewed in the new medium referred to as "virtual reality".

Technology is enabling Fire Service Officers to be effectively and efficiently trained in a most cost effective manner yet achieving as good, if not better results than traditional methods.

Consideration by GMCFS is already being given to developing other simulators to deal with Fire Investigation and Hot Fire Training.

The future is very exciting, and it is the intention of the GMCFS to play an active role in in creating and controlling the future that we want rather than simply being carried along with the flow.

7 References

1. Employment Department (1992) ICCARUS Learning Technologies Project report, HMSO, London.

2. Desmond Fennel OBE QC (1988) Investigation into King's Cross Underground Fire, HMSO, London.

44 TRAINING AND RESEARCH IN FIRE SAFETY IN MOSCOW STATE UNIVERSITY OF CIVIL ENGINEERING (MSUCE)

A.V. ZABEGAYEV and A.N. BARATOV
Moscow State University of Civil Engineering, Moscow, Russia

ESSAY

An information about the Program of training engineers, scientists and teachers in the field of engineering safety in construction in the whole and fire-explosive safety in particular is proposed. Note, that training includes lecture course, practical studies and laboratory studies. It is showed, that MSUCE is realising a broad complex of scientific works on different aspects of fire-explosive safety in construction. The report aguaints with results of the investigations to create the basis of fire-explosive standardisation, new means and methods of fire-explosive protection, elaboration of fire-explosive standards.

Keywords: training, program, engineering safety, fire-explosive safety in construction, basis of standardisation, aerosols extinguish composition, fire-explosive safety standards.

1. INTRODUCTION

Nowadays MSUCE is realising a broad training program for specialists in the field of engineering safety in construction. This program has been under supervision of Russian Academy of Sciences and Russian Government and consists of four block as follows:
- general aspects of safety;
- safety of structures and buildins;
- safe technology of buildibg production;
- management, economy and psychology (hyman factor).

The first block of general aspects of safety includes examination of hazardouse situations, catastrophes and their consequences, both general and applied to construction industry, special chapters of high mathematics, risk assessment, principles of accident's modelling and control, special psychology.

Fire Engineering and Emergency Planning. Edited by R. Barham.
Published in 1996 by E & FN Spon. ISBN 0 419 20180 7.

The second block is related to design of structures subjected to accidental loads and actions. It includes studing of problems conserning a resistance of structures and buildigns to intense dynamic loadings (earthquakes, blasts, explosions, impacts, hurricanes etc.), temperature actions (fires, frosts and thaws, heat radiation ets.), nuclear radiation, aggressive enviroments, as well as rehabilitation of damaged structures, safe planning of buildings, potentially subjected to accidental actions, ets.

The third block of disciplines includes resque work's techniques, special equipment as well as labour protection.

Economical aspects of safety, project management in liquidation of hazardouse situations and catastrophes' consequences etc.) legislation and selection of personnel make up the forth block.

Most of knowledge taught within the courses is based on original research results obtained in MSUCE. The researches are supervised by scholars of authority, members of international scientific organisations (NFPA, IFSAC, IAFSS, ICE, SECED etc).

As exposed, training of specialists is being developed on a broad range of subjects of engineering safety so that graduates of MSUCE are highly appreciated as qualified specialists not only in the field of construction safety, but in other branches of industry and moreover they are able to investigate the problems of engineering safety themselves.

In the present paper an information is given about those parts of training and scientific work in MSUCE which are dedicated to ensuring fire and explosive safety in construction.

2. Training Program of Ensuring the Fire and Explosive Safety in Construction.

The training includes lecture course, practical studies and laboratory studies.

2.1. Lecture Course.

This course consists of the following parts:

2.1.1. Information About Combustion, Fire and Explosion.

Definitions of these processes, ideas of thermal and chain mechanisms of combustion, thermodynamics and kinetic of reactions in flame and critical conditions of starting the combustion and fire-hazardously properties of substances and materials (taking into account their aggregate state) are given. Ideas of combustibility of materials and structures, methods and criterion of estimation of fire-hazardous properties of substances and materials are given; real conditions of their using (influence of localisation of the explosion hazard volume, pressure, temperature, etc.) are taken into account.

2.1.2. Fire Prophylactic.

In this part the following is considered: fire-protective standardisation, assess ment of fire resistance and measures to raise it by fireproofing, planning decisions and fire-protective brakes, smoke removal and peoples evacuation, fire signalling.

2.1.3. Extinguishing.

Principles and methods of extinguishing, surface and volume extinguishing, means of extinguishing and classes of fire, standard indexes and optimal conditions of extinguishing, waterworks for extinguishing, fire-extinguishing equipment and devices and a basis for their design, fire automatics are presented in this part.

2.1.4. Protection Aganst Explosions.

This division includes explosion stability of buildings and structures development of explosions in gases and dust mixtures with air allowing for a combustion intensification by flame turbulisation on obstacles, methods of explosion protection and principles of calculation and design of protecting devices, discovery of explosion hazard accumulation of gases and vapours, inertization of explosion hazard mixtures.

2.1.5. Organisation of Fire Service in Russia. Fire Inspection. Fire Legislation.

Public organisations: voluntary fire societies, voluntary fire ganges.

2.2. Practical Studies.

They include case-studies on practical problems **of** different aspects of fire safety. In particular, design estimates of main indexes of fire-explosive hazard of substances and materials are done (flammability limits, flash points, conditions of materials' self ignition during storage, ets.), thermal regime of fire, fire resistance of structures. categories of fire-explosive hazards, required quantity of extinguishing substances, ways of evacuation of people, area of explosive protective devices, vent,etc.

2.3. Laboratory Studies.

They include an experimental determination of indexes of fire-explosive hazard of substances and materials, heating of structures according to an assigned temperature regime.

3. Results of Some Investigations Carried out in MSUCE in the Field of Fire-Explosive Safety of Buildings and Constructions.

As mentioned above, MSUCE is carrying out a broad complex of investigations in the field of ensuring the fire-explosive safety of buildings and structures. It is impossible to inform about all these works. In the limits of the short report the present report presents a piece of information about the results only of several investigations, which are, to our mind, especially important in the observed field.

3.1. Elaboration of Fire-Explosive Standardisation Principles.

MSUCE took part in elaborating two standards fundamental for Russia: "Standards of Technological Design ONTP 24-86" [1] and the State Standard 12.1.004-85 "Fire Safety. General Requirements" [2]. The first document is intended for a design of industrial and storage buildings and based on a determinant approach, dividing the objects into categories. Their limits are defined by fire-explosive properties of substances disposed in the buildings, their quantity and possibility of the explosive hazard mixtures formation in the volumes.

There are totally five categories required by these documents:"A"- explosive fire hazardouse (rooms with combustible gases, liquids with flash point $28^{0}C$ and lower, pirophoric substances), "Б" - explosive fire hazard (rooms with liquids having flash point higher then $28^{0}C$ and up to $61^{0}C$ or with flash point higher then $61^{0}C$, but are heated higher then the flash point or forming an explosive hazardouse aerosol during their leaking from an equipment under pressure, as well as rooms containing explosive hazard dusts), "В" - fire hazardouse (rooms where a hazard of formation of the explosive hazardouse mixtures is absent, but they contain combustible materials, capable to create thermal loading higher then 50 (MDj/m^{2}), "Г" - non fire hazardous, but with ignition sources (rooms of boiler-houses, with open flame, for welding, etc.),"Д" - non fire hazardous.

A determination of the categories is made according to the folloving formula:

$$\Delta P = \frac{m \cdot H_T \cdot P_O \cdot Z}{V_{room} \cdot C_a \cdot P_a \cdot T_O \cdot K_{un}} \qquad (1)$$

where ΔP = superfluous pressure developed in rooms during the combustion of explosive hazardous mixture, which can form because in accidents of technological equipment (kPa);

m = maximum possible (taking into account possible emergency situations) mass of combustible substance, which forms the air-gas or air-dust mixtures in the room (kg);

H_T = combustion heat of substance, forming explosive hazardous mixtures (kDj/kg);

P_0 = atmospheric pressure (101 kPa);

Z = coefficient, allowing for a portion of explosive hazardous substances, got out to the room and taking part in the forming of the explosive hazardous mixture (Z=0.1+0.5);

C_a, P_a= heat capacity of the air (kDj/kg·K) and density of the air (kg/m^3) in the room;

V_{room}= free cubic capacity of the room (m^3);

T_0= air temperature in the room (K);

K_{un} = coefficient, taking into account non-hermetisation of the room (usually equal to 3);

If ΔP < 5kPa category "A" or "Б" is prescribed depending on properties of substances, if $\Delta P \leqslant 5kPa$ - the category "B" , "Г" or "Д" depending on thermal loading and ignition sources is given .

The requirements of fire-resistance for constructions, number of storeys, spatial and engineering decisions are formulated according to abovementioned categories.

The second document is based on the concept of permissible risk and foresees complex of engineering measures, meeting the requirements that a probability of the people's defeat in fire or explosion at the present object does not exeeded 10^{-6} per year.

The probablility of the striking action of fire or explosion is calculated by the formula:

$$Q_L = Q_p \, (1-P_p)(1-P_a) \qquad\qquad (2)$$

where $Q_p = Q_c \cdot Q_o \cdot Q_i$ = a probability of occureing the fire
or explosion;

Q_c = a possibility of occuring of the comburant amount
enough for forming explosive mixture;

Q_o = a possibility of occuring the oxidant;

Q_i = a possibility of occuring the ignition sources;

P_p = a reliability of prophylactic measures;

P_a = a reliability of active measures (extinguishing,
inertisation, explosion opression, etc.);

The difficulty of the last standard's utilisation is that a knowledge of statistical data about emergency situations, recorded for a number of years, is needed for the practical use.

3.2. Working Out the Fire Protection Means.

In collaboration with "The Gabar" firm and other organisations investigations and elaboration of a new volume fire extinguishing mean have been carried out^ resulting in the aerosol extinguishing composition (AEC), obtained by burning down a solid-fuel composition (SFC) consisting of a non-organic oxidiser and organic re-establisher. The AEC is a mixture of potassium and smallest (about 1 μ) solid particles of carbonates, chlorides and K-oxides [4]. This composition possesses the highest extinguishing ability $(0,03-0,007 kg/m^3)$, which is 5-8 times more effective then halon, and is characterised by a lack of ozone depletion potential (zero O.D.P.), low corrosion activity and low toxic. It is intended for volume extinguishing and keeps advantage over the ecologically dangerous halons.

An absense of pressure vessels and pipe-lines, low cost aerosol systems (AEC) in general comparing to other extinguishing stationary systems (water, foam, gas, etc.) expand merits of AEC.

Mecanism of a fire extinguishing action of the AEC is defined by the same processes and conformities as the usual dry pow-

ders. Since the size of the solid particles in AEC is signifiant-
ly smaller then these of the usual dry powders, the effectiveness
of AEC is much higher. As AEC is formed at the moment of fire,
the danger of sticking and pieceing is absent.

The facts that creating of AEC is accompanied by an open
flame and AEC by itself have the temperature higher $1000^{0}C$ may be
reffered to as demerits. The first circumstance does not allow to
utilise AEC in the explosion hazardous compartments, the second
reduces an effectiveness of AEC, because the nighly heated aerosol floats up to the ceiling and starts to propogate in the volume while cooling occurs only.

To come over these demerits special generators of extinguishing aerosol (GEA) of the "Gabar" type were developed, by means
of which an effective and reliable extinguishing (including the
reservoirs with oil) may be provided.

Numerous successful tests under real conditions has confirmed the forgoing. The generators have got through different commission tests and allowed in Russia for fire protection of the
objects of all categories.

3.3. The Main Results of Investigations, Havig Conducted in MSUCE in the Field of Explosion Protections of Buildings and Constructions.

Safety of buildings and structures, subjected to accidental
exposions, has been investigating at a special laboratory of MSU-
CE, having existed for about 25 years and being the leading organisation in Russia in the field mentioned. Among a broad complex
of projects fulfilled, particular attention showld be drawn to
the results of investigations on intensification of gas clouds'
combustion and, based on this effect, elaboration a method of
analysis of the areas of precautionary devices, providing timely
throwing off the surplus pressure caused by deflagrating explosion.

It was founded that the intensification of gas combustion is
influenced by turbulization of flame and gas flow, formed at meeting of the flame with several obstacles and by the gas flow
through an outlet or a canal with reduced cross-section in compa-

rison with primary conditions.

As a result of the investigtions a standard document was developed [5], according to which required section of precautionary devices (S_r, m^2) is calculated as follows:

$$S = \frac{0.105 \cdot \alpha \cdot U_n \cdot (\varepsilon-1) \cdot \beta_m \cdot K_{c.p} \cdot V_p^{2/3} \cdot \rho_g^{1/2}}{\Delta P_p^{1/2}} \qquad (3)$$

where

α = a coefficient of intensification varying from 3 to 30 and depending on a nature of fuel and specific conditions;

U_n = a normal speed of flame spreading, m/s;

ε = design intensity of the combustion products' compression during burning down in the confined volume

β_m = a coefficient, allowing for a degree of filling the volume with the explosive hazardous mixture;

$K_{c.p}$ = a coefficient, allowing for the influence of combustion products on ΔP_p;

V_p = the volume of the room, m^3;

ρ_g = a gas milieu density, kg/m^3;

ΔP_p = a permissiable surplus pressure, kPa;

Besides that, data concerning dynamic effects of the explosive loadings on the bearing structures at falwre of the window glasses were obtained.

As a conclusion , we shall notice once more that the given information is far from an explicite one .

Based on a profound experimental background elaborated advanced engineering methods of analysis and protective measures gives advantages to users because of their simlicity, reliability, easy shift to time-saving software etc.

4. References

1. Standards of Technological Designing, ONTP 24-86, VNIPO, 1986, 25 pp.
2. State Standard 12.1.004-85 "Fire Safety. General Requirements", Gosstandard, 1985, 130 pp.
3. Baratov A.N., Kravets O.P. "Explosive Safety of the Organic Heat-carrier", Problems of Engineering Safety , No 6, 1990, p.68-85.
4. Baratov A.N. "Modern Means and Methods of Extinguishing", "Fire Explosive Safety", No 2, 1992, p.56-60.
5. Construction Standards and Rules. "Protection Against Explosions", SNIP 2.01.11-94.

PUBLIC INFORMATION SYSTEMS

45 HAZARD MANAGEMENT SYSTEMS: INFORMATION MANAGEMENT AT MAJOR HAZARD INCIDENTS

D.T. DAVIS, B. CHEUNG and E. MORRIS
Cheshire Fire Brigade, Chester, UK

ABSTRACT

The provision of information for fire-fighters and the public at major incidents is a crucial action in seeking to achieve a safe and speedy resolution to events. The area of Cheshire within the North West of the United Kingdom, contains many major hazard sites. The Cheshire Fire Brigade has attended several major incidents in recent years and has researched and sought to develop technological and media solution to these issues. The paper outlines the use of computer aided data storage and retrieval systems whilst exploring the interaction to other technologies and agencies. It also outlines the development of a public information initiative designed to promote the concept of shelter rather than evacuation.

KEYWORDS: Hazard Management Systems, Public Information Systems, Fire-fighting Information Systems, Emergency Planning Information.

1. Background

Cheshire is set in North West England and is predominantly a semi-urbanised region which has a significant number of major petrochemical sites and large numbers of transportation movements due to the North West's motorway network within the Fire Brigade's operational area. In the United Kingdom the mix of chemicals to other industries is 90:10, in Cheshire it is 50:50. Because of this particular environment, combined with the experience gained over many years of handling major incidents, the Brigade has consciously and routinely investigated the likelihood of an incident involving a major escape of toxic or flammable gas or similar major emergency. Recently the Brigade has conducted a review of gathering and disseminating of information to both the

Fire Engineering and Emergency Planning. Edited by R. Barham.
Published in 1996 by E & FN Spon. ISBN 0 419 20180 7.

public and fire-fighters and investigated the equipment support systems required by fire-fighters. There are of course many existing provisions within UK legislation which enable the Fire Service (1) to gather information on hazardous materials, the design and construction of buildings and chemical manufacturing plants. The legislative base, whilst helpful, in clearly establishing the need for the public and fire-fighters to have access to information that they will require at the time of emergencies unfotunately fails to recognise that today's firefighting and rescue service needs for information are complex, and due to the dynamics of the fire itself, require rapid translation into useable fireground information and subsequently translation into action.

The Hazard Management System (HMS) project commenced in February 1994 with the intention of researching and providing a technological solution to the identified difficulties. During the project the preliminary rcview focused on how to secure, access and retrieve usable risk information for fire-fighters at the time of any major emergency.

Concurrent with the HMS project the Brigade also introduced research into how best to assist public dissemination of information. One major aspect of its work was the introduction of a public information leaflet baser around the concept to shelter, rather than evacuate, and the use of existing media rather than creating new stand-alone systems.

2. Lessons learned from previous incidents
Major incidents have occurred in Cheshire on a frequent basis. Some involve flow line systems of products whilst several others are operated by batch processing to final chemical substance development. Batch processing adds difficulty to understanding. The nature of the product both in its constituent forms prior to batch mixing and at the conclusion of the process in its new refined form may be well understood and clearly identifiable both in terms of human and environmental risk (2). However during the batch process itself, whilst undergoing catalytic or pressure and temperature changes, the product itself may present other alternative risks. It is usual in such circumstances for fire-fighters to seek the assistance of the on site manufacturer and identify clearly at which stage the batch process is actually in process at the time of the incident. Such a process would seem to be robust and capable of supplying all information needs. Unfortunately occurrences do happen which destroy monitoring equipment and may also injure or kill the relevant site personnel who have the technical knowledge to assist the fire-fighters.

Another example envisaged and encountered is one where the core product will react differently both in its released state and when subjected to fire. The principal risk whilst involved in a spill situation may be the chemicals' own flammability, that flammability may result in a fire and the consequent product evolved during combustion may be considerably different in risk terms to the product before combustion. Again it is important that technical assistance be made available and invariably there will be situations where technical assistance might not be forthcoming due to the circumstances of the actual incident.

In a similar way transport emergencies may put distance between the technical assistance and the fire-fighters who themselves may be either at personal risk or at least seeking to minimise risk to other persons and the environment. With these problems in mind it becomes necessary to achieve a higher level of technical understanding both of the process and the products and the various stages during manufacture (3).

Moving away from the product substance hazard itself, there is also a difficulty at any chemical plant in accurately determining the exact nature of inventories, where they are located, how they are protected, and what action should be taken to minimise the overall risk. Also there is a real need, on the grounds of personal safety alone, to ensure all possible solutions for environmental monitoring, containment and the good practices, i.e. incident management are investigated to help protect fire-fighters (4). Whilst a great deal can be achieved through pre-planning (1) to assist in this regard, when a Brigade, such as Cheshire, is confronted by a large number of such hazardous sites it becomes extremely difficult, if not unmanageable, to hold this new improved archive and recover the detail, in a presentable form, to aid the fire-fighter who, at present, uses hard copy at the incident.

3. The HMS project

Cheshire's HMS project seeks to address this need. All the aforementioned points are seen as extremely important since the concept has to be that fire fighting in itself is an intervention, which managed well with considered judgements and decisions can minimise potential hazards. Conversely if managed badly with poor judgements and decisions, it could in fact escalate the situation and certainly place fire-fighters, if not others and the environment, in extreme danger. Technology is seen as a major tool which can aid this process, both by making an interactive system within the improved risk information and central archive and providing the means to relay or convey that information to a mobile appliance some distance away.

To try and assist in this preplanning and judgmental process the HMS project envisages the use of video and information technology to both record the information, of any particular chemical plant or process, and then to allow that information, utilising a lap top computer (S), to be made available in a rapidly accessible and understandable form to fire-fighters, who may be attending an incident on a particular chemical plant or within a specific building or site for the first time in their career. The assumption made is that it is impossible to train every fire-fighter for every known risk. Skilled fire-fighters still need good information and the better the information and the wider the information sources given, the more likely we are to achieve good decisions. Against this background the provision of risk information in the way outlined in the project is seen as essentially a safeguard for the general public, and more importantly, the fire-fighters, who are called to deal with the incident.

Cheshire Fire Brigade has therefore invested in attempting to move forward these ideas. The HMS project involves the participation of a team of fire-fighters to research a lap-top interface with video, CAD drawings, improved risk information, weather and toxicological monitoring equipment for on and off site activity and electronic delivery of that information to the incident. This equipment is being trialled at one of the Brigade's Fire Stations so that the end users may start to review the concept and consider the practical difficulties and user requirements. *(Details on the project and the current status of the trial are available from Cheshire Fire Brigades.)*

4. Networking and technical development

The project scope has concentrated upon meeting the needs for fire-fighters to have good information and environmental equipment for dealing with all types of incidents. The Brigade is also active in introducing electronic ailing facilities between those agencies involved in the management of the offsite effects arising from major hazard incidents. These major accidents require considerable skill and judgement not only at the incident scene, but at those supporting emergency headquarters attempting to resolve the offsite impacts. The use of modem technology, i.e. cloud plume monitoring, lap top computers and close circuit television in this area is seen as greatly aiding the tactical management of the incident and reflects considerable research already undertaken. The electronic mailing of information concerning actions taken, in real time, together with regular bulletins for more general distribution will, it is believed, prove of considerable benefit to those other participants who have either a legitimate demand for information (7), like local municipal authorities, or who need to interact with tactical decision making, like Police or Health Authorities. Electronic mailing is seen as a positive way of creating this broader network.

5. Major incidents tactical response

The safety record of Cheshire's industry is good and whilst the Brigade's operational Cloudburst Plan (6) for major toxic gas emissions is aimed at all eventualities, in the vast majority of the emergency services response, it results in immediate mitigation following an accident. In most cases the risk remains fairly low being contained on site. The Fire Brigade's part in this response is therefore well tested and proven and historically it has been seen that in the response to Cloudburst incidents there is a need to vary the degree of activity in relation to the situation being confronted. This has led to proposals for a graded response to incidents. This would allow every participant if they so wish to predetermine the weight of response relevant to the potential risk. It would also allow industrial operatives to exercise a professional judgement at the source without over compensating the essential need for safety. The resources deployed would have to be sufficient to meet the risk and allow for expansion if the potential increased.

Principal concerns are the need to:

a. Relieve public anxiety by mobilising only those fire appliances or other vehicles to meet the risk.
b. Give more accurate public information relevant to the risk.
c. Ensure a basic level of resources is available at all incidents.
d. Allow professional decisions to be made at the source.
e. Permit the overall Emergency Services and Local Authorities response to reflect the incident activity.

It is therefore proposed to have three initiation levels of response:

Level 1 On site risk with limited potential to off site
Level 2 Off site with no major potential to harm
Level 3 Off site with a potential to harm

After the initial notification from an industrial site indicating say a Level 1 response, the Police and Fire Brigade will upon responding immediately assess the extent and nature of the incident in conjunction with the companies, specialists and scientific advice to confirm that the initial level notified is correct. Cloud plume prediction and gas monitoring are focused on upgrading existing Cloudburst sector maps by applying radii parameters at 500m, 1 000m, 1 500m with grid map references and local land marks, together with Draeger tube sampling equipment for fire-fighters to go off site gas monitoring. This improved procedure will allow the Fire Brigade to liaise with the Company's off site monitoring teams and confirm the level of categorisation by sampling off site toxicity levels.

The existing communications for a Cloudburst incident between companies and all other emergency agencies are by telephone contact with established links in their on site Emergency Works HQ to the off site Emergency Services Reinforcement Base. Other emergency service agencies communications are radio schemes, cell phones and foxes which are not networked and incident information is not centralised. During an incident there is a plethora of information from numerous sources from which various agencies have to act and respond. In order to respond effectively and make balanced decisions and judgements it is essential that this information is firstly shared with all agencies involved and then, in appropriate ways used to inform politicians, the media and the general public. With the range of technical experts involved in an incident it is imperative that they see the incident unfolding in front of them. Only by doing this will we be able to maximise available knowledge and expertise which will allow the agencies involved to have an effect on the outcome of the incident.

The safety for fire-fighters and ultimately all those directly affected depends on effective command and control at the incident scene. The HMS project is designed to enhance and support that given by technical advisers. Clear tactical decisions need to be made at the scene by the Fire Service. Inevitably this impacts on other agencies which are then required in some cases to act rapidly. They therefore need to receive live information, in an explicit and understandable way. The process of feedback of occurrences and decisions taken at the scene is therefore vital to creating the effective team approach which will aid a successful outcome to an incident.

The current view as suggested above is that the need exists to create an efficient method of joint logging of key messages and decisions so that all five principal players, Company, Police, Fire, Ambulance and Local Authority are involved. In the long term as better technology becomes available it may also allow the strategic decision making process to move further away from the scene and for the Fire Service to, the Fire Brigade Control Room. This would allow the tactical decision making to take place at the scene of the incident and the strategic response to come from the centre where there is access to the widest information sources. In turn technology would also allow the constant sharing of information with other services through their controls or key locations such as District Off Site Emergency Centres. In this way the strategic decisions will be shared and immediately disseminated. This is a logical evolution which has tremendous resilience and strength.

6. Off site monitoring
Local communities invariably turn to their emergency service for assistance to prevent the effects of damage upon the environment. Time and equipment are a significant factor because a measured response can be constructed involving internal and external expertise to mitigate the effect of say the accidental release of chemicals into the environment that

may have a significant effect on water and air pollution.

Personnel training and scientific advice are two issues, but equally there is the question of equipment. National data sources will provide information for physical and chemical properties and to the products of combustion in some cases. What they cannot do is identify the local weather and micro climate related to an incident. It is extremely unlikely that existing fixed sampling or weather equipment that has been strategically placed throughout the area for other purposes could be so fortuitously sited that it would assist in assessing emissions in an emergency, although wind data might be used to help estimate say plume dispersion.

The Brigade has therefore purchased three portable gas samplers which determine oxygen deficient or flammable atmosphere at the source of emission, together with three portable computers for local weather monitoring with additional gas samplers (Draeger) to determine the level of contamination off site by fire fighting teams. This equipment is made available on the Fire Service Operational Support Units (OSUs). To mitigate the effect of water pollution, consultation with the National Rivers Authority has resulted in the offer of absorbent packages to render first stage prevention. Should the incident escalate then pneumatic seals, neutralisers and booms will be made available on Fire Service OSUs to prevent entry of contaminants into the water causeways.

7. Public information

It is accepted that the better informed we are the more likely we are to take sensible actions. In an emergency that is particularly important. Finding the right platform at a railway station may aid a successful journey. Taking the right exit from a collapsing building can be crucial to very survival.

The avoidance of panic, the safe exit from a dangerous situation and a more sensible reaction do depend heavily upon how much information is available to the individual making the decision (8). In emergency situations we need to be able to use those daily systems and other means of accessing information. Visually explicit warning signs or exit route markings are simple examples but the process is more complex.

Quite often the decision has to be made suddenly, unexpectedly, and the ability to analyse the information being delivered to the brain may well be competing with a natural desire to run or hide from danger.

It helps to think about the problem in three phases. Before an event, during the emergency, and after an accident or injury has occurred.

Before the Event

Fire awareness is a good example of pre incident education. Different strategies work better with adults, who accept the event as credible to those who do not. Recognising such credibility issues can provide a focus on delivering a fire awareness message to an adult. By contrast a child may accept from an adult the fact a risk of injury exists but their interest and concentration may be limited or the threat of injury may make the subject under discussion frightening. Using role models or characters, in Cheshire the use of an elephant by the Fire Service (9), makes the subject interesting, funny and non threatening.

Messages about what to do in earthquakes in the Pacific rim area pose similar needs as to forest fires in France or hurricanes in America. The earthquake message may be to stay put and hide under a table, the forest fire may need evacuation along designated highways and hurricanes introduce the need to take shelter. Overall the message needs to be conveyed before the event to increase probabilities of survival.

During the Emergency

Rational thinking during any emergency is difficult. To those in the intervention service, trained in trauma situations, the coping strategies are high. To the public they are all threatening and coping strategies are not necessarily available.

Panic can be avoided as shown in human behavioural studies (10) if information is given. This is highly relevant in crowd situations. Again practical examples may be as simple as announcements telling people how to leave a building threatened by terrorists. In a complex building the use of zoned alarms may aid phased evacuation - an essential need in hospitals or high rise buildings. Flood situations demand advice on preparedness and options as to when and which route to chose when it is time to leave. Hurricane and earthquake situations develop a strong sense of threat which can be greatly eased by knowing the strength of seismic activity or how storm cloud formation is moving. Even at sporting events good information to the crowd can greatly aid and indeed avoid crush situations developing.

In town centres when mothers and children are separated by an accident occurring from say an overturned vehicles blocking a highway information on where to go to find each other will provide a massive reassurance to mother and child. Similarly the avoidance of injury to health from risks to the environment like radiation or toxicity require extremely positive and unambiguous public announcements.

After the Accident

The emergency having occurred then recovery and restatement of appropriate actions are a vital aspect of the public information system. Advising people when it is safe to go home is obvious. Less obvious is the need to let individuals know how to obtain water or basic help following a major natural disaster or how to obtain medical information following a toxicological event. These processes can further be extended through real tragedies to show how survival might have been an outcome if previous advice had been followed. Again by way of example we have used in UK the fire deaths through foam furniture to promote a change in furniture standards and to mount highly successful schemes to fit smoke detection.

8. Public information - a practical response
Cheshire Fire Brigade employ a full-time Media and Education Officer. Part of the individuals job is to respond to major incidents to liaise with the media and reflect the role that the Fire Service is undertaking. As a public servant, the officer has a duty to ensure that accurate timely information is communicated regarding incidents attended.

Experience has shown that in responding to the needs of the media there is a need to balance the needs and to adequately take care of public information.

Increasingly individuals and organisations are being held accountable for their actions whilst dealing with a disaster on an emergency. In the aftermath agencies disseminate actions and information and if these are found not to have been effective or in the best public interest, then the question of "accountability" is invariably raised.

The media has an editorial need and the Emergency Services have the need to communicate effectively. These two needs do not necessarily satisfy each other but they can it is believed, help each other and subsequently ensure more accurate reporting and concise information.

One incident, used to provide further thinking of the public information need, involved a very serious fire with toxic potential, at which the media could not gain access to the site - They were required to stay at the Emergency Services Reinforcement Base (ESRB) which was a Fire Station approximately one kilometre away. Here a media briefing centre was established where regular briefing sessions took place. Interviews were conducted and in addition hundreds of media calls were taken over a twelve hour period.

On the night of the incident copies of the Cloudburst plan was issued to correspondents which explained the Fire Service's approach to this type of incident and the Media Officer assisted in producing feature - two to three minute items for television - which accompanied the national and local news highlighting the approach to incidents and how, by pre-planning and training, the Fire Service sought to manage these types of incidents successfully

Whilst this information provision helped, the following day newspaper articles appeared which conveyed contrary information and raised public concern and questions in the UK Parliament (12).

As a result of these concerns a review of the educative needs of the public was undertaken by a working party drawn from the Emergency Services, Industry and Media. Following the first meeting the very common messages was a clear feeling from all those involved that the level of anxiety was above that of the perceived risk.

The working party observed there were needs to be addressed. For example, creating a better flow of accurate information to the media and improving the confidence with the community in relation to emergency response.

After the first meeting views were summarised and designed into a leaflet which was subsequently called "In case of emergency" (13). Interestingly the industrial members on the working party asked if the emergency services logos could appear on the cover as a form of endorsement. It was also considered that a simple action message needed to be incorporated.

The Fire Service for many years has used "Get Out, Get the Fire Brigade out, and Stay Out". What was required needed to fit the chemical incident and would promote the policy of "shelter", consequently, Go in - Stay in -Tune in - was decided upon together with an established telephone answer line based on existing one stop shop information network provided by County Council level Information Points. The points are on a "searching" group so that a call will consistently search until it finds a vacant line, in addition they are connected by an electronic mail system. A maximum of 25 lines can be staffed by professional information givers.

The leaflet also makes the case that the chemical industry aids the public stating we all need modem drugs and use consumer items and transportation - if we expect these goods we should accept the industries which produce them.

Indeed in Cheshire the industry employs 25,000 people and contributes £1.26 billion towards the wealth of the Country. Such quotations are seen as important in emphasising balance.

Information in the leaflet headed "Working together for your Safety" informs people that if they live close to an industrial hazard then industry, by European law, has to inform them of certain issues. The section goes on to explain that plans formulated in a co-operative way and that the emergency services and emergency planners constantly work and exercise their plans to ensure effectiveness.

Having given a positive picture the leaflet goes onto a step by step approach to explain "What would happen in an emergency"

This information page includes eight local radio station logos who have broadcast areas into Cheshire. Over a period of a year every news editor, both radio and television, has discussed the information leaflet with the Brigade. The level of support received has been most encouraging and following each of the meetings confirmation in writing of what was discussed and agreed as definite guidelines has been issued. The objective throughout, recognises that perhaps in the past there had been some inaccuracies, however, the media can help and the proposal should not only get the "public information" message right, it should also furnish the media with timely information which should greatly assist in the accurate reporting of the event.

The leaflet's importance is that it is a policy statement of those who have been involved in incidents in the past and therefore have a great deal of experience. The agreement of the words used in this public leaflet by the agencies involved bonds the common approach of shelter, not evacuate, and thereby provides a first step in a major educational initiative to help ensure all the community, the chemical industry and emergency response agencies are working along a common path of understanding, allaying fears and helping put industrial hazard risk in perspective.

One of the problems historically is the use of TV graphics supporting news items. Graphics intended to show this gas cloud for the incident quoted appeared on television. Working out the scale of the cloud by using the width of a river shows that at one point the cloud actually stretched to a town some 30 kilometres away from the site.

In reality gas monitoring taken at the incident showed safe readings some few hundred metres down wind from the site. The result of such graphics can cause anxiety to thousands of people. Having discussed this particular problem with TV editors the Brigade has now agreed to fax copies of Fire Service operational sector maps to TV stations.

Combined with scaled response and environmental monitoring this will allow the Brigade to give very accurate information regarding potential public risk. For example, if the Fire Service are registering gas readings at certain operational sectors then this information will be shaded in on sector maps and faxed to TV stations.

The final page of the leaflet again highlights the team approach and a quick response provided by the emergency services. It is intended to send the leaflet to every household in Cheshire and to launch it using the media as well as recognising them as key players in this communications process. The Brigade are now also encouraging industry to use the leaflet as part of the information they send out to householders in their legal notification of hazard zones.

Telephone lines will also be available for emergency information response 24 hours a day. In the day time they will operate from information points in public libraries which are staffed by trained librarians. Out of office hours the system can be brought on line in 20-30 minutes of call out from a special room at the administrative Headquarters for the County. In this room a team of three people will be responsible for formulating information and faxing it to radio and TV stations. This will be done utilising electronic mail fox systems. The target is to issue "public information bulletins" every 20 minutes

Public bulletins will have a pen picture of none speculative information regarding the incident plus a public information statement based on basic pre-determined guidelines such as :-

We are attending an incident at ...... involving a fire/leak. The wind is blowing from .... towards .... We are asking the residents of .... as a precaution to stay indoors and close doors and windows and stay tuned/switched on to .... radio/TV to listen for further information.

It should be noted that it is not intended to give wind direction as westerly or whatever. Similarly locations are stated by name such as a village or town which communities can identify with directly.

9. Conclusions

There is considerable need within areas having high hazard sites to develop public and fire fighting service information systems which are robust and capable of delivering high quality and unambiguous data which both aids successful tactical decisions and avoids necessary public anxiety. Cheshire Fire Brigade has sought to use technology to help in

References

(1)   Control of Industrial Major Accident Hazards Regulations 1984
       Fire Services Act 1947
       Control of Substances Hazardous to Health Regulations 1988
       Notification of Installation Handling Hazardous Substances Regulations 1982.
(2)   Emergency Planning for Industrial Hazards H.Gow, R.Kay
(3)   Environmental Protection "Fire and its Effects on the Environment" Seminar
       Documentation 27 January 1993 (IBC Technical Services Ltd)
(4)   Monitoring Options for Local Authorities by Sean Beevers (by IEMO March
       1992) On-Site Meterological Program Guidance for Regulatory Modelling
       Applications Paper by US Environmental Protection Agency. June 1987.
       Shell Research Papers: H G System Gas Dispersion Model, H G System II
       Dispersion Model, Development of Open Path Systems for Omission
       Rate Measurements.
       Cheshire Fire Brigade Operations Group Order 19/1 -Incidents Involving Large
       Scale Toxic/Flammable Gas Emissions.
       Manual of Firemanship HMSO
(5)   Cheshire County Council Management Consultancy Report (July 1994).
(6)   Cheshire County Council Cloudburst Procedure for Major Toxic Gas Emissions.
(7)   Dealing with Disasters HMSO Publication.
(8)   Quarantelli EL Evacuation Behaviour and Problems: Findings and Implications
       from the Research Literature.

(9)   Welephant: A character developed and used by UK Fire Brigades to promote fire safety awareness.

(10)   Sime J D - Human Behaviour in Fires: Summary Report. JCFR. FROG Report 45.1992.

(11)   The Associated Octel Fire 1994 - Cheshire Fire Brigade Report.

(12)   Hansard Emergency Debate - Associated Octel - February 1994.

(13)   In Case of Emergency Leaflet - Cheshire Fire Brigade.

46 AN EMERGENCY PLANNING TEAM AS A SOURCE OF PUBLIC INFORMATION

S.M. CHAMBERS
Get Science Write, Calbourne, UK

Abstract
When the public become aware of an incident, they want information. This is especially true if questions about health and safety arise. Information may be gained from a variety of sources such as the mass media, gossip or experience. For those in local government with a statutory obligation to provide information [1], being believed and taken seriously by the public, are important.

Credibility, trustworthiness, familiarity and competence, as perceived by the public, are significant to anyone providing information. Public opinion and perception of the source of information influence the effectiveness of communication.

The Isle of Wight Emergency Planning Team is addressing these issues. The Team is collaborating with a public communication specialist in an attempt to produce a coordinated information response as an integral part of their emergency plans.

Quality of information is increasingly a priority. Focusing on information style and content is supplemented by raising the Team's public profile. This paper looks at issues raised by the research, case studies of relevant issues, and the methods the Team is using to improve its communication with the public.

Keywords: Credibility, Emergency Planning, expertness, familiarity, *Get Science Write*, information source, risk, trustworthiness.

1 The use of a specialist communication consultant

The Isle of Wight Emergency Planning Team recognises the need for effective communication with the public during an incident. The Team has started to use a public communication consultancy called *Get Science Write*. *Get Science Write* has advised the Team on the preparation of public information.

Get Science Write is based on a unique combination of scientific training and research, coupled with communication expertise. The services provided by the consultancy mean

Fire Engineering and Emergency Planning. Edited by R. Barham.
Published in 1996 by E & FN Spon. ISBN 0 419 20180 7.

that information received by the Team regarding an incident can be interpreted and then 'translated' into a form suitable for the public. Information may come from Central Government or the Team's own advisors.

From late 1994 to date *Get Science Write* has been advising the Team on public information policy. The Team and *Get Science Write* have started to coordinate policy to place public information prominently in emergency plans.

Collaboration began with the consultancy researching the field of public understanding of science. This lead to research into the public perception of risk, risk assessment, public perception of information sources, the role of the mass media in opinion forming and media interpretation of risk issues. The research highlighted areas relevant to the work of the Team. A meeting in September 1994 allowed the points raised by the research to be discussed with other agencies involved in County Emergency Plans.

Research findings showed that a trustworthy, expert source is needed for the public to react positively. Information had to be accurate and unambiguous. Among those providing the information, there must be agreement about what is being said, a 'singing from the same hymn sheet' approach. Information must be put across in a style easily understood by the public. Problems may arise when the public perceive a risk when they are not in danger. All these points have relevance to the public communication role of the Emergency Planning Team.

Three examples, taken from the range of work researched, highlight the relevant points. First is public reaction to information provided following the explosion at Chernobyl in 1986. The second is a study of problems faced by the authorities in New South Wales, Australia over the siting of a hazardous waste incineration facility. Third is work studying public perception of risk.

2   Chernobyl

The explosion at Chernobyl in 1986, provided those studying public reactions with an excellent research opportunity. The work done by Hans Peter Peters, in the then West Germany [2], describes the 'information disaster' which followed the Chernobyl accident. In his analysis Peters examined the role of sources of information, the public perception of these sources and source credibility.

This 'information disaster' is just the type of mix up that the Emergency Planning Team is looking to avoid. An example is as follows,

"While the Minister of the Interior 'absolutely' ruled out any danger for the West Germany public in a television interview, the Government, at the same time, asked farmers not to let their cattle graze outside and imposed strict controls on incoming food supplies at the eastern boundaries of West Germany."

Peters points out that even the Federal Government did not have a clear or comprehensive picture of the preceding events and the risks posed to the public. The responsibility for those charged with informing the public is to tell them what is known and to be consistent. If things are unclear then say so. Any short term outcry is nothing compared with the damage done by later exposure of cover-up or incompetence.

What result did the Chernobyl disaster have on the public perception of information sources? Peters studied public reaction to seven information sources; Federal Government, Political Opposition, Nuclear Research Centres, the Nuclear Industry, an Ecological

Institute, action/pressure groups and journalists. Peters looked at four factors in his analysis; familiarity, public interest orientation, competence and credibility. Peters used a study group of around thirteen hundred members of the public.

2.1   Familiarity
The public were asked how familiar they were with certain organisations and institutions. Very familiar were the Federal Government and the Political Opposition. Unfamiliar were the Nuclear Research Centres and the Ecological Institute.

2.2   Public interest orientation - 'trustworthiness'
Do the public view an organisation as having its own interests at heart? Or does the organisation have the public interest at heart? The public believed the Federal Government and Political Opposition had the public interest as their primary concern. Contrasting this was the nuclear industry and journalists. Public interest orientation can also be called 'trustworthiness'.

2.3   Competence - 'expertness'
Peters asked those surveyed how competent they felt the sources were to give advice. Highly competent were the Nuclear Industry and the Nuclear Research Centres. The Government and Opposition were close joint second. Least competent were journalists. Competence can also be called 'expertness'.

2.4   Credibility
Those surveyed were asked to assess the credibility of the seven organisations as a source of information. The Federal Government and Opposition came out on top followed by Nuclear Research Centres, action groups, journalists, Ecological Institute and last the Nuclear Industry. The credibility of a source comes from a combination of 'expertness' and 'trustworthiness'.

3   New South Wales

In the paper 'Siting a hazardous waste facility: The tangled web of risk communication' [3], Beder and Shortland discuss aspects such as source credibility and perceived risk. Their discussion centres around the siting of a hazardous waste incinerator in rural New South Wales, Australia. The work of Beder and Shortland is an interesting study for those involved in public communication.

Beder and Shortland looked at two players in this case study, the Federal and Local government's Joint Taskforce on Intractable Wastes (the authorities) and Greenpeace Australia.

Neither of these groups lied, yet they used their own view of the technology of waste disposal to support their stance. The information battle between the two is important to the Emergency Planning Team as it highlights how the authorities lost credibility by their actions during the long running public debate.

In their conclusion Beder and Shortland identify four flaws in the authorities communication process. These four examples are of particular interest to Emergency Planners.

1 The authorities portrayed the ideal technology operating in the ideal world. This didn't wash with the public and the authorities lost credibility.
2 The efforts made and the lengths to which the authorities went to reassure the public were seen as salesmanship.
3 'Gaps' in information and the lack of answers to some public questions lead to the communication of the 'opposite' message, from the authorities to the public.
4 The authorities failed to consult the public. This destroyed the idea that the authorities were acting in the public's best interests.

3.1 Lost credibility

The authorities took one extreme and Greenpeace Australia the other. The authorities presented the hazardous waste plant operating in an ideal, human error free world. Greenpeace Australia took the other extreme and put forward the 'worst case scenario'. Beder and Shortland suggest that,

"...such polarized models of technological systems can be found in many technological controversies."

The authorities must take up some stand point but this extreme view of simplicity was not credible and was attacked by Greenpeace Australia. The public sided with these attacks, as the position of the authorities was not credible in the 'real world'.

3.2 Perceived salesmanship

The authorities, in the face of the Greenpeace campaign had to go to great lengths to try to reassure the communities that were short-listed as sites for the new facility. But, as Beder and Shortland point out,

"There is also some evidence that messages of reassurance inadvertently communicate insincerity and dishonesty. The contractions and incongruities that arise from the need to reassure rather than openly inform...are easily picked up by those who are most likely to be affected and are amplified by opponents."

3.3 Unintended, opposite messages

"If this facility is as safe as the authorities claim, why is it 100 kilometres from Sydney?" The unintended message from the authorities gave those living near proposed sites the feeling of being 'sacrificial lambs'.

The authorities assured the public that the new facility would 'be away from any environmentally sensitive areas such as wetlands, national parks and significant streams and lakes.' One proposed site for the facility was only two kilometres from the Murray river. The Murray is one of Australia's major waterways, used for drinking water and irrigation in three states. At public meetings the public asked, 'Is the Murray not a significant waterway?' As Beder and Shortland note,

"The failure of those officials to give what locals considered to be an adequate answer to this and other questions communicated more to the audience than all the purposeful, reassuring statements they made all evening." Beder and Shortland also note that in this situation, "Learning to say, "I don't know" may be one of the most difficult communication lessons."

3.4   Failure to consult

The combination of lost credibility, reassurance rather than information and unintended messages combined to leave the communities affected with the feeling of being left out. Indeed the whole campaign by the authorities appeared to the public to be centred around the idea that, 'Have faith; we are in charge.'

This simplistic approach from a source lacking credibility showed how poor handling of a vital communication programme can lead to great difficulties in resolving an issue.

4   Risk

Whether an incident occurs or not and whether any danger to the public occurs can be irrelevant. The problem of risk and its perception by the public should be a major consideration to those involved in providing public information. There are two relevant aspects of risk, actual risk and perceived risk.

4.1   Actual risk

Actual risk places the public in physical danger. The idea of an odourless, tasteless, invisible yet physically damaging force is very difficult to understand. Yet this could be a description of ionising radiation. If another incident such as Chernobyl occurs, where the public may be in physical danger, they must understand or be made to understand that they are in danger. Public behaviour can be contradictory. If an incident occurs, no risks may exist, yet panic may follow. When there is real physical danger the public may be very slow to accept that anything is wrong. To overcome this possibly fatal contradiction, effective communication of risk is needed.

Everyday examples should be used to communicate risks. Is it possible to compare the levels of exposure to having a dental X-ray or sunbathing? The public are not experts, but they are not stupid and do not need patronising. It is not a good or successful idea to tell the public they are being exposed to becquerels or millisieverts of radiation. Such descriptions mean nothing to the layman.

4.2   Risk as a social construct

Susanna Hornig studied the public perception of risk. Her work brings a new aspect to the consideration of how the public react to the risks [4]. Hornig states,

"Risks are socially constructed; they are interpreted (whether by the lay public or by the scientific elite) in a particular social and cultural context."

This idea is of particular interest to those involved in public communication. The public consider risk not in the ideal world in which risk calculations are made but in the real world in which they live. The public are aware of, and place importance on, factors that may not be considered by those wishing to describe the levels of risk.

Public perceptions cannot be viewed as right or wrong, they are arrived at by using a different set of values from those used within, say, the scientific community. 'Irrational' is a word used by some scientists to describe public reaction to risks. Hornig suggests

"The public thinking may be 'irrational' to the extent that it differs from scientific thinking."

So where does this leave the Emergency Planning Team and those others involved in communication of scientific information and risk? Hornig points out

"No amount of information on probabilities of harm - however phrased - will serve to create a favourable climate of public opinion unless social context issues are also addressed."

People may be more concerned about how a new technology or risk fits into, and can be controlled by, existing social systems. The public seem to find issues of ethics and regulation of new risks or technology more important than data and statistics. Whereas the scientific community will accept new risks if 'adequate testing and monitoring' has been carried out, the public view the risk in a much wider context.

5 Putting research findings put into practice

Source credibility is very important to the Team. They have a statutory duty to provide public information. The work necessary is now being reviewed in the light of *Get Science Write*'s research findings. Fulfilling statutory obligations to a minimum level, is not enough. The research findings point out steps that can be taken to provide effective public communication. So where has this research lead the Team?

A public exhibition has been commissioned from, and produced by *Get Science Write*. Familiarity, expertness, trustworthiness and credibility, issues raised during the research, are addressed by the style and content of the exhibition. This portable exhibition tries to set local Emergency Planning, its work, future projects and resources in context. The intention is to use the exhibition at public meetings and events.

By taking information to the public the Team hopes to avoid the 'salesmanship' perception. The public are drawn into the planning process by offering information and opportunities for involvement. Letting the public get involved with planning tries to prevent any ill-feeling over a failure to consult.

Bearing all the points in mind and testing them during exercises should ensure that no unintended messages are sent to the public. By combining all the findings of the research the Team will be in a strong position. A familiar and credible source is much more likely to be listened to and understood by the public. When the issue of actual or perceived risk arises, the Team can produce information that should satisfy the public. The use of a specialist to produce suitable information coupled with their credibility as a source will maximise the effectiveness of the Team.

In practical terms the Team now have an exhibition to show the public. As a first step to providing information the Team have information sheets on file ready to use. Currently, the prepared information covers a radiation emergency. These sheets only need specific details of an incident to be added and they can be copied and distributed. The future of the Team's information campaign involves having information prepared and ready to amend as soon as details are confirmed.

Information needs to be controlled during an incident. This is not to stop people finding out the facts, but the quantity of information given out should reflect the seriousness of an incident. A constant flow of information feeding saturation media coverage may not be appropriate.

Now that the Team have theoretical confirmation of the main points of public communication, they are ready to monitor future events in a new light. *Get Science Write* is also continuing to research public communication. Accidents and disasters do not all happen in one place, so *Get Science Write* is now gathering more information on the response strategies of other Emergency Planning Teams.

The use of a specialist has raised a series of points that the Team can now work on. Progress has been made. Future work will involve the production of more prepared information that can provide a quick, accurate response to a demand for information. The media's role and agenda are also topics for future research. *Get Science Write* and the Team are at the start of a long term effort to produce an effective information system able to cope with any emergency. Further study of prepared information will take place in the future. Exercises scheduled for early 1995 will provide an opportunity to test this prepared work.

Principle references

1 The Public Information and Radiation Emergency Regulations (PIRER) (1993) Statutory Instrument No. 2997 1992.
2 Peters, H. P. (1992) The credibility of information sources in West Germany after the Chernobyl disaster. *Public Understanding of Science* Vol. 1 pp. 325 - 343.
3 Beder, S., Shortland, M. (1992) Siting a hazardous waste facility: The tangled web of risk communication. *Public Understanding of Science* Vol. 1 pp. 139 - 160.
4 Hornig, S. (1993) Reading risk: public response to media accounts of technological risk. *Public Understanding of Science* Vol. 2 pp. 95 - 109.

Other references

Barker, F. (1992) South Yorkshire Fire and Civil Defence Authority, Public information in the event of a radiation emergency. Assessment of the research literature and proposals for future work.

Barker, F. (1993) South Yorkshire Fire and Civil Defence Authority, Public information in the event of a radiation emergency. Responding to public inquiries.

Carver, S., Myers A. (1993) The Role of Public Perception in the Response Planning for Major Incidents: Public Perception and Memory Retention Questionnaire Survey.

County Emergency Planning Officer, (1993) Public information and guidance on accidents involving radioactive incidents. Isle of Wight County Council.

Dakin, J. (1994) A media master plan is a must. *Civil Protection.* Issue 31 Summer pp. 12 - 13.

Dunwoody, S., Paters, H. P. (1992) Mass media coverage of technological and environmental risks: A survey of research in the United States and Germany. *Public Understanding of Science* Vol. 1 pp. 199 - 230.

Frewer L. J., Shepherd, R. (1994) Attributing information to different sources: effects on the perceived qualities of information, on the perceived relevance of information, and on attitude formation. *Public Understanding of Science* Vol. 3 pp. 385 - 401.

Handmer, J., Penning-Rowsell, E. (1990) *Hazards and the communication of risk.* Gower Technical ISBN 0566 02784 4.

Health and Safety Executive, *Arrangements for responding to nuclear emergencies*, HMSO, ISBN 0 7176 0828 X

Home Office, (1992) *Dealing with Disaster*, HMSO.

Knorre, H. (1992) "The star called Wormwood": the cause and effect of the Chernobyl catastrophe. *Public Understanding of Science*, Vol. 1 pp. 241 - 249.

The Royal Society, 1985, *The Public Understanding of Science.* ISBN 0 85403 2576

Warner, F. (1992) Calculated Risk, *Science and Public Affairs,* Winter edition.

Waterhouse, R. (1994) The balance of power, Feature article, *The Independent on Sunday Review Section.* pp. 10 - 12.

Wilkinson, J. (1992) Channels of communication, *Science and Public Affairs,* Spring edition.

Wynne, B. (1992) Misunderstood Understanding: social identities and public uptake of science. *The Public Understanding of Science.* Vol. 1 pp. 281 - 304.

47 A CRITICAL REVIEW OF HUMAN BEHAVIOUR IN SHOPPING MALLS

P. HUMPAGE
West Midlands Fire Service, UK

Abstract

Over the last two decades numerous surveys have indicated that people have died or been injured as a result of inadequate provision of means of escape within buildings of all types, particularly those used by the general public. as a consequence legislation within the U.K has been amended or issued with pressure from the general public, fire authorities, and insurance companies and is incorporated within the present standards of the building regulations.

The effects by external influences on people in a mall within a shopping centre are complex, but has the area of the actions of these people within the mall been fully explored. In that the area of emergency exits provided within bear a true representation of those that will be used within an emergency. Rather than those provisions within the regulations.

The findings of this study is based upon a structured questionnaire of five hundred randomly selected shoppers within the Merry Hill Centre Birmingham.

The result of the survey strongly indicate that the exits specifically designed to be used during an emergency will not be fully utilised, and the normal (entrance/exit) will be over subscribed. The survey however does not take into account the interaction of other variables, and as such should not be taken in isolation.

Human behaviour, Methodology, Survey, fire alarm, fire exit.

Introduction

The hypothesis put forward in my discussion is that in an emergency the occupants that are expected to use the emergency exits will be below that which they are designed for, and that more people will exit in the direction by which they entered creating problems in design. In addition to the evacuation of the occupants the introduction of emergency exits has implications at the design stage, and in the financial aspect of reducing the retailing area available as a consequence of usable floor space being taken up with emergency exits.

Fire Engineering and Emergency Planning. Edited by R. Barham.
Published in 1996 by E & FN Spon. ISBN 0 419 20180 7.

Methodology

In order to investigate the possible actions of individuals within the mall. I devised a structured questionnaire which attempted to identify factors that contribute to the actions of individuals within a shopping mall in the scenario of an emergency situation. With the questionnaire covering a sample of 500 people.

Survey Analysis

To gain an understanding as to peoples reactions the survey identified three areas.

1 Analysis of seeing a fire
2 Analysis of fire alarm sounding
3 Analysis of method of exit

For the purpose of my study this provided the natural progression of events for people within the mall and identifies the factors that influence the actions of occupants with a shopping mall.

The questionnaire provided useful data I could compare to already established ideas on human behaviour and consequently areas of the building regulations in relation to the number of emergency exits and there requirement within a shopping mall.

To provide some continuity through the actions and events that may take place in the given situation of a fire. Then the area of research comes in the last part - that of the direction chosen. Before this however is the identification of a fire, the reaction of the fire alarm sounding and finally the exit used.

Analysis of action of seeing a fire

The difference between male and female in their actions in seeing a fire are well documented in that. It would normally be expected for females to tell people in authority or people around them, needing conformation of what they could see and as is the case where the female is with her partner, informing them.

Males on the other hand should be more inclined to raise the alarm and tell people around them. The results of the survey showed up some striking similarities in the options available to the individuals concerned as opposed to what would be expected.

Fig.1 Analysis of action of seeing a fire

		Find someone in authority	Tell people about you	Raise the alarm
Female	269	31.97%	20.45%	47.58%
Male	231	29.00%	19.91%	51.08%
Totals	500	30.60%	20.20%	49.20%

The difference between male and female in the answer to tell people around you was very similar at 20% and 19% respectively fig.1.

More surprising was the number of males that would look for someone in authority in comparison to the females with only two percentage points between the two. This indicates that either the males used for the survey are of a lower self esteem than that of the females or, assuming that the intelligence of both groups is spread over the full spectrum, the actions and percep- tions of individuals within the sexes has changed in that more males would look for someone in authority has been previously recognised.

It follows that if this is the case then individuals are going to more inclined in a situation within an enclosed shopping centre to take notice of security personnel. This can be used to the advantage of controlling the people in getting them to move in a direction by the staff.

But by far the most striking thing was the number of females who were prepared to bring attention to themselves and raise the alarm, it was expected that this would be a high number among the males, but the results show that the females were only 4% behind the males, in taking this course of action.

If it is that the individuals used for this survey is bal- anced, then there has been a dramatic change in perceptions of individuals, in comparison to previous surveys, especially that of females, in there actions. Perhaps the reasons behind this are of a social nature. What is highlighted is that overall only 20% of people questioned would be prepared to stand and shout, to make others aware of the situation. Fig.1

Having looked at the reactions to the discovery of a fire, natural progression to this would be the sounding of the alarm within the ESC and the reactions of individuals within the mall, both in local to the fire,and more importantly unaffected areas of the mall.

Analysis of action in fire alarm sounding

As with the previous section, the use of sexes was the most important, and as such Fig.2 shows the results in that the dif- ference between the males and females, within each choice, is minimal.

By far the largest group were those who elected to leave the mall, 63% in total, those remaining being split between ignoring the danger, at 24%, and carrying on shopping at 13%. Fig 2

Fig.2

		Ignore danger until identified	Carry on shopping	Go out
Female	269	23%	14 %	63%
Male	231	25%	11%	64%
Totals	500	24%	13%	63%

It would appear from first impressions that the latter two are the same response in that neither would leave the mall. This is incorrect in that the individuals who said that they would ignore it until identified were aware that there was some form of problem but were not prepared to take any action until they could perceptibly identify the danger, and if it was going to affect there plans that were already in motion.

Of the individuals who said they would carry on shopping were not prepared to have obstacles put in their path from what they wanted to do, (This was graphically illustrated in the fire at Bradford Football Stadium, and Hendersons in that although there was an identifiable danger the individual perception was one of continuing the designed routine).

These results show that although 63% of people are going to leave immediately there is still going to be substantial number of people who are going to stay within the centre. The implications of this on the escape time that are required to empty the mall are that it will take longer than envisaged. Especially if there is a movement of people from the shops on to the mall as there undoubtedly would be.

Allied to their reaction in the event of the fire alarm sounding was what does the fire alarm sound like? 27% of the survey were unable to select from either of the options given to them. This means that if correct, the fire alarm sounding then a quarter of the people would take no notice. Of this quarter, only 19% of the respondents said that they would leave if the fore alarm were to sound. If this is correct then the number of people evacuating the mall immediately would only be 44%

This has implications on the movement of people within the mall with regards to the smoke control methods and fire suppression.

Analysis of method of exit

This is the main area for the project research, the results of which are shown in Fig.3

Rather than split the results by sex it creates more meaning by age as the data provided gives some insight into the behavioural actions of groups with different responsibilities, and abilities.

To obtain theses results and to give some relevance to the results The survey was in sight of a normal means of exiting the building, with only a third of the distance to two emergency exits. It was hoped by giving individuals a selection of exits to choose from they would take time to, look around before deciding on there action.

Fig.3

Age	Sample	Way they came in	Nearest Exit	Don't Know
0-16	32	46.75%	53.25%	0%
16-21	53	49.06%	47.17%	3.77%
21-35	161	44.72%	54.04%	1.24%
35-50	142	57.75%	42.25%	0%
50 +	112	48.21%	49.11%	2.68%
Totals	500	49.60%	49.00%	1.40%

In comparing the results obtained against age group there are some striking differences between the ages of 21 - 35 and 36 -50. In the other three age groups the split between the two answers was nearly the same. Yet in these two groups the deferential is in excess of 10% Why should this be so? I can only surmise that reasons behind this could be anything from education standing, greater awareness,or perhaps ethnic origin of the individuals within the groups.

The results as a whole show a split of nearly half to the two questions available (Fig.3. This means that if it were a real emergency then 50% of the occupants would exit the way that they entered, or more importantly choose an exit that offers an normal escape from the building.

Before looking at the implications of this the results high-light some other interesting facts, in that I had expected ever-yone, given the choice of two, to select one of them. Surpris-ingly, there was a percentage that had no idea what their actions would be. The group with the highest incidence of this was the group 16 - 21 at 3.77% with overall of the sample at 1.4%. It could be that this group is tied into the fact that 4% of the survey were unable to decide how the would find an emergency exit. If this is the case then this indicates that if these individuals are not going to make a decision as to there choice of direction that they are going to follow everyone else.

Another major factor that is going to effect an individuals selection of route is there familiarity with the building. What if any effect this has on the results cannot be measured, as this was one area within the questionnaire that was not covered.

Perhaps the reason for such a high percentage of people choos-ing the way that they came in was the unknown of what lay behind an emergency exit when comparing it to the knowledge of the ground between them and the way that they came in. Another summation would be that people would know where to find there vehicles with reasonable ease. The guidelines for the means of escape from buildings as been well researched within areas of the shops. These same principles have been adopted to the mall areas of the ESC but as the results have shown the direction that individuals will take is different in this situation of that of a shop area.

Implications on means of escape

The implications of these results on the means of escape are far reaching in the legislation that covers the area of the fire exits and the normal means of exit from the building taken to-gether when assessing the number of emergency exits that are required.

Using the basis with the building regulations of the area of the mall divided by 0.75 to give the number of occupants that the exits have to accommodate. It therefore follows that the number and size of emergency exits is going to have to be significantly larger in actual size that the normal means of exit in that as a rough guideline to shopping malls the split is in the region of 80% of the fire exits with the remaining 20% taken up with the normal means of exit.

It means that if in this case of the questionnaire, and the mall were at a capacity which is allowed for them, the normal means of exiting the building would be over subscribed and the emergency exits under subscribed.

It is interesting to note that of disabled people there were in fact two that I found to question. In both cases their answers were the same, that of going out the way that they came in although this sample figure of two is too small to provide any conclusions of this group of people, it does indicate that the disabled are going to exit the way that they came in. The reasons for this are perhaps ease of movement.

The implications of these results cover more than one area of concern and have effects on not only the ESC but also other public buildings. Under present legislation a developer of an ESC has to set aside a proportion of the available floor space adjacent to the mall for the purpose of emergency exits. The cost, not only the developer but also the local government, the infrastructure of the area in creating jobs and revenue, can in the very worst scenario be jeopardised by this loss of space.

If conditions were to change the availability of space could be put to far more productive use rather than them being set aside for the public to use as emergency exits. This may appear on the surface as a cavalier attitude towards the safety of the occupants of the mall. If it is that my results are accurate to +\- 15% then they have some validity. That is not to say that in the actual event peoples reactions are going to be as stated.

With the introduction of external elements including smoke and all its hazards, a level of panic, and the increase in areas of heat, and the locality of the fire are all going to affect the decisions of individuals and their interactions with others.

If it is that the regulations, as my survey suggests, are out of sync with the actual reactions of individuals then the solution to the problem of evacuating the occupants needs to be looked at from a different direction.

An idea of this is given in some of the other results that I obtained from my survey.

The alarm

Is there a need for there to be an audible alarm affecting the whole area forcing people from the shops into the mall? If we upgrade the detection system with a greater use of video surveillance and silent alarms to which security staff would to attend to assess the situation before deciding on whether there is any need for the movement of people. For a system of this type to be effective then there are several areas that need to be addressed in that the control of smoke and its prevention of its entering the mall is paramount even to the point of not utilising the roof areas and its prevention of its entering the malls as smoke reservoirs.

This reduction in the role that the mall plays within the smoke control of the ESC will mean that on of the factors affecting the human behaviour within the mall has been cancelled out. This solution is possible with modern forms of design, the implications are obvious in that the initial cost of construction will be greater in that the expunging of the smoke would have to be through ducting within the shops, a higher level of training the personnel within the ESC would also be required to :-

* Assess situations
* Take the relevant action on there own intuitive

All of these changes will mean that the action of the people within the mall are going to match the results from the question-naire if the situation to remain as it is now.

There are limitations to the questionnaire. The results obtained were over one day. Variations in the results could occur through repeating the questionnaire on a different day and different place.

To prove conclusively the finding of this project, a large scale investigation would be required covering a number of ESC's throughout the UK. Therefore the reuse of the areas of emergency exits into shops creating greater revenue, coupled with the redesigned use of the malls as area of total safety will ean that if it is required, than the total evacuation from the malls can be made in a far more orderly manner than would otherwise be the case. This situation can only be catered for the majority of circumstances, and there is a downside in that there may be an event, such as an explosion, that would have catastrophic reper-cussions. As we cannot plan for every eventuality, the risk of a situation such as this arising is that of minimal. As to make the cost of incorporating extra forms of safety into the mall as being beyond economic reasoning.

Summary

For a system such as I have described to work there have to be changes with the shops in addition to those already mentioned in that there are two other areas that could be changed.

The first is the introduction of tighter legislation on the types of materials used in the use of items produced for selling. This lowers the combustibility of the shops and reducing the possibility of smoke being produced.

Secondly the greater drive by the construction industry and all the satellite operations involved around it in developing new forms of building materials and more efficient, effective and reliable forms of extraction systems of smoke from the shopping area.

CONCLUSION

This project set out to identify whether the present legisla-tion covering the egress of people from the mall was appropriate in its assumption that people within the mall area or entering the mall from the adjacent shops in the event of a fire will use the emergency exits in the numbers that have dictated the space that needs to be provided for them. Or go out through one of the exits by which they entered. In addition to the main area the survey provided information closely connected to this in that:-

* Action of the individual if the fire alarm were to sound
* Action in the event of finding a fire

The reasons behind the selection of exit that the individual takes can be due to any or a number of the following, the sex, the age of the individual, the ethnic origin, level of intelli-gence, panic of others, the visual hazard of smoke, noise of fire alarms, changes in perception due to smoke inhalation, and loca-tion of the car. The results of the survey show that under normal conditions 50% will exit by the way that they came in, the other 50% said that they would leave by the nearest exit. 4% of the latter did not know what the alarm sounded like, so how are these people going to know that they are to leave?

The survey also shows that at least 37% of people are going to fail to leave immediately. Therefore, is the sounding of a general alarm the best method of evacuating? If it is, a more subtle approach is used to convey to people to leave the mall in the event of fire, of for any other reason. If, as the results show the mall areas are able to be kept clear, then the orderly evacuation of the occupants can take place at a leisurely pace meaning that the exits provided for normal movement in and out of the mall can be used without the need for emergency exits to be fitted within the mall.

The options available are:-

1 Improvements on fire legislation with regard to materials used within the structure of the shops and changes to the building regulations dealing with electrical installations.
2 No smoking in the malls or shops.
3 Better training for the security staff and centre management.
4 Public awareness of the dangers of fire.

Not only do we need to develop subtle methods of evacuating the mall, but also the training given to security within the centre, coupled with improvements in technology and the understanding of human behaviour will enable the safety of the occupants of the Enclosed Shopping Centres to be improved.

It therefore follows that by keeping the variables to a minimum wherever possible the findings of the survey can be used to design and implement better ways of emptying the mall through the greater use of information technology, and an investment in the security with the ESC.

Limitations

This project is limited in scope and further research is required to obtain a more meaningful conclusion. Perhaps comparing the views of shoppers of various ESC's so that a greater weight can be added to the results.

48 ASSESSING OCCUPANT RESPONSE TIME: A KEY ISSUE FOR FIRE ENGINEERING

J.D. SIME
Jonathan Sime Associates (JSA), Research Consultants,
Godalming, UK

Abstract
Research indicates that Occupant Response (OR) time should figure prominently in predictions of escape times from a variety of settings (eg underground station, Channel Tunnel train, stadium, shopping complex, office, hotel, hospital). This paper reviews a prototype procedure for assigning the pre-movement phase of OR a design value and prioritising OR time as a key fire engineering performance criterion. Tpre is derived from a matrix of tpre best (b.p.s.), average (av.p.s.) and worst possible scenarios in response to alternative warning systems (w1 = alarm bell, w2 = non-directive prerecorded fire warning message, w3 = live directive public address + CCTV). The tpre av.p.s. is adopted, or the tpre b.p.s. is multiplied by an OR efficiency weighting, Weff, derived from ratings of eight tpre parameters B - I defined in the paper. This tpre adjusted design value for the occupancy is then compared with the benchmark criterion of the tpre av.p.s. = tpre b.p.s. x safety factor of 2. Further research is needed to refine, validate and calibrate the methodology through comparison of ratings, post-occupancy OR measures and fire scenarios. OR measures of occupancy risk (population profile), occupancy movement prior to direct escape and wayfinding design also need to be included in fire engineering calculations.
Keywords: occupant response, performance, warning, pre-movement, time, efficiency, post-occupancy.

1 Introduction

One of the most striking features of fire codes is the fact that there are no predictions of escape time in relation to warning systems. Escape times are assumed to be a function of population size, travel distance and exit widths. Research indicates that delays in warning the public of a fire threat are a consistent feature of fire disasters. An early response gives people more time to cover travel distances and pass through exits. There is a gradual realisation that the timing and pattern of Occupant Response (OR) needs serious attention. Since the time available for people to avoid a danger has to be matched against the time they need to respond and escape, OR is a key issue for fire engineering. The aim of this paper is to outline a method of calculating OR pre-movement time (tpre) for inclusion in fire engineering calculations of escape time.

Fire Engineering and Emergency Planning. Edited by R. Barham.
Published in 1996 by E & FN Spon. ISBN 0 419 20180 7.

2 Occupant Response (OR)

2.1 Definition
The term Occupant Response (OR) has begun to appear explicitly or implicitly in draft fire engineering code documents and related research in Australia [1], Canada [2] and the UK [3]. Reference [1] defines 'occupant response' as 'the actions of an occupant in interpreting, investigating and validating a fire cue'. A wider definition of OR includes a range of OR movement which also occurs prior to and during attempts to escape (eg wayfinding). One way to think of OR is as the aspects of pre-movement and movement time ignored by a traditional physical-science, fire code and engineering model of human behaviour. In contrast to people equated with objects emptying a space immediately in response to an alarm bell, research of human behaviour indicates that the time it takes people to respond and escape is influenced by social, psychological, organisational, as well as physical factors [3]. One way to characterise this is to compare the following definitions of T = Time required to escape:

T = travel time + flow time via exits on escape routes (physical-science model)   (1)

T = trec + tcope + tesc (psychological model)   (2)

Here, trec = recognition time from people being first alerted by a cue to recognising there is a fire (includes actions such as investigate); tcope = coping time (sometimes called the gathering phase which includes actions after recognition, such as warn others and fight fire); tesc = escape time (escape behaviour). A further distinction can be made in terms of patterns of movement [3]:

T = t1 + t2   (3)

Here, ti = the time to start to move or pre-movement time (tpre) and t2 = movement time. Equation (1) is equivalent to T = t2. The relationship between equations (2) (behaviour) and (3) (movement) need to be considered carefully. In some occupancies and situations t1 = trec and t2 = tesc is more likely (eg an audience moving from seats in a theatre); in others, movement occurs during trec, tcope and tesc. T = t2, equation (1), is an ideal, minimum escape time, rather than an assured reality. This paper is concerned with tpre, rather than other important OR patterns of movement as well.

2.2 Occupant Response and Fire Engineering Codes
Occupant Response (OR) receives minimal attention in some fire engineering codes and is represented more directly in others. For example, whilst the EC Construction Products Directive [4] defines the fire engineering performance of products in terms of specified 'mechanical, thermal and/or environmental actions', thereby ignoring human actions, the Australian draft Code [1] addresses OR. This Code [1] includes a prototype Occupant Classification and Rating, Occupant Communication and Response and Occupant Avoidance submodels and efficiency ratings, addressing the same phases of response (Tme = Tr Response + Tp Preparation + Tm Avoidance) as equation (2).

A primary question is what probabilities, predictive times or safety margins to include for the different types of warning system and phases of OR? In New Zealand reference [5] defines tev (calculated evacuation time from ignition) as td (time from ignition to detection) + ta (time from detection to alarm) + to +ti +tt; to = trec, ti = tcope and tt = tesc; to and ti are represented by minimum times of 30 secs each, tev being multiplied by a safety factor (SF) of x 2 (or more for occupants who are young, or have a disability) to allow for uncertainties in calculating the likely times. Reference [5] is similar to the British draft Fire Engineering Code [6] in concentrating on engineering performance criteria relating primarily to fire scenarios, rather than OR scenarios as well. The draft Code [6], unlike [1], assumes that once people are moving they are escaping. According to [6], escape time (tesc) = tdet (detection time) + tpre (pre-movement time) + tflow (flow time). Trec and tcope movement are excluded from consideration (except perhaps by default in terms of a safety margin of tflow x 2 for public settings which are large, complex and 'unfamiliar').

2.3 Occupant Response Research: Examples

Research indicates that alarm bells are generally less efficient than Informative Warning (IFW) voice and visual display alarms [7,8]. In a series of CCTV video monitored evacuations of a relatively crowded open plan underground station [9], the following times were recorded in response to the same alarm initially sounding: alarm bell only (approx. 15 mins), alarm bell + 2 staff on site (8 mins, with passengers misdirected into the high risk concourse), alarm bell + repeated non-directive Public Address (P.A.) announcement (equivalent to pre-recorded voice alarm) (approx. 11 mins); alarm bell + live, directive P.A. announcements with and without 2 staff on site (approx. 6 - 7 mins). This illustrates that the warning system in place can radically influence the overall evacuation time achieved in the same setting.

A more recent evacuation study of four buildings with 'mixed abilities occupants', conducted as data input to a Fire Risk Evaluation and Cost Assessment Model (FIRECAM) [2], has indicated that the 'times to start' can be wide for a population separated in different units (apartments). Again, the major contribution to the overall evacuation times was the tpre. In this case, tpre was measured on video as the elapsed time between the fire alarm sounding and the moment the person left his/her unit. The times from the alarm to exit from each building were *on average* as follows: Building 1 (B1) = 3:05 mins, B2 = 9:36 mins, B3 = 10:57 mins, B4 = 4:38 mins. The last person exited at B1 >15 mins, B2 >23 mins, B3 >25 mins and B4 >12 mins.

Research of this kind [2, 9] indicates that a suggested 'extreme' of 1 - 2 mins, for people 'in considering an evacuation' (reducable by an effective IFW) [7] may be optimistic, representing a best rather than worst possible tpre scenario. In some fire disasters there has been as much as 20 mins delay between staff first discovering a fire and people starting to move [3].

Bearing in mind the fact that communication and public warning systems have to increasingly instruct people in alternative life safety strategies (eg defend in place and/or move to an area of refuge/phased vis-à-vis simultaneous evacuations) warning systems differ primarily in the degree to which they effectively prompt people to start to evacuate (tpre). The remainder of the paper addresses this important issue. The methodology evolved out of an invitation to the author to provide a matrix of occupancy specific tpre times for inclusion in the British draft Code [6].

3.   Method of Assessing Occupant Response (OR) Pre-Movement Time

3.1   Communications: Minimum Baseline Pre-Movement Times

The first step in the assessment of tpre OR for a particular occupancy is to consider:

(A) COMMUNICATIONS: What kind of warning system alternatives are there, ranging from an alarm bell (system), 'non-directive' (prerecorded) Public Address (P.A.) announcements, Informative Warning (IFW) visual displays, and/or live 'directive' P.A. from a Control Room (using CCTV)?

According to the rationale of OR, movement is as much a communications as a physical issue. Application of the OR assessment procedure may itself influence the decision as to which warning system should be provided. A particular type of warning system may be a minimum requirement or more feasible in certain types of occupancy. No assessment procedure for the content of warning system messages is outlined here. The procedure presented is concerned primarily with the type of warning system.

The assessment of Communications should be conducted at a pre-design and design stage and also in Post-Occupancy Evaluation (POE) checks. The POE is a means of checking the validity of the original predictions, and whether adjustments are necessary in the design, operation and management of the fire safety engineering solution. The assessment procedure can be applied to new and existing buildings.

Table 1 lists 3 broad types of COMMUNICATIONS (warning) systems under the headings: w1, w2 or w3. The assessment method may eventually incorporate further gradations in terms of combinations of auditory and visual systems. At present the pre-movement (tpre) assessment involves the following initial decision:

- *Either,* adopt the average possible scenario (av.p.s.) baseline design values of w1 (6 mins), w2 (4 mins) or w3 (2 mins) in Table 1 as the design values in a fire safety engineering appraisal (irrespective of the occupancy type),
- *Or,* enter the best possible scenario (b.p.s.) baseline design value of w1 (3 mins), w2 (2 mins) or w3 (1 min) derived from Table 1 into the following formula:

$$\text{tpre adjusted} = \text{w1, w2 or w3 tpre b.p.s. x Weff} \tag{4}$$

Weff = the pre-movement Occupant Response Efficiency Weighting

Table 1. Matrix of baseline estimates of tpre (pre-movement times)

warning system		Pre-Movement Time (tpre)		
		best scenario (mins)	average scenario (mins)	worst scenario (mins)
w1	alarm bell	<3	6	>9
w2	non-directive pre-recorded P.A. and/or IFW	<2	4	>6
w3	live directive P.A +CCTV	<1	2	>3

3.2   Calculation of Pre-Movement Efficiency Weighting (Weff)
This section outlines the method of calculating Weff derived from factors which OR research indicates influence tpre [3] and represented by the following formula:

$$\text{Weff} = 5 + \text{Average B - I ratings} \tag{5}$$

B to I are defined as follows:

(B) ALERTNESS: How likely is it that people will be awake or asleep?
(C) MOBILITY: What are the sensory (eg hearing, vision) and mobility abilities and disabilities of the range of people likely to be?
(D) SOCIAL AFFILIATION: Are individuals most likely to be alone, separated from or in a primary social group (eg a family) in the setting when first alerted?
(E) ROLE: What is the ratio of public to staff in the setting?

(F) POSITION: How likely is it that people in the setting will be lying down, sitting, standing or moving at the time when first alerted?

(G) COMMITMENT: To what degree is the setting characterised by activities which people will be committed to finish (such as queuing to obtain a ticket) before recognising the need to evacuate?

(H) FOCAL POINT: To what degree does the setting have a particular focal point in terms of the direction of attention (eg a theatre)?

(I) FAMILIARITY: How familiar are the majority of people likely to be with different areas, entry and exit routes from the setting?

(B) to (I), whilst also relevant to movement, are used in the present context to assess tpre only. The ratings of B - I are on 5 point rating scales represented by Table 2.

Table 2. Occupant Response Pre-Movement Efficiency (Weff) Rating Scales

			Efficiency Weighting Factors				
B alert-ness	C mobility	D social affiliation	E role	F position	G commit-ment	H focal point	I familiarity
asleep *	low *	group *	public *	lying	high *	none *	unfamiliar *
**	**	**	**	sitting	**	**	**
***	***	***	***		***	***	***
****	****	****	****	standing	****	****	****
*****	*****	*****	*****		*****	*****	*****
awake	high	alone	staff	moving	low	focussed	familiar

In applying the assessment method to a large-scale complex, with multiple functions on the same and/or different floors, separate tpre Weff ratings for the each area should be made and compared. Tables 3 and 4 provide an *illustrative* example of the steps in the Weff calculation listed below.

1   Rate the occupancy, each floor, or each functionally different space or compartment, on each of the B to I scales 1 to 5 (* to *****) in Table 2. (Each rating scale ranges from 1, for the least efficient or slowest tpre response, to 5 for the most efficient or fastest tpre response likely).

2   Sum the B to I ratings (the row totals of Table 3 are given in column 2 of Table 4).

3   Calculate the average B - I rating (ie B - I/8; column 3 in Table 4).

4   Calculate Weff based on formula (5) above (column 4 of Table 4; this reverses the scale so that the higher the Weff score the slower or longer the tpre expected).

5-7 Enter Weff into formula (4) above to establish the tpre adjusted design values in mins for w1, w2, w3 (in Table 4, columns 5, 6 and 7, the tpre values for w1 = 3, w2 = 2 and w3 = 1 have been multiplied by the Weff for each row to give the corresponding illustrative tpre adjusted times in mins).

Asterisks (*) on a 5 point scale, rather than numerical ratings have been used in Tables 2 and 3. This is to indicate that the ratings and derived values in Table 4 are *illustrative* of the kinds of rating one might expect for the generic occupancy type headings listed. These are NOT definitive ratings or ones necessarily applicable to a particular setting. In Table 3, occupancies with less asterisks and therefore a smaller B - I aggregate score (column 2, Table 4) are those in which a slower tpre is expected.

Table 3. Illustrative Occupant Response pre-movement efficiency ratings

Occupancy	Efficiency Weighting Factors							
	B	C	D	E	F	G	H	I
Hospitals	*	*	****	****	*	**	*	**
Residential Buildings	***	***	*	*	**	****	*	*****
Nursing Homes	**	*	****	***	**	****	*	****
Hotels	***	****	****	***	**	****	*	*
Places of Assembly	*****	****	***	**	**	*	*****	**
Sports Stadiums	*****	*****	***	**	**	*	****	**
Shopping Complexes	*****	****	***	***	****	***	**	*
Shops	*****	****	***	***	***	***	***	**
Underground Stations	*****	****	.****	***	*****	**	**	**
Offices	*****	*****	****	*****	**	**	**	****

Table 4. Illustrative calculation of tpre adjusted for different occupancies

1	2	3	4	5	6	7
Occupancy	Sum B to I	Avg B to I	Weff = 5/avg	w1 alarm bell 3 mins x Weff	w2 non-dir PA 2 mins x Weff	w3 directive PA 1 min x Weff
Hospitals	16	2.0	2.5	8	5	3
Residential Buildings	20	2.5	2.0	6	4	2
Nursing Homes	21	2.6	1.9	6	4	2
Hotels	22	2.8	1.8	5	4	2
Places of Assembly	24	3.0	1.7	5	3	2
Sports Stadiums	24	3.0	1.7	5	3	2
Shopping Complexes	25	3.1	1.6	5	3	2
Shops	26	3.3	1.5	5	3	2
Underground Stations	27	3.4	1.5	4	3	1
Offices	29	3.6	1.4	4	3	1

In Table 4 some of the tpre adjusted figures are higher and some are lower than the hypothetical average (of w1 = 6 mins, w2 = 4 mins and w3 = 2 mins) in Table 1. Thus, in some instances application of the tpre adjusted method might produce a more stringent safety margin and in others a relaxation. Calculation of Weff for a particular setting is recommended, since the process highlights aspects of OR which should be addressed in an integrated fire engineering, architectural, management solution.

5.   Further Development and Application of the Methodology

Further research, refinement, calibration and validation of the parameters and ratings for a range of occupancies, OR movement and fire scenarios are needed before more definitive times or probabilistic predictions of response times for different types of occupancy could be derived.   Predictions also need to be checked against a Post-Occupancy Evaluation (POE) database of distributions of response times within and between occupancy types and settings.   Indeed, POE measures of tpre OR should be a fire engineering performance requirement, thereby also contributing to the database for tpre OR which is needed.

At present the tpre av.s. in Table 1 suggests a normal distribution for all occupancies or a particular evacuation, whereas in reality this may not be the case. Tpre distributions between and within settings could be 'normal' (as assumed here) or skewed to the left (eg the average being faster than is assumed here, but with a tail-end of late starters and evacuees) as in [2].   At present the times in Table 1 and the Weff formula are broadly in line with current OR research.

There needs to be consideration of issues such as the relative contribution to tpre of the different factors in different types of setting, their interaction in relation to different superordinate goals (eg reduce tpre or trec, tcope, tesc), the feasibility, validity, appropriateness or not of aggregating or averaging Weff scores or ratings for functionally different areas of a setting.   The tpre methodology needs to be reviewed in relation to alternative OR movement and fire scenarios and OR measures such as occupancy risk (population profiles) and a wayfinding design index, yet to be devised.

The inclusion of OR in fire engineering performance calculations should direct attention to the crucial relationship between communications and escape times in a variety of settings.   This includes transport systems, such as the Channel Tunnel where language differences could be important.   An OR review by this author, of evacuation trials in the Channel Tunnel several years ago, recommended that car and coach passengers in wagons should be presented with an instruction video at the beginning of the journey (comparable with international flights) as well as an IFW and directive P.A. messages in the event of an emergency.   The content and likely effectiveness of warning messages should form part of pre-design and POE fire engineering performance assessments.   For further discussion of communications and warning systems which allow 'democratic' distributed intelligence (prompt and accurate public warnings), rather than 'autocratic' centralised intelligence (delay in warning the public) see [9, 10].

The current draft British Code [6] has taken up the issue of tpre, reproducing the illustrative data in columns 5, 6 and 7 of Table 3 of this paper as 'Design values for pre-movement time' (Table 16.2 in the draft Code).   This, unfortunately, gives the impression of a definitive data set, rather than an illustrative calculation as intended. Table 1 seems to provide a reasonable set of tpre safety margins in keeping with current research knowledge and has been reproduced in [11].   Table 4 should be withdrawn from the text of [6].   As originally recommended by this author, Table 4 might be included in an illustrative fashion in an appendix of [6] and/or supportive document together with the necessary explanatory text.

In conclusion, the aim of this paper has been to introduce a prototype tpre OR assessment method. The method needs further definition and refinement. A more substantive tpre OR database in relation to different kinds of warning systems is undoubtedly needed. Research to date indicates that OR is a key feature of the time required by people to avoid a danger. It is hoped that this paper promotes the inclusion of tpre and other aspects of OR in fire engineering performance calculations.

Acknowledgement

This paper has evolved out of a consultancy report in 1993, on behalf of Warrington Fire Research Consultants Ltd and the British Standards Institution, as a contribution to [6].

References

1. Johnston, P.J. and MacLennan, H. (1991) Occupant Communication and Response Submodel/Occupant Avoidance Submodel. In V. Beck et al Appendix A Draft National Building Fire Safety Systems Code. C. Eaton (ed) *Microeconomic Reform: Fire Regulation.* Building Regulation Review Task Force. Australia.
2. Proulx, G., Latour, J. and MacLaurin, J. (1994) *Housing Evacuation of Mixed Abilities Occupants.* NRC-CNRC. Internal Report No. 661. Ottawa: National Research Council Canada.
3. Sime, J.D. (1994) Escape behaviour in fires and evacuations. In P. Stollard and L. Johnston (eds) *Design Against Fire.* London: Chapman and Hall, Ch. 5. pp. 56 - 87.
4. EC (1994) Interpretative Documents of Council Directive 89/106/EEC, Construction Products. Safety in Case of Fire. *Official Journal of the European Communities,* 28 Feb.
5. Buchanon, A, H. (1994) *Fire Engineering Design Guide.* Christchurch, NZ: University of Canterbury.
6. BSI (1994) *Draft Code of Practice for the Application of Fire Safety Engineering Principles to Fire Safety in Buildings.* Milton Keynes: British Standards Institution.
7. Canter, D., Powell, J. and Booker, K. (1988) *Psychological aspects of informative warning systems.* Building Research Establishment Report BR 127, BRE, Fire Research Station, Borehamwood.
8. Technica Ltd (1990) *Experimental programme to investigate fire warning characteristics for motivating fast evacuation.* Fire Research Station. Building Research Establishment Report BR 172. Watford: BRE. 1990.
9. Proulx, G. and Sime, J.D. (1991) To prevent 'panic' in an underground emergency: why not tell people the truth? In G. Cox and B. Langford (eds) *Fire Safety Science: Proceedings of the Third International Symposium.* London: Elsevier Applied Science. pp. 843 - 852.
10. Sime, J. D. (1994) Intelligent buildings for intelligent people. In D. Boyd (ed) *Intelligent Buildings and Management.* Aldershot: Ashgate Publishing Ltd/Gower Press. pp. 223 - 235.
11. CIBSE (1995) *CIBSE Guide: Fire Engineering.* London: The Chartered Institution of Building Services Engineers (to be published).

RISK ASSESSMENT – PROBABILITIES AND PUBLIC PERCEPTIONS

49 REALITY AND THE PERCEPTION OF RISK – RISK ASSESSMENT FOR THE FIRE SERVICE

R.A. KLEIN
CFRS, Huntingdon, UK and Institute for Physiological
Chemistry, University of Bonn, Germany

Abstract
This paper discusses some of the important components of risk assessment in the UK Fire Service and makes the point that a realistic perception of risk requires complete identification of all the hazards present, as well as estimation of the potential for exposure in terms of the risk environment. Four risk environments, some of which are specific for the Fire Service, are considered: maintenance and support activities; training; operational (tactical) assessment; and contingency or emergency (strategic) planning.
Keywords: reality, perception, risk assessment, emergency services

1. Introduction

Maintenance of a balanced view of the potential and actual risks in our environment is essential in order that scarce resources, both financial and in terms of manpower or equipment, are not diverted unnecessarily to deal with a high-profile but low probability risk at the expense of risks that are perceived as "ordinary" but which have a high probability of occurring with serious consequences.

The gap between reality, as objectively expressed in quantitative risk assessment, and the perceived risk needs to be addressed. The impossibility of living in a world entirely free from risk must become accepted more openly by the public at large. These attitudes can only be changed by education and the availability of information which is acknowledged as unbiased and accurate.

The unattainability of zero risk is especially true of those professions and occupations in which risk is, by definition, part and parcel of what is being done, e.g., the Emergency Services, mining or deep-sea fishing. In situations where risk is inherent, it must be assessed and controlled in some acceptable way.

The essence, therefore, of risk assessment is in identifying particular situations affecting particular populations and the harm that is likely to ensue. Population in this sense can mean either large groups, for example members of the public within a predetermined distance of a major hazard, or much smaller groups such as those

Fire Engineering and Emergency Planning. Edited by R. Barham.
© R.A. Klein and CFRS Consultancy Services 1995. Published in 1996 by E & FN Spon.
ISBN 0 419 20180 7.

Emergency Services personnel attending an incident. Risk assessment is, however, critically dependent upon actual circumstance. One risk assessment for a given hazard may be totally inappropriate under different conditions for the same hazard.

Having analysed the risks, control measures can be put into place to reduce the risk, or even to eliminate it, in line with the philosophy behind the COSHH Regulations 1988.

This reduction of risk involves, in general, application of the "ALARP" principle - As Low As Reasonably Practicable - a concept developed as a means of balancing the perceived risk against the costs of controlling it. ALARP is a peculiarly British principle embodying inter alia the concept of how a reasonable man would behave and view the situation. ALARP is not understood by other European countries with legal systems based on the Napoleonic Code; such a principle is philosophically inconceivable within this codified framework. This is one of the more basic reasons why it is so difficult to achieve European harmonisation in the area of safety-related legislation.

Establishing best practice requires taking account of information and opinions from many different sources, both inside and outside of the Fire Service. As such, best practice must (i) be Fire Service specific and relevant; (ii) conform to legislative requirements and the current interpretative philosophy of the enforcing authority; and (iii) be consistent with those systems and procedures developed by specialised sectors of industry, including the Fire Service itself, based on accumulated experience.

2.   Historical background within the Service

Risk assessment in the UK Fire Service is not new. It has been carried out in some form or other at least since, if not before, the Fire Services Act 1947 imposed requirements on Chief Fire Officers to acquire information relevant to Brigade operations within the Fire Authority's area. The exclusivity to fire-fighting was removed by the Fire Services Act 1959. Risk assessment, with a slightly different emphasis, has been carried out as a means of preventing fire and its consequences through powers conferred under the Fire Precautions Act 1971. Within the Service, inspections carried out under Section 1(I)(d) of the Fire Services Act 1947 have traditionally provided the basic hazard data on which assessments of risk are founded.

What has changed within the intervening half-century, particularly since the Health and Safety at Work etc Act 1974, together with its derived legislation, is the legal basis for the requirement to carry out formalized risk assessment. This now includes requirements for the introduction of control measures, surveillance, monitoring and auditing, as laid out in the Health and Safety Executive's document entitled "Successful Health and Safety Management" HS(G)65 1993, based on the Approved Code of Practice "Management of Health and Safety at Work" 1992. Because the management of risk assessment and related health and safety matters now requires many different areas to be addressed, we refer now to integrated risk assessment. In particular the Approved Code of Practice underlines that "...*a suitable and sufficient risk assessment will reflect what is reasonably practicable to expect employers to know about hazards in their workplace and should involve a systematic general examination of the work activity...*".

The background to risk assessment within the Fire Service is explained, together with discussion of some of the problem areas, with the intention of suggesting ways of approaching a solution suited for the Brigade in question. There is no one solution, no one way of doing things that can be said to be 'right' and applied across the board to all Fire Services within the United Kingdom.

It should be clearly recognised that the Service has been carrying out risk assessment for a long time. Probably around eighty percent, or more, of the necessary procedures are already in place. What is now required is that the remaining twenty or so percent are addressed in the formal ways required by recent legislation. This is very different to having to establish a whole new system.

3. Legal background

One of the main provisions of the Fire Services Act 1947 was "*...to transfer fire fighting functions from the National Fire Service to fire brigades maintained by the councils of counties and county boroughs...*", thus establishing the United Kingdom Fire Service in its present day form. Restrictions on the employment of fire brigades and equipment for purposes other than fire fighting were removed by the Fire Services Act 1959.

In practical terms the responsibilities of fire brigades have been extended greatly over the intervening decades since the 1947 Act came into force. Dealing with "fires" would now be taken to include a wide range of incidents such as road traffic accidents, chemical and biological spills, civil disasters such as air and rail crashes, flooding or nuclear accidents, as well as a range of rescue and cleaning-up activities. The significant component of general rescue activities is reflected in many brigades being called "fire and rescue services".

In the context of risk assessment, Section 1 of the 1947 Act imposes a number of specific relevant duties on fire authorities for the provision of fire services, including securing

- *the services for their area of such a fire brigade and such equipment as may be necessary to meet efficiently all normal requirements;*
- *the efficient training of the members of the fire brigade;*
- *efficient arrangements for obtaining, by inspection or otherwise, information required for fire fighting purposes with respect to the character of the buildings and other property in the area of the fire authority, the available water supplies and the means of access thereto, and other material local circumstances;*
- *efficient arrangements for ensuring that reasonable steps are taken to prevent or mitigate damage to property resulting from measures taken in dealing with fires in the area of the fire authority;*
- *efficient arrangements for the giving, when requested, of advice in respect of buildings and other property in the area of the fire authority as to fire prevention, restricting the spread of fires, and means of escape in case of fire.*

The duty to acquire information required for fire fighting purposes, whether by inspection or otherwise, has a long history in the fire service and forms an essential part of the risk assessment process, that of hazard identification. Risk assessment has been carried out by fire officers at all levels, perhaps under other names, as a fundamental and necessary part of their job. What then has changed?

As a result of the Health and Safety at Work etc Act 1974, together with legislation derived from this Act over the last twenty years, there is now a requirement to formalise procedures used in risk assessment. In many cases this means committing to paper accepted procedures which have been in use for many years. In others it may mean developing new procedures to comply with the legislation.

The role of the firefighter, however, remains the same. As succinctly put by Sir Kenneth Holland, the then Chief Inspector of Fire Services, in a letter to all Chief Fire Officers (DCOL 15/1978) - "...*The introduction of the above legislation has not as far as I am aware changed the traditional role of a fireman to protect life and property to the best of his ability and to accept reasonable, calculated risks to himself in doing so; or the responsibility of the officer in charge of an incident to ensure that safe practices are followed in fire fighting and other activities and that so far as is reasonably practicable in the circumstances risks to the personnel under his command are if possible eliminated, or if not, reduced to the minimum commensurate with the needs of the task...*".

The Lyme Bay canoe disaster, in which school children were drowned off the coast between Lyme Regis and Charmouth in Dorset, is an important milestone in health and safety history. This disaster has resulted in an historic legal judgement at Winchester Crown Court (see the legal report in The Times, 13 December 1994), as well as a the Young Persons Safety Bill being presented to Parliament.

Until 1925, based on a judgement in 1701, a corporation was not indictable but its members were; it was not possible for a corporation to "appear" and plead through a representative. Prosecutions involving loss of life through industrial accidents have, until recently, been brought under the Health and Safety at Work etc Act 1974 and associated safety legislation, if at all. The precedent for indicting a company for manslaughter was established in 1965. Apart from P&O European Ferries, no company has ever stood trial for manslaughter, although even this trial was stopped before its completion.

The judgement that OLL Ltd. was guilty of manslaughter establishes that a company can be held criminally responsible. There are, however, still legal difficulties involving technicalities such as "aggregation of guilt" and the concept of the "controlling mind", *mens rea*, as highlighted by the author of the report.

Whatever the legal difficulties, however, the precedent is set in case law. It is possible for an organisation, such as a Fire Brigade, to be held criminally responsible for its corporate actions, whether these be sins of omission or commission.

3.1 Health and Safety Executive Improvement Notices

During the past two or three years the Health and Safety Executive has served, under the provisions of the Health and Safety at Work etc Act 1974, improvement notices on a shire county brigade and on a large metropolitan fire authority, which have considerable significance for the management of health and safety in the Fire Service as a whole.

Two of the notices highlighted first the need to improve communication of the results of inspections carried out by fire safety and operational departments and secondly to improve training specifically, in this instance, as regards breathing apparatus entry control procedures.

Two other notices, served on a large metropolitan brigade, were more generally concerned with management systems to ensure the effective monitoring of occupational health and safety and with personnel receiving adequate and documented operational training. A third improvement notice, served in February 1994, and following on from the first two, required implementation of general organisational arrangements for "...*the effective systematic management of health and safety and the training which supports it as specified in Section 2(2) of the above Act and the above Regulations...*" (the HASAW Act 1974 and the MHSW Regulations 1992).

The detailed requirements of this third HSE improvement notice, contained in a Schedule, are of sufficient importance to the way in which the Service approaches its management of health and safety, including risk assessment, that extracts are quoted here.

In particular:

A. *Provide an information system to support the management of health and safety. The system should:*
i. *provide the information needed by managers at all levels and the Authority, in order that they can discharge their responsibilities in relation to health and safety and associated training and resource issues.*
ii. *develop methods for ascertaining the training provided to all ranks, particularly in relation to the health and safety aspects of operational performance, and recording this to accurately reflect the subjects covered, competence reached and the dates it was carried out.*

B. *Ensure that the development of risk assessments concentrates upon risks arising from:-*
(1) *The system of command;*
(2) *The system of tactical firefighting and*
(3) *The use of personal protective equipment including the support systems for breathing apparatus.*

Highlighting their requirements in the form of a summary the Health and Safety Executive specified, in particular:
► a management information system designed to support the development of systematic management of health and safety.
► a health and safety policy, organisation and control system which clearly articulates health and safety objectives and provides sufficient performance standards and measures.
► appropriate methods of hazard identification and risk assessment of operational incidents and training.

These requirements are far reaching. Coming, as they do, from the enforcing authority, they set the standards by which the Fire Service should approach the whole range of health and safety management, including risk assessment.

In the remainder of this contribution I should like to consider briefly two areas which have critical importance to the way in which reality and the risks posed are perceived within Fire Service operations, namely the identification of hazards and the various risk environments in which these hazards are found.

Risk assessment has to do with determining the likelihood, in a particular situation, that harm will be caused. For this one needs to know the hazards and the potential for exposure, i.e., the risk environment. Risk can be defined simply as the severity of the hazard multiplied by the potential for exposure.

4.   Hazard identification

The Management of Health and Safety at Work regulations 1992 contain a number of detailed requirements related to the organisation and arrangements for managing health and safety at work which reflect the implied requirements of Section 2 of the Health and Safety at Work etc. Act 1974. At the heart of these regulations, as with other more recent, "modern" regulations post Robens, is the concept of risk assessment. In principle this concept requires precautions to be taken which are commensurate with the risk and are consistent reasonable practicability. Risk assessments are also required under both the Ionising Radiations Regulations 1985 (IRR 1985) and the Control of Substances Hazardous to Health regulations 1988 (COSHH 1988) but the principles remain the same.

The purposes of risk assessment are:

▶   to identify hazards and the risk environment in which they are found;
▶   to focus attention on measures for the reduction or control of risk to acceptable levels - the ALARP (as low as reasonably practicable) principle;
▶   to help employers decide on priorities for action based upon an objective cost-benefit analysis;
▶   to assist employers in discharging their legal obligations under the regulations.

Risk assessment, which the Health and Safety Executive sees as a practical process concentrating on real risks rather than a paper exercise, must address each specific risk environment in which a hazard occurs. Although generic risk assessments are acceptable for similar risk environments, this approach is unsafe if applied to the hazard(s) themselves rather than their environments. The same hazard may pose a very different risk depending upon circumstance.

4.1 Identification and classification of hazards

A hazard exists or it does not. There is no halfway house. Recognition or identification of hazards is not so clear cut; it requires, essentially, the collection of good intelligence.

Identification of hazards is primarily an operational exercise undertaken by trained personnel. In industrial or research environments this may be carried out by Safety Officers or their deputies, or by members of line management. In the Fire Service it is usually carried out by the members of the Watch at Station level. Identification and classification of more complex hazards, e.g., CIMAH sites, or those which are considered marginal, will require decisions to be made by more senior specialized members of staff, such as the HAZMAT officer or specialist scientific adviser for the Brigade. Problems of hazard identification may arise with apparently low risk sites, such as agricultural or domestic premises, when these contain unexpected or unusual hazards.

One of the purposes of an inspection under Section 1(I)(d) of the Fire Services Act 1947 is to identify potentially hazardous situations within the Fire Authority's geographical area of responsibility. Recent Improvement Notices served by the Health & Safety Executive have underlined the need to collate information gained from Section 1(I)(d) inspections and those carried out as part of Fire Prevention. Another valuable source of information for identifying hazards comes from inspections carried out under the Petroleum Spirits Regulations.

Within the Local Authority procedures should exist to ensure that information received by the Chief Executive from bodies such as the Department of the Environment, the Directorate of Pollution or the Health and Safety Executive, which is relevant to fire brigade operations, is passed with minimum delay to the Chief Fire Officer.

These inspections are data collection processes requiring fairly crude guidelines as to what constitutes a hazard, more often than not guided by common sense and experience. The data then needs to be stored in a retrievable form. It is of no use whatsoever if this information cannot be retrieved or manipulated when required.

What guidelines for hazard identification exist for Fire Service use? As a result of recent legal action, a real practical problem exists for UK Fire Services. Down to what level should we look for hazards? Should we inspect every chicken-house, smallholding or garage for hazards, particularly in rural areas? Unfortunately the answer is probably, yes! But this has appalling manpower implications for the Service.

The way in which hazards are identified and then classified will depend upon the use to which these data are put.

4.2 Criteria for selection of premises

Criteria that may be applied for considering whether premises are selected for inspection and at what frequency, have been dealt with in Fire Service Circular (FSC) 16/1989. The suggested selection criteria for determining whether or not premises are inspected and the type of records to be maintained, include in particular:

▸ an exceptional occupancy and high risk to life;
▸ unusual or complex (building) structures;
▸ the storage or processing (or handling) of hazardous materials;
▸ and high potential loss (human, financial or social).

This represents a logical dilemma. Unless premises have been inspected, it may not be apparent whether any or all of these criteria apply. It seems appropriate, therefore, to recommend that a fairly crude form of hazard identification be applied using a classification based on a limited number of major premises categories, supported by an operationally relevant questionnaire distributed to all occupiers.

The questions asked of the occupier of the premises must be chosen to answer questions of operational relevance for fire service activities. For instance, it may be necessary to know whether compressed gas cylinders are being stored on agricultural premises, quite apart from chemical compounds such as ammonium nitrate or cyanamid. Industrial premises may be holding larger quantities of flammable/toxic materials than thought, e.g., solvents, particularly as byproducts, or smoke detectors containing radioactive sources.

A small business may have large quantities of paint as aerosol cans. Small engineering works may have gamma radiography sources on site. The possibilities are legion. An efficient questionnaire must be constructed based on experience and a certain cynicism.

In determining the nature and frequency of an inspection - that is the complexity and thoroughness of the inspection - brigades are advised to apply the test of whether basic fire fighting techniques would be sufficient to cope with the majority of incidents that could be foreseen as occurring at the premises.

4.3 Pre-planning for inspections

Reliable risk assessment needs to be based upon good hazard identification for the area served by the fire authority. If hazard identification is patchy then risk assessment will be less than realistic and may be positively dangerous and misleading for operational personnel. Efficient hazard identification has, as a prerequisite, the acquisition of complete, up to date and accurate information by means of inspections, or otherwise, i.e., through letters and questionnaires.

Planning inspections need not involve a great deal of work. One cannot over-stress the importance to the end result - effective hazard identification - of indicating the type of information required and producing this information in a standard format for recording purposes.

Two-way consultation is also an essential part of the collection of hazard information and personal contact with local safety officers or fire officers is an efficient method of maintaining liaison and cooperation.

In order not to waste resources, financial or manpower, premises will have to be classified as to the types of hazard to be expected so that the appropriate information may be sought. In the first instance, seven key areas of operational significance should be considered:

- ► occupancy and use;
- ► access, both to and within the premises;
- ► structural features and layout;
- ► the presence of hazardous materials or processes;
- ► fire fighting resources and fixed installations already present;
- ► control of incoming sources of power, i.e., gas, electricity and high-pressure steam.
- ► proximity and density of the neighbouring civilian population.

4.4 Classification of premises

For the efficient identification of all significant hazards in the fire authority's area, some form of initial screening is necessary when inspections are carried out. It would be pointless and extraordinarily expensive to apply the same criteria to domestic premises as to a site classified under the CIMAH regulations! Although a fanciful example, this makes the point.

The information gathered through inspections, or otherwise, on the identity of hazards in the fire authority's area has a number of purposes which may require a lesser or greater degree of "digestion" before use. The various purposes for which the information may be used include:

- ► conversion into operational information for crews making up the attendance at an

incident;
▶ more detailed information may be required at Incident Command and Control level to enable longer term decisions to be taken by more senior officers;
▶ risk assessment as part of pre-planning and production of the Risk Cards, normally done at or with the assistance of headquarters staff;
▶ formal contingency and worst-case planning required by regulation, such as for CIMAH classified sites;
▶ post-incident analysis and monitoring;
▶ surveillance of the occupational health of crews.

5. Risk environments

There are four different categories for risk assessment in the Fire Service which have certain distinguishing features.

5.1 Office and support services
The employment of civilian, non-uniformed staff in offices, kitchens and as cleaning staff can pose problems not encountered in the uniformed part of the Service. Although the hazards and their associated risks appear at first sight to be at least an order of magnitude less, injury or even death must be considered with electrical or mechanical accidents.

Moreover, because this part of the workforce is non-uniformed and not used to Service-type discipline, the presumption that instructions will be obeyed and that levels of training are adequate may not be valid. All the usual problems of modifications to electrical appliances such as extension leads, storage of large amounts of paper or other combustible waste, inappropriate footwear associated with falls, slippery floors, poor working positions with computers, and minor injuries, form part of the accepted "administrative building" safety picture.

5.2 Maintenance activities
Maintenance operations, i.e., Brigade workshops, encompass fairly standard factory- type risks in terms of risk assessment. Mechanical and electrical hazards predominate, although the hazards associated with petroleum spirits should not be forgotten. Other maintenance activities, such as BA cylinder filling, may involve very specific hazards, in this case that of high pressures (approx. 200 Bars) or of contaminating air that will be later breathed by operational personnel. The hazards associated with repairing appliances, i.e., ladders, may be those of using the equipment itself.

Other maintenance jobs on Station, particularly those involving civilian staff or outside contractors, should be subject to "Permits to Work" and to adequate internal codes of practice. Care should be taken that financial stringency does not result in lax or positively dangerous systems of work.

5.3 Operational assessments
This area is the one normally considered when talking about risk assessment in the Fire Service. Risk assessment has been carried out over the years, both pre-incident and on the fireground, as part of the job long before the term 'risk assessment' came to be used.

Changes that have come as the legislation has developed, include formal pre-planning, notifications, and requirements for medical health surveillance of personnel.

5.4 Worst-case and contingency planning

Worst case and contingency planning are really extensions of operational risk assessment with an implied loss of control.

The assumption is made under normal operational conditions that managerial and organisational infrastructures remain intact and functional. Likewise it is assumed that control systems for plant, whether electronic or mechanical, will remain working. Disciplined services assume control over the available manpower. The behaviour of a civilian workforce or members of the general public cannot be guaranteed under emergency conditions.

5.5 Training and the "safe and competent" person

Risk assessment is part of a tetrad comprising assessment, planning, implementation and evaluation. Within this tetrad, training has an important role especially for the Fire Service. Training crosses all the boundaries and includes operational (tactical) as well as planning (strategic) matters and procedures, apart from maintenance and support activities.

The purpose of training within the Fire Service is to produce a safe and competent person as a means of controlling risk.   What do we mean by a safe and competent person?

It is clear that an inherently hazardous operation is inherently safer, i.e., associated with less or reduced risk, when carried out by someone who is experienced, well informed and well trained, than by someone who is not. This is, in principle and as a matter of commonsense, what we mean by a safe and competent person.

6.   Models for risk assessment within the UK Fire Service

The two diagrams which follow illustrate the various features of risk assessment in the Fire Service, including monitoring, surveillance and auditing.

References

Integrated Risk Assessment, eds. R.A. Klein and M.R. Pallister (1994) Proceedings of the European Conference on Integrated Risk Assessment,   21-22 March 1994, Cambridge, England. CFRS Consultancy Services, Huntingdon.

Fire Services Act 1947 and 1959.

Health and Safety at Work etc Act 1974.

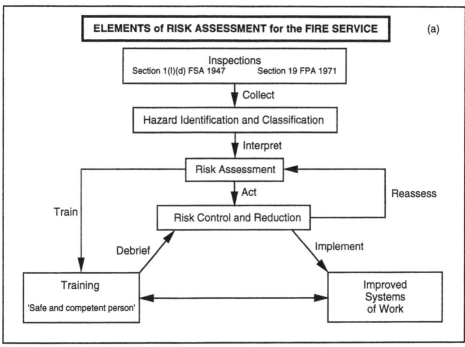

ELEMENTS of RISK ASSESSMENT for the FIRE SERVICE (a)

Inspections
Section 1(I)(d) FSA 1947   Section 19 FPA 1971

Collect

Hazard Identification and Classification

Interpret

Risk Assessment

Act

Reassess

Train

Risk Control and Reduction

Debrief   Implement

Training

'Safe and competent person'

Improved Systems of Work

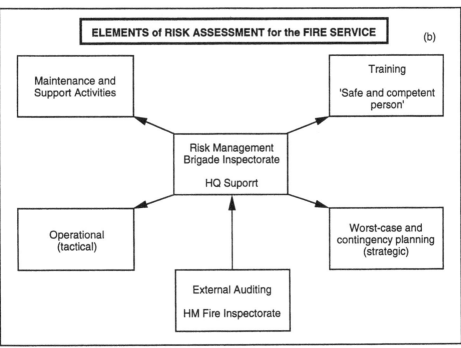

ELEMENTS of RISK ASSESSMENT for the FIRE SERVICE (b)

Maintenance and Support Activities

Training

'Safe and competent person'

Risk Management Brigade Inspectorate

HQ Suporrt

Operational (tactical)

Worst-case and contingency planning (strategic)

External Auditing

HM Fire Inspectorate

50 PUBLIC PERCEPTIONS AND RISK ASSESSMENT: LESSONS FROM INDUSTRIAL HAZARDS IN DEVELOPING COUNTRIES

P.S. FLORA
University of Central Lancashire, UK

Abstract

Industrial hazards, like natural disasters can be understood in terms of their occurrence in time and place, how they affect social units and how these units take responsive actions to mitigate disaster consequences. To some extent the causes of industrial crisis in developing countries may be rooted in the rapid pace of industrialisation following independence from their former colonial rulers. Between the years 1960 to 1982 most developing countries showed a significant increase in their GDP derived from Industrial activities [1] with the transfer of technology being determined by intergovernmental contracts and multinational corporations. Such technology transfers brought new investments, jobs but only in localised urban areas, whilst the masses remained relatively unaffected in rural areas. The necessary infrastructure to support these technologies for a sustainable development could not keep up with these developments due to chronic fiscal problems and policy and planning failures. The quality of the infrastructure eg. water, energy, public health systems, transportation, communications, educational instituitions and the labour force all influence the probability of the occurrence of accidents and also the escalation into a crisis.

This paper will discuss the discrepancies in perception of the major players in identifying and assessing the causes of two industrial accidents one chemical, the Bhopal tragedy [2,3] and the other due to a fire at a nuclear power plant. These accidents, exemplify many issues pertaining to the decision making processes of the different participants involved in the process of mitigating circumstances in developing countries.

Introduction

The effects from industrial accidents and hazards such as Fire or Natural disasters often lead to the loss of human life, property and financial damage and considerable environmental pollution. The severity of disasters occuring in developing countries is exacerbated due to high levels of poverty, illitracy, inadequate housing structures, under-resourced health services, lack of expertise in training and planning for risks and hazards and of course the necessary support structures and finance from relevant organisations for the rehabilitation of victims. Both preventing and mitigating effects of crisis depend upon the improvements made in the above areas.

In South-Asian countries, the unplanned rapid pace of urbanisation outstrip growth in infrastructure development which also have consequences for the environment in terms of pollution abatement from toxic wastes, water, air pollution and other hazards. The poor not having any resources to adequately protect themselves are more vulnerable and exposed to such hazards. Even today, some 10 years after the Bhopal crisis, liberal government policies although encouraging foreign investment are not directly addressing the needs of infra-structure development, for instance in the power,

Fire Engineering and Emergency Planning. Edited by R. Barham.
Published in 1996 by E & FN Spon. ISBN 0 419 20180 7.

transport and water industries. The following statistics for the capital perhaps typifies conditions of urban life [4].

A 10 million population of Delhi use a meagre 6000 public and private transport buses. One sixth of this population lives in slums, called Jhuggis. More than 50% of the medical facilities are only available in the private sector. A 10 year comparison (from 1981 to 1991) of the number of occupied residential homes shows an increase of nearly 70%. More than 30% of Delhi residents suffer from respiratory ailments. There is a boom in unauthorised constructions and flat conversions with rents skyrocketing. Taps run dry every summer, with hundreds of people drinking contaminated water. Electricity supplies are erratic, with voltage fluctuations and power cuts frequently occurring.

Risk assessment

A full assessment of risk requires studies and data collection of population movements in and out of towns, social patterns such as sleeping hours, peak congestions, availability and response timing of emergency services, corporation organisational structures, governmental administrative services etc. Computer programs assessing risks are useful from a design and control point of view however they often lack detailed data and prior to Bhopal incidence such techniques were not available or utilised only scarcely in order to satisfy minimum regulations. Subsequent to Bhopal many authors have constructed mathematical simulation models to assess a variety of different scenarios of possible accidents [5, 6]. The non-availability of data on the above social factors and how they relate to each other is a complex issue for evaluation. Qualitative assessments however may be made on the basis of historical events.

Chemical Hazards-The Bhopal Incident

The city of Bhopal reflected very similar trends with the slums and shantytowns occuring near the industrial plant. The slum dwellers served the officers and staff at the plant who employed cheap forms of domestic labour. This consequently led to a high population density near the plant site. In terms of quantifying chemical hazards, the Bhopal case may be taken to be the worst case of an industrial chemical disaster with a figure of above 6800 deaths and several hundreds of thousands of people injured with long term health effects [7].

The technical aspects of the Carbide's safety measures have been covered by many authors [5, 8]. An aftermath of the essential issues such as the choice of technology, siting of the industry, design of the plant, operation and maintenance system and public information dissemination channels should have been considered when first evaluating risk factors. There are many other technical authors who are also very forthright in isolating the cause to a simple operator error and have totally missed out the inadequacy of egronomic design and safety features of the plant.

The US. chemical giant Union Carbide responsible for the manufacture of Methyl isocyanate (MIC) used as a pesticide at the Bhopal site was faced with multimillion dollar lawsuits (which still has not been paid to all the victims of the disaster) whilst all over the world, governments, businesses and industries worked on developing informative programs on hazards and risks for their local residents. Questions on could it happen here were frequently asked or is it merely a problem that happens over there were vigourously debated. The fall out from such a crisis to some extent jolted some governments and industries into an assessment of their measures for the prevention of major industrial accidents [9], eg. in Canada the Transport of Dangerous Goods (TDG) was implemented, in the UK the Control of Industrial Major Accident Hazards (CIMAH) regulation of 1984 was passed whilst in India, three acts dealing with industrial hazards (Factories Act (1948), The Water Act (1974 and 1977) and The Air Act (1981)) were amended and a comprehensive new law called The Environmental Protection Act of 1986 was passed.

Fire-Hazard-The Narora Incident

At 3.31am of March 31st 1993, a fire broke out after two blades in a turbine generator of the Narora Atomic Power Plant, Uttar Pradesh snapped. The broken blades sliced other blades and caused the turbine rotor to vibrate excessively, thereby igniting hydrogen gas cooling pipes. The resulting fire spread rapidly and also burnt out back up power supply cables. although, staff were able to avert a major catastrophe by following a 'crash cooling sequence' of the reactor core.

Governmental response

Subsequent to the Narora incidence, the Government ordered its watchdog on nuclear safety (AERB) to form an expert committee for an investigation of the incident. Some of the committees findings are given below [10]:

-Hairline cracks in four other blades of turbines.
-Erroneous interpretation of control panel instrumentation highlighting eight times more than normal axial vibration of the turbine.
-Four back up emergency power supply cables using the same duct, without any fire-resistant material in between. The fire spread from one cable to another, thus there was no available backup power supplies.
-Although it was said that there was no radioactive release, containment integrity of the dome had been lost during the power break down.

On discovering the design faults, the AERB ordered a reluctant NPC (public sector corporation in charge of building and running Indian nuclear plants) to close down other plants with the same turbines and recommended modifications to the design. The turbine blade technology had been transferred by the UK based company GEC who had warned much earlier about the possibilities of structural failures of blades and suggested design modifications for blades that had completed between 10,000 to 35,000 running cycles. The turbine blade at Nagai had completed 16,000 cycles. NPC says that it had no knowledge of GEC's warning.

The complete findings of the expert committee's as can be expected are held confidential even to other closely related sections of the government's departments and furthermore praises the effectiveness of its existing safety procedures, thereby allowing the NPC to continue with a 'pat on its back' in averting a disaster that could have been a nuclear disaster.

In the case of the Bhopal tragedy, the Government was passing through the midst of its political campaign and was trying very hard to influence public opinion. Negative publicity could have ruined its chances for re-election. It therefore reacted to control all information regarding the accident and releasing it gradually to control desired specific events which could increase its own popularity. Voluntary organisations were thus prevented from entering government relief camps and some were closed down while there was still an urgent need for them [11].

Relief efforts by the government were heavily publicised and Union Carbide's negligible contribution to the relief fund further enhanced the governments public image. By easily passing a Gas Disaster Bill through its national parliament, the government became representative of the victims interests and thus sued Union Carbide in the US.

Corporation Response

In the case of the Narora incident, the NPC reported a loss of $32 million. On being requested by the government to shut down its other plants with similar turbines, the NPC complained about further huge losses. Further, it denied knowledge of earlier recommendations for the design modifications to the turbine blades.

The Bhopal accident triggered many conflicting issues which were debated in the media across many parts of the globe and threatened the closure of Union Carbide. Communities, Governments, Cities voiced their objections to transporting, producing and storing Union Carbide chemicals. In Breziers, France the local community objected to the reopening of a Union Carbide plant after it was shut down following the Bhopal accident. Chemical companies around the world reviewed their own safety operations and emergency response procedures and also re-evaluated the risks of operating in developing countries. Union Carbide's perspective subsequent to the accident were three fold; initiate damage control and investigate causes, coordinate with government agencies and handle legal affairs.

In March of 1985, Union Carbide released its own internal report on the Bhopal accident. This suggested that the accident was caused by sabotage, local operator errors and local management lapses. The report highlighted the narrow thinking involved by the corporation to defend itself against all charges by denying liability and shifting the blame on to 'local effects' and even suggest blame on to a shadowy Sikh terrorist group, 'The Black June Movement', and simplify the technical complexities to a one simple action of water entry into a storage tank. This theory however lacked credibility when its source became known: some people in one Punjab city were reported to have seen a poster from this Black June Movement claiming responsibility for the disaster.

In terms of rescue and relief efforts, Union Carbide donated a $1.1 million to the Prime Minister's standing relief fund and $10,000 for medical supplies to local hospitals. Judge Keenan, appointed by Reagan ordered the company to pay $5 million for interim relief which reached Bhopal after a time lapse of a year. Union Carbide's attempts to help government agencies in relief operations were rejected on the grounds that the government did not want Union Carbide to exploit this situation for public relation purposes. The company also claimed to have offered $20 million to mitigate effects of the accident, but there was a condition that information on the health of the victims had to be known which then also could be used in the Union Carbide's legal defense. Other offers by the company to their insurance limit of $200 million were also seen to have been insincere and merely generated negative publicity about the company.

Voluntary Agencies

The role of voluntary agencies vary considerably, but generally as non-profit and independent organisations they are better able to inform and assist people particularly in rural areas. Their analysis of events such as by surveys, death counts and subjective evidence may lack the expertise and rationality of government and corporation methods, who have vested personal interests and often only implement community demands no more than is required for satisfying minimum policy requirements. Other factors such as religion and cultural tradition may also dominate perceptions of some voluntary organisations thereby influencing their risk assessment methods. Nonetheless, their approach will reflect a bottom up and a grass-roots level strategy which in some areas of rehabilitation and relief work suit local needs.

There are also many other organisations affiliated directly to humanitarian and environmental groups that have some technical expertise and resources to effectively provide a rational picture of the events. Although their interests sometimes may be in direct conflict with both government and corporation interests, they are well able to spearhead and mobilise governmental actions.

Agencies such as the voluntary anti-nuclear organisation in Gujarat, revealed that the rate of congenital deformities and unformed limbs was 3.5 times higher than in villages far away from the plant. This was refuted by an alternative survey carried out by the government, however, the government also revealed that the dosage of radiation received by Indian workers in nuclear plants was eight times higher than that of the world average when comparing similar power plants.

The Bhopal incident brought together many voluntary, environmental groups and activists to focus on political issues. Responses and the feelings of victims were sometimes effectively channeled into demonstrations which had the effect of forcing the Government and Union carbide to be more responsible. Although limited by resources, their ability to coordinate mass movements proved effective in lobbying these larger instuitions.

The Role of The Fire Emergency Services

All accidents or natural disasters can lead to multicomponent events where damage of property and life can be at risk whether by fire, radiation, water, gas or other natural hazards. The role of the Fire services as in most countries extends beyond that of dealing with fire-fighting to cover general emergency security and protection. The rapid construction of high rise buildings, hotels, factories, offices, shops, industries and many unauthorised dwellings all lead to an increase in risk of fatalities. Sharma [12] gives a fatality risk value due to fire of 13% higher in India than in other European countries with an estimated material damage being 0.3% of GDP. Prior to the Indian 1986 Fire Prevention Act, a Delhi survey revealed that 194 sky scrapers had virtually no or inadequate fire safety provisions.

This increase in demand due to the effects of rapid urbanisation have opened up areas for Fire protection such as in education, training, research, manufacturing and employment. An excellent account by S.K.Dheri [13] of Delhi's Fire Services since 1862 provides an insight into how India is meeting the demands of its infrastructure development. The recent conference, 'Fire India '94' provided many references to collaborative organsational, technical, research and training programmes between the UK, India and the US in the growing area of Fire Engineering. Such initiatives in the development of emergency services infrastructure should be valued, developed and expanded for the mutual benefit of both the developing and the developed world.

Conclusions

This paper analysed two industrial incidents one of which developed into a major crisis. From such an analysis it can be concluded that many aspects leading to an increased risks of industrial accidents need to be addressed for developing countries as well as effective measures for mitigating crisis situations. Aside from addressing careful engineering design and manufacturing processes to suit local conditions, there still remains the problems of financing adequate infrastructure in training, communication, education and placing a value on human life in developing countries.

It is suggested that industries regularly assess their hazards to their local communities in terms of manufacturing, storing, transporting or disposing hazardous materials. In addition, by actively taking part in the infrastructure development of their immediate surrounding will in the long turn bring higher benefits by increased efficiency and profitibility. Further, by involving the local community in its decision making structures will allow an informed assessment of the risks which also take into account social patterns of their community as well as increasing awareness of the risks for preventative measures. In the case of Bhopal, medical practitioners were not informed of the toxicity of the gases, and therefore could not be effective in reducing casualties.

The expansion of domestic industries in the States of Gujarat, Maharastra and Punjab have conveniently neglected effects of their unchecked hazardous dye wastes (which sometimes catch fire) on their local residents, inspite of governmental guidelines on hazardous waste handling [14]. Pending enforcement of government policies and guidelines, it is up to the responsible and informed industrialists to invest in the understanding, training and disposing of industrial wastes for a better future.

References

[1] World Bank (1984) *World Development Report* New York, Oxford University Press

[2] Shrivastava P (1992) *Bhopal -Anatomy of a Crisis,* Chapman, London

[3] Jones T. (1988) *Corporate Killing Bhopals Will Happen,* Free Associations, London

[4] The Hindustan Times, New Delhi, 7th August 1994, pg 13

[5] Slater D.H.(1986) Chemical Engineering in Australia Vol ChE11, No 1 pgs 12- 16

[6] Psomas S (1988) *Planning for Chemical Disasters* in Disasters Vol 14 Num 4 pgs.301-308

[7] India Weekly, 12-18th of August 1994, pg 15

[8] Bowander B.(1985) *The Bhopal Incident: Implications for Developing Countries* The Environmentalist, Vol 5, Num. 2 pgs 89-103

[9] Swick L.A. (1988) *Bhopal: Lessons for Canada.*in Natural and Man-Made Hazards, pgs 799-804 eds M.I. El-Sabh and T.S. Murty

[10] India Today June 30th 1994 pgs 55-60.

[11] Health and Safety at work, May 1985 pg 41 & pg. 56

[12] Sharma T.P. (1988) Interflam'88 pgs. 261-268

[13] Dheri S.K. (1994) Fire International No 142 pgs 23-26

[14] Down to Earth, June 30th 1994, pgs. 26-36

51 ANALYTICAL APPROACHES TO EMERGENCY PLANNING AND ASSESSMENT OF EMERGENCY RESPONSE

K. ODY and B.A. LEATHLEY
Risk Management Consultants, Four Elements Ltd,
London, UK

Abstract

Emergency planning is crucial to the effectiveness of an emergency response. Advancements in technology and the ability to deal with a large variety of emergencies, has enabled rapid response to threats. Arrival at an emergency by an emergency decision maker, is not the overall aim of the emergency plan; the design of the emergency plan also has a significant impact on the efficiency of the ensuing response. It is argued that many emergency plans illustrate an organisation's ability to run an emergency exercise, not to ensure adequate response to an emergency. This paper discusses techniques which address three problem areas:

(1)   Ensuring that an emergency plan addresses an organisation's goals, e.g. via critical task analysis.

(2)   Ensuring training is effective, using data capture techniques to analyse decision maker attitudes and perceptions.

(3)   Ensuring that emergency plans are tested for feasibility, and that analysis of emergency simulations and of real incidents is carried out effectively.

It is argued that a structured analysis of responses will make assessment and comparison of emergency plans possible - previously made through subjective attribution of success, based on little more than how rapidly the threat was 'made to go away'.

Key words: Emergency planning, crisis management, analysis techniques, decision making, methodical approaches, critical task analysis, testing.

Fire Engineering and Emergency Planning. Edited by R. Barham.
Published in 1996 by E & FN Spon. ISBN 0 419 20180 7.

1 INTRODUCTION

This paper argues that emergency planning is currently approached in a manner which is unsystematic, and proposes techniques for the methodical analysis of planning needs and plan effectiveness.

Research into real international crisis events, and full scale emergency exercises has shown that each stage of a crisis (i.e an unstructured event; outside the scope of regular procedures) demands specific considerations and an individually tailored response. Thus, an emergency plan which anticipates all possible eventualities is practically impossible to design. A good plan therefore needs to be flexible, coping with the unforeseeable, and supporting the decision maker in anticipation, perception, and control of each evolving event - be it fire, toxic release, or bomb threat, etc.

This paper outlines three areas of emergency planning where analytical techniques may be used:

1) At the conceptual stage of the plan.
2) To assess the effectiveness of training by understanding and comparing decision maker assumptions.
3) In the recording and analysis of response activities.

2 BACKGROUND TO THE WORK

Four Elements have been involved in a three year CEC project, partly funded by the Health and Safety Executive, to develop recommendations for the support of the decision maker in crisis. Extensive research has involved modelling of real events, such as the King's Cross Underground Station fire (Fennel, 1988 and Ody, 1993a) to establish the core decision maker activities that lead to a quality response. It is revealed that during a crisis, information reaches the incident decision maker in a relatively random fashion. This information is used by the individual to form a mental representation of the event. Following the development of this representation the decision maker is reluctant to reassess it, or the decisions that are based on it, as any reassessment will put yet more demands on already overloaded people.

The mental representation held by the decision maker will influence their aims for the response, who the decision maker chooses to interact with, and what information is requested. This in turn affects the accuracy of the mental representation as it develops. A decision maker's representation rapidly becomes outdated if all relevant data is not continually integrated as the incident continues. Even if the original mental representation is the correct one, an incident may change from a bomb threat, to an evacuation, to a fire in a matter of minutes. It is therefore critical to identify the elements of training and support which can aid accurate diagnosis on arrival, and the continued collection of data and the regular re-diagnosis of the situational demands.

3   TECHNIQUES FOR ANALYSIS OF EMERGENCY PLANS

3.1   Assessment of Emergency plans

The role of the decision maker is ill-defined and transient, involving many varied responsibilities and tasks. Emergency planning and decision maker research has revealed that the common approach to emergency planning (in this context the identification of threats, the identification of response goals, and the analysis of an emergency response), is presently an ad hoc affair with developments based rarely on systematic assessment. Emergency plans can help the decision maker to improve the effectiveness of response in a crisis.

The key elements in an emergency plan, shown to influence the effectiveness of a response, are:

1.   Anticipate the likely threats and worst case scenarios.
2.   Prompt an accurate definition of crisis on alert.
3.   Place communication within and across agencies as a high priority.
4.   Consider and integrate the potential involvement of other agencies.

The first requirement, anticipation of likely threats and worst case scenarios, is often provided for through risk assessment techniques (Cox & Bellamy, 1992).

The remaining requirements for definition of crisis, communication support, and integration and co-ordination of multi-agency decision makers, can also be improved applying Human Factors techniques already available, such as task analysis and function analysis. Critical tasks central to the success of an emergency plan can be identified by addressing potential risks, potential threats, facilities available, and the organisational priorities. In this way an organisation may establish which central tasks pre-planning needs to cover, and where to allocate resources to ensure completion of these tasks.

By breaking an emergency plan down into its basic elements, and identifying each critical task, time line assessments and barrier analysis can be employed to assess whether each critical task can be achieved. Critical task analysis combined with an emergency goal checklist can provide a rapid means of identifying the critical elements within a particular emergency plan, the goals, and the roles decision makers are required to fulfil. For example: "What does the organisation see as first priority - protection of personnel? return to production? or alert of head office?" The type and shape of a response developed will be dependent on the goal. Table 1 shows an example barrier analysis for assessing part of an emergency plan.

It *is* possible to establish the goals and aims of the emergency plan through discussion. However, a technical approach such as critical task analysis will provide a structure within which to identify, in an exhaustive manner, the key needs of a tailor made emergency plan.

Table 1: Use of a Barrier Approach to Assessment of an emergency plan

BARRIER		BARRIER FAILURE	
Function	Type	Design features and assumptions	Human errors
Receive alert from Train chief Crew member required to obtain more information if location of incident is unclear	Procedural	Alarm alert is in a clearly understood form, containing information that will not be confusing or stressful	Fail to understand message Fail to receive message
Receive alert from Train chief via Personal radios Standard message format	Physical	Radios have open channel Assumes can be heard above ambient noise Assumes message can be received in all areas	Fail to carry radio Fail to switch on radio Message unintelligible
Receive alert from Train chief via public address system Crew acknowledge receipt of alarm and confirm that they are on their way	Physical	Messages can be heard in all locations	Fail to attend to public address system

3.2 Analysis of emergency response training success

To make a valid assessment of the success of training, a good understanding of both organisational and training goals is needed (eg. return to full production rapidly, or reach threat limitation through a co-ordinated response; gain understanding of the importance of cross-service co-ordination) as well as a decomposition of the "optimum" decision-maker's mental model.

Comparison of the mental representation held by different decision makers following a training exercise will give some idea of the success of that exercise. One psychological technique which can be used to capture and compare the mental representations of individuals is 'repertory grid technique' (Fransella & Bannister 1990) . There is not space to describe the process in detail, but in this context, the training goals provide the 'elements' or objects on which people provide their 'constructs' or ideas. The 'grid' provides a map of each individuals representation of the incident, which allows a systematic comparison of experts with novices, observers with participants, and different emergency services with each other. The technique has already been used successfully to compare the mental representations of members of control room teams (Leathley & Tinline, 1995 in press).

The use of this approach enables comparative assessment of different training techniques and emergency plan designs, by comparing the completed grids from decision makers.

In addition, the structured analysis of an emergency exercise will aid decision makers by providing a tool with which to compare views of the exercise. Such work will enable decision makers to understand more about their own decision making, and hence improve performance in future exercises, and in real incidents. Such a technique of structured analysis of emergency exercises and incidents is described in Section 3.3.

3.3 Analysis of the emergency exercises and incidents

It is crucial to response success to test a plan fully, at least once in its life, to identify the key elements that have been forgotten or that require redesign. For example, during the King's Cross fire, it was discovered that police radios would not function underground, so that and all radio communication was made by travelling to the surface, and that no emergency services were in possession of maps outlining the layout of the station (Ody, 1993a). An emergency exercise would have revealed these facts before tragedy struck.

This section presents a technique which can be used once an exercise has been completed, or following a real incident, to capture weaknesses in order to improve an emergency plan.

Figure 1:
Summary of warehouse fire flow-chart description

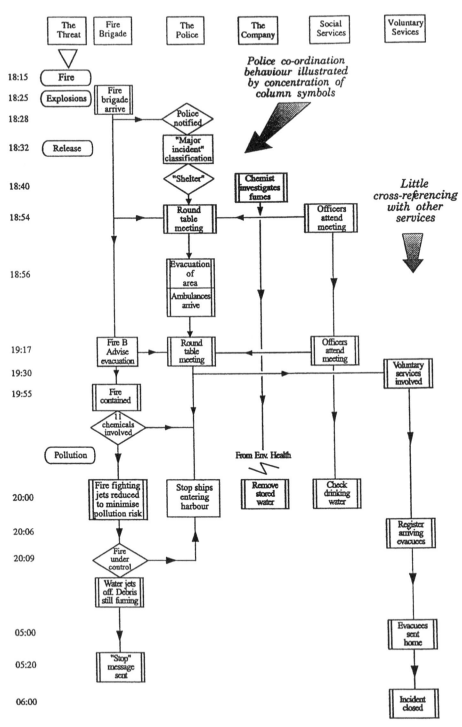

Flow-chart modelling of an emergency response

Debriefings of emergency exercises currently make use of structured discussions and "prompt words". However, as yet no analytical techniques are used to compare the efficiency of one response against another, and therefore to judge just how effective the exercise has been. Similarly, de-briefings following real incidents could benefit from a more structured approach. In the UK, incidents are assessed in "hot-debriefings" soon after an event, and "cold-debriefings" a day or so later, attended by the key decision makers. However, psychological research has shown eye-witness testimony of this sort, gathered as anecdotes, to be, (a) unreliable and, (b) easily biased (Cohen et al, 1993, Glietman, 1986).

Four Elements have used a flow chart analysis technique to illustrate an event graphically, displaying activities as information-decision-action links. The flow chart model is produced by plotting actions and communications, and illustrates stages of emergency response from initial alert, through the response processes as the event develops. Figure 1 shows a summarised analysis of Poole BDH Warehouse fire (Ody, 1993b), using the flow chart approach.

This approach has clear advantages over the conventional debriefing currently employed:

- Flow-chart analysis enables the systematic correlation of reports, from varying sources, into one visual illustration of events, indicating when data collection may not be exhaustive (via cross analysis of communications and requests for action), and provides a direct comparison across different scenarios.

- The assessment technique produces similar results across different modellers. Therefore, through flow chart assessment it is possible to develop a library of plotted and comparable emergency responses, whoever has modelled the process.

- Plotting the event by time and actor provides a clear indication of the time lapses involved, and decision time taken. Quantitative measures can be taken, e.g. time taken before co-ordination between decision makers started, also introducing a visual means of measuring effectiveness of response.

- Process modelling enables identification of the culture of each group of players, through patterns of decision-making, and the resulting success of their response. For example, Figure 2 illustrates a culture which encourages reference to authority, where information goes in one direction only. This "waterfall" pattern is characteristic of such a culture. Alternatively, where "loops" can be seen in the flowchart (see Figure 3) this indicates a culture where information is passed in both directions, enabling more rapid response to the emergency. The first of these examples is based on an analysis carried out of the King's Cross disaster.

Figure 2: Example "waterfall" flowchart, showing continuous reference to authority

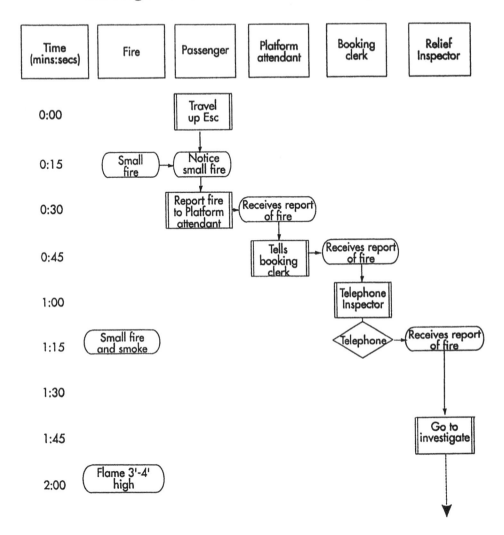

Figure 3: Example "looped" flowchart, showing feedback and support

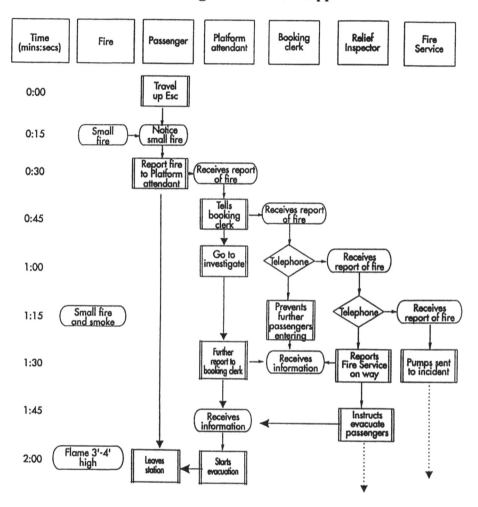

- A display of information-decision-action links means that the information requested and used by decision-makers can be traced backwards from that decision. That is, the illustration will provide an indication of why a decision was taken, by tracing the data the decision-maker had acquired and who the decision-maker had talked to.

The philosophy behind process flow technique holds that by establishing the understanding of key decision makers taking part in an event, and the received information that produced that understanding, the logic behind actions taken may be assessed and understood (Drieu C & Durand C, 1992). The core principle holds that decision-makers must have (1) an accurate understanding of events, as well as (2) access to the important players from other emergency services. Only when this is the case can an effective and efficient emergency response be achieved.

4   CONCLUSIONS

There is obviously a need for emergency plans to cope with the unexpected crisis and emergency event. However, as each emergency is, by definition exceptional, a plan can only be designed on a generic basis, to cope with the threats identified through analysis of the risks.

Therefore, the testing of a plan is crucial to:

(1)   Ensure that the structural elements such as efficient co-ordination and communication have been designed in.

(2)   Ensure those involved in carrying out the plan understand their role and can carry out their responsibilities.

The goal of a generic plan is to be flexible, and to enable identification of changes, and a rapid response. Emergency plans are difficult to test, with high demands on time, and the extensive allocation of resources. There is therefore the tendency to stop short of analysing the effectiveness of a plan following testing - being satisfied that the exercise itself was carried out.

Use of techniques such as critical task analysis, data capture, and response modelling enable the emergency planner to record the results of actual incidents and emergency response simulations, in a way which allows a comparative study of the effectiveness of training for decision makers and of emergency plans.

REFERENCES

1. Major Civil Emergency Planning (1992), *Major Civil Emergency Planning Seminar,* London, 23 November 1992, IBC Technical Seminar, Bedford

2. Baldwin, R., 1994, Training for and Management of Major Emergencies. *J. of Disaster Prevention and Management,* 3 (1), 16-23

3. Cohen G, Kiss G, & LeVoi M, 1993, Memory: Current Issues, Second Edition, *Open Guides to Psychology,* Open University Press.

4. Cox, RA & Bellamy, LJ (1992) The use of QRA and Human Factors Science in Emergency Preparedness & Crisis Management. Risk Analysis & Crisis Management, the Interface, 22-23 September, 1992, London. Organised by BPP Technical Services / Cremer & Warner.

5. Borodzicz, E., Ody, KJ., and Pidgeon, N., 1994, Birkbeck/Four Elements Comparison and Synthesis of National Case Studies: Using Case Study Categories, 3 June 1994, *Revision 2, Internal work for the CEC.*

6. Dreiu C & Durand C, 1992, Crise de Nantes, 29.10.87, *Modele Organisationnel de Traitement,* Gemini Consulting; (STEP Project internal report).

7. Glietman, H. 1986, *Psychology (2nd Edition),* Norton & Company,

8. Fennel D, 1988, *Investigation into the King's Cross Fire,* Department of Transport, Her Majesty's Stationary Office.

9. Four Elements Case Study of the Emergency Response to Poole Warehouse Fire, Dorset, UK 1988. Rev 02, 10.9.93 *(Unpublished report for the CEC, STEP Project 90-0094)*

10. Fransella F, & Bannister D, 1990, *A Manual for Repertory Grid Technique,* Academic Press Ltd.

11. Leathley, BA & Tinline, G, 1995 (in press)

12. Ody K, 1993a, *Four Elements Assessment of a Flow-Chart Technique for Crisis* Assessment Using the Kings Cross Fire Disaster.

13. Ody K, 1993b, *The Information Needs of a Decision Maker During Crisis,* Internal study for the CEC (CEC STEP Project 90-0094).

14. Sills D L, Wolf C P, & Shelanski, V B, 1979, *Accident at Three Mile Island: The Human Dimensions,* Westview Press/Boulder, Colorado

52 AIRBORNE HAZARDS: MOVE OR STAY PUT?

J.R. STEALEY
John Stealey & Associates, Kent, UK

Abstract
Materials released into the atmosphere may be dispersed by meteorological effects and deposited over considerable distances from the site of release. If this material is toxic the potential problems for injury to humans and contamination of the environment could be severe.

An understanding of dispersion mechanisms plays a fundamental rôle in pre-event disaster limitation and contingency planning. Should a release occur, prediction of downwind spread and subsequent monitoring are essential for post-event consequence management including decisions on public protection.

Therefore, it is important for strategic planners, emergency responders and operational decision makers to consider ways in which materials released into the atmosphere spread downwind and the consequences which might result. This paper addresses aspects of such releases particularly those occurring close to the ground.

Keywords: airborne hazards, consequence management, evacuation, decision making, dispersion, meteorology, public protection shelter.

1. Introduction

Cloud emissions from power station chimneys or chemical processing sites are a common sight and people living nearby will report that the shape of these clouds and their plumes changes dramatically with variations in the local meteorology.

On a larger scale, the explosive and thermal release of radioactive material into the atmosphere from the nuclear reactor at Chernobyl graphically demonstrated how materials can be transported and deposited across oceans and continents. The disaster at Bhopal was a trigger for an international response and legislation on the hazards and risks from the spread of toxic clouds.

Fire Engineering and Emergency Planning. Edited by R. Barham.
Published in 1996 by E & FN Spon. ISBN 0 419 20180 7.

Even accidental releases from relatively small events such as venting from a factory chimney, a large scale fire at a tank farm or perhaps the spillage from a road or railway tanker can give rise to the airborne dispersion of hazardous material.

Of course, some atmospheric releases occur as part of normal site operating procedures where environmental contamination and human burden, in terms of toxicological risk management, should have been assessed and calculated as part of the licensing and operating conditions.

The important events are accidental releases which continue to occur [1, 2],and which by definition, are uncontrolled and undefined and where the downwind effects on humans and the environment are unknown. The immediate requirement is to warn the public and provide advice on protection; this is likely to be a choice between stay-put or evacuation. Consequently, decision making requires a knowledge of the way in which firstly, materials are spread and then their biological effects; an inappropriate decision between these two alternatives could place many people potentially at greater risk.

2. Release processes and sources

In general terms releases might be characterised as:--

- Active or jetting
- Neutrally buoyant
- Positively buoyant
- Negatively buoyant

Other effects of the accident, such as heat loading from fires, might also cause initial upward displacement and dispersion.

Neutrally buoyant material has a density close to that of the surrounding atmosphere and its mixing process with that atmosphere is random. Movement away from point of release is dominated by the wind speed and the stability of the atmosphere.

By contrast, a negatively buoyant material is denser than its surroundings such that aerosol and vapour clouds slump to the ground and spread out in a radially from the point of deposition sometimes demonstrating unusual qualities such as the ability to "flow up hill"; however, in spreading, they increase their surface area which in turn can affect the rate of evaporation.

Positively buoyant materials will rise into the atmosphere on release until they reach a height where upward momentum is overcome by wind speed and direction. They then tend to behave like neutrally buoyant materials.

Active jetting is the release of material under pressure at a speed in excess of the wind speed. Once the velocity of release falls below the wind speed meteorological factors tend to dominate the continuing dispersion process with deposition rates dependent upon whether the material is neutral, positively or negatively buoyant.

Two other factors should also be taken into account. The first is time related such that releases can be: continuous (plume), instantaneous (puff) or time varying. In addition, the material may be released as a gas, an aerosol, a liquid or a mix of all phases.

3. Meteorology and dispersion

Once material has been released conditions in the atmosphere influence any further dispersion. Firstly, wind speed determines how quickly the material moves away from the point of release and secondly, the stability of the atmosphere affects the vertical and horizontal spread of material and hence the dilution rate.

The physical principles and characteristics relating to atmospheric stability have been well documented and a number of text books are available on the subject [3]. The important factor to note is that stability is related primarily to the temperature of the air in the lower layers. This can vary significantly on a diurnal basis due principally to changes in the amount of sunlight absorbed by the ground during the day and re-radiation of heat from the ground at night.

Unstable air is characterised predominantly by rapid vertical mixing of entrained material which is quickly dispersed and diluted. Alternatively, stable air shows little vertical (or horizontal) mixing and the concentration of material entrained remains relatively constant for a long time. Under the influence of light winds, also a feature of stable air, the mass of material will move slowly away from the site of release with little diminution in concentration. Under certain meteorological conditions a temperature inversion can form a "lid" some tens, or perhaps hundreds of metres above the ground. Material released below the inversion is trapped into a lower mixing layer and tends to fumigate areas downwind.

Atmospheric stability is often categorised into one of six classes [4]. Simple rules can be devised to estimate stability class based on the amount of incoming sunlight, cloud cover and wind speed (Table 1).

Table 1 Estimating Stability Classes

Wind Speed m/s	DAYTIME Incoming Solar Radiation			NIGHT TIME	
	Strong	Moderate	Slight	Thin overcast or >4/8 cloud	<3/8 cloud
<2	A	A - B	B		
2 - 3	A - B	B	C	E	F
3 - 4	B	B - C	C	D	E
4 - 6	C	C - D	D	D	D
>6	C	D	D	D	D

A	Extremely unstable	D	Neutral
B	Moderately unstable	E	Slightly stable
C	Slightly unstable	F	Stable

From a large number of experimental releases it has been shown that material in the plume follows a classical Gaussian bell shaped distribution where the concentration is at a peak in the centre and falls off towards the outside edge. It should be stressed that much of the experimental data on dispersion have been obtained from flow over "smooth" surfaces and open country. The heat island effect of urban areas and disturbances caused by tall buildings and built up residential areas are exceedingly difficult to describe and model; some empirical data exist but calculations on downwind dispersion and plume definition can only be approximations. Vortex flow and stagnation are two factors which modify concentrations in urban areas very markedly. Therefore, in the event of an accident in built areas, real time decision making based on computer simulations and models of airborne dispersion should be treated with caution unless the user has full confidence in the quality of the input data and assumptions.

4. Dose and dosage

Injury to people by toxic material can be caused by inhalation, absorption or ingestion. Inhalation of a vapour is a very efficient way of introducing a substance whereas absorption of the same substance through the skin or ingestion through food is less effective. There are two measures which quantify uptake; the dose and the dosage.

Dose can be defined as the mass of a substance taken into the body, often measured in milligrams per kilogram of body weight. However, without reference to the time over which the amount is received is not an immediate measure of the effectiveness.

The term dosage is often used to overcome this apparent problem and refers to the "integral of concentration with time" or concentration:time product (Ct), expressed in units of mg.min/m^3 . On this basis it is argued that exposure to a low concentration for a long time could result in the same dosage as exposure to a high concentration for a short time. "Half as much for twice as long is just as bad".

However, other work [5] suggests that a linear relationship may not be an appropriate measure of harm caused by exposure to a toxic material. Instead, the term "toxic load" has been used and defined by the equation $L=C^n t$ where C is a constant gas concentration raised to a positive power (n) which has a value greater than 1. The UK Health & Safety Executive [6] have suggested that n equals 2 for chlorine; the Institute of Chemical Engineers [7] give the same value for ammonia. Toxicity evaluation for a number of other chemicals [8] suggests that the value for the exponent n may lie in the range of 1.5 to 3.5. Consequently, toxic loads, with exponents greater than 1, will be very sensitive to changes and fluctuations in peak concentration of toxic material.

A variety of terms are used to relate the amounts of different chemicals or substances required to give the same effect: the LD_{50} is the dose of a particular required to kill 50% of an exposed population and the ID_{50} that required to incapacitate the same number. Similarly, the LCt_{50}, can be defined as the dosage required to kill half of an exposed population. However, differences in biological responses between individuals mean that members of a population do not all behave in exactly the same way to the same substance. Refined statistical techniques can be used but accuracy can not always be justified from the available experimental toxicological data. A conservative "safer" approach might be to use the simple Ct relationship on the basis that the dose response curve is reasonably steep and that the LCt_{50} is a useful representation of an "all or none response".

Dosages can be further modified by other factors such as wind speed and the rate of a person's activity, exertion and breathing rate at the time of exposure.

Therefore, exposure and subsequent effects depend on a number of inter-related mechanisms all of which result in a complex process of evaluation in judging risks and options for protection. This complexity can be simplified to assist real time decision making; however, in these circumstances an understanding of the underlying principles of dispersion and toxicology is required in order to make value judgements when managing an accident and its consequences. Conversely, in any programme of pre-planning and risk analysis it is important to consider the widest range of possible factors which might contribute to accidental releases and evaluate the downwind spread of material and its effects upon people and the environment. It is in this pre-planning phase that the toxic load concept can be evaluated through sensitivity analysis using ranges of values for the exponent n.

5. Protecting the population

In the event of accidental release of materials into the atmosphere the first consideration is the protection of the population living downwind from the site of release Two options might be considered: stay-put or move.

Whilst there maybe a natural tendency to try and escape from an emergency area particularly if it is thought to present an airborne toxic hazard there have been a number of major reviews [9] of evacuation procedures, many based on real events, provides support to the proposal that keeping people indoors is a justifiable first option. Firstly, it is unlikely that any evacuation process could be initiated in less than 30-45 minutes after first notification. Thereafter, the time taken and complexity to achieve an evacuation is likely to be proportional to the number of people to be moved but not necessarily in a linear relationship; published data suggest that it might take up to one hour to move 200 people and perhaps up to three hours to move 1000 [10, 11, 12].

In addition, moving people into the open presents two fundamental problems. Firstly, they would be exposed fully to the toxic cloud; secondly, they are vulnerable to any associated and subsequent effects such as secondary explosions or fires.

There are a number of other reasons to suggest that staying indoors is preferable to moving outside. Ventilation rates for UK houses (determined from air penetration rates [13]) indicate that the number of air changes per hour for a house with doors and windows shut might vary between 0.2 and 1.5 per hour depending on the age, style of construction, double glazing and secondary draught exclusion. Other factors which influence ventilation rate include size of openings, wind speed and direction and the temperature gradient between the air space inside the building and the outside air.

Therefore, as a cloud of airborne material passes the outside of a building a certain amount will enter and the concentration inside the building will rise, but slowly. If the release is continuous at source and the external concentration remains constant (an idealised assumption) then eventually the inside concentration will approximate to that outside. However, once the outside concentration has fallen, because say the release was an instaneous puff, or the cloud has meandered with fluctuations in peak concentration or the wind direction has changed, then there is no further ingress to the building space and the inside concentration can no longer rise.

But equally, unless doors and windows are opened after passage of the cloud the inside concentration will only decay very slowly as material is deposited onto surfaces, absorbed or ventilated outwards. Consequently, exposure inside a building to a low concentration for a long time might eventually be as harmful as exposure to the high concentration outside but which passed by relatively quickly. Therefore, information about the characteristics of the release are required for once a cloud of material has cleared the best advice to occupants is to move outside, open all windows and doors and ventilate the space as quickly and efficiently as possible.

6. Evacuation

If an operational decision is made that a population be moved then knowledge of the release, its likely duration, toxicity and physical properties must be known and evaluated. In addition, details of meteorological conditions, now and for the next few hours, particularly changes in wind speed and direction, must be determined.

To a first approximation, the best direction in which to move is at an angle acute to the downwind centre line of the plume across its cone. Protection gained by moving directly away from the site of release, down the centre line, applies to those damaging effects which spread radially, such as blast and thermal radiation. However, advice to move across the plume holds good only if based on sound meteorological data.

Furthermore, the physical exertions and changes in breathing rate brought about by movement itself and the stress or anxiety of evacuation can alter the toxic load or dosage parameters.

Finally, the possibility of subsequent or secondary effects, such as blast or fire, may influence significantly the decision to move or stay-put: once a population has been brought out into the open it is at its most vulnerable and all possible damaging effects must be evaluated and a relative risk assessment carried out.

7. Particulate releases

The increased use of new technology, such as biotechnology and genetically modified organisms (GMOs) in a variety of biological and pharmaceutical industries raises particular concerns about downwind spread of particulates such as biological organisms. UK Regulations on GMOs [14] now require an assessment of risk and preparation of an emergency plan which in certain circumstances extends off-site.

Particles tend to behave like neutrally buoyant vapour clouds, and under favourable conditions, their downwind spread can occur over very considerable distances . Impaction on buildings and other structures can cause dilution (re-suspension into the air is unlikely) and the same principles apply to particulate ingress into buildings as do those for chemical vapours or aerosols. Partitioning between external and internal concentrations can be approximated to that for chemical hazards. Particulate deposition on vegetation off-site can lead to a potential hazard through uptake and passage into the food chain.

8. Managing the consequences of airborne releases

8.1 Pre-planning

Analysing the downwind spread of materials from accidental release scenarios should form part of a comprehensive site evaluation, emergency plan and response capability evaluation. Whilst it is often argued that emergency planning should be flexible enough to cope with all eventualities scenario led analysis can set bounds on the problem such that limitations in the response capability can be examined.

In the event of an accidental release of material to the atmosphere the first priority will be to warn people downwind to stay indoors, close doors and windows and listen for further advice. Such action suggests a fundamental need for a rapid and reliable public warning or alert system which is properly designed, tried and tested by operators and understood by the local communities. A siren system may provide this capability; however, pre-planning should assess and evaluate a range of warning options to define the most appropriate to the circumstances.

Analysis could be carried out on the efficacy of remote sensing devices to detect and monitor the movement of material in the atmosphere. These might also help to confirm whether a release is a puff, a plume or an intermittent source.

Prediction of downwind dispersion requires a capability to measure local meteorological conditions. Equipment estimating wind speed and direction, and wet and dry bulb temperatures could be valuable but require good training in their use and application to the evaluation of basic meteorological data.

All planning functions require evaluation and review. Exercising operational decision making on a range of airborne hazard scenarios would be of great benefit.

8.2 Post-Release

On site assessment of the release source and material(s) are valuable first steps. It is probably impractical to try and analyse and identify the mix or composition of materials released; certainly, in the event of associated fires the combustion products will vary greatly according to the source of material, the temperature at the scene, rate of release and the amount of oxygen available during the combustion process. These conditions are likely to change during the release process such that meaningful data on the mix of materials and their toxicology are not valid.

Determining material dispersion downwind requires measurement of the local meteorological conditions although these can vary considerably over very short distances and their use in interpreting the direction and flow of clouds in built up areas should be treated with caution.

The UK Meteorological Office provides a service, CHEMET [15], to local authorities, police, fire and other emergency services to help in the management of accidental toxic hazard releases. This service, based on advice from a professional weather forecaster using basic site data provides a number of generalised templates depicting likely downwind threat areas which can be used as decision support tools for evacuation or stay put. It also advises on the possible changes in local wind conditions in the next few hours which might influence decisions on directions for evacuation and locations for sheltering evacuees.

9. Environmental impact

This paper has stressed the impact on people living in the path of accidental airborne releases and whilst this will always be the over-riding operational consideration there is a pressing need to consider the effect of materials on the environment and the possible short and long term damage they might cause. Deposition on land and into water courses, the possible uptake by vegetation, livestock and food crops and how this might pass into the food chain, are important assessments. Many other agencies will be involved with this process and they too will need information on the type of release, the likely mix of materials, the meteorological conditions and nature of downwind spread.

Conclusions

The advice to populations to stay-put in the event of an accidental release of toxic or hazardous materials is well founded on experimental evidence and should be the preferred option for advice on protecting populations in downwind areas. However, in evaluating the impact of such events it is very helpful to understand the mechanisms which influence the downwind dispersion of material in order that a sensible, balanced and authoritative assessment of hazard, risk and consequence can be made. This then leads to robust judgements in emergency and contingency planning as well as decision making and consequence management should such an accident occur.

References

1. Health and Safety Executive Report (1993) *The Fire at Allied Colloids Ltd.* HMSO.
2. Davis, D. (May 1994). *Disaster Prevention and Limitation in Industry Seminar.* Cleveland County Fire Brigade.
3. Meteorological Office. (1978) *A Course in Elementary Meteorology* HMSO.
4. Pasquill, F and Smith, F. B. (1983) *Atmospheric Diffusion* 3rd Edition Halstead Press (John Wiley & Son).
5. Models for Predicting Outdoor and Indoor Exposure Hazards from Toxic Gas Releases. (1990) University of Alberta.
6. Health & Safety Executive Report. (1990) *Toxicology of Substances in Relation to Major Hazards - Chlorine.*
7. Institute of Chemical Engineers - *Ammonia Toxicity Monograph.*
8. ten Berge, W.F., Zwart, and Appleman, L.M., (1986) Concentration - Time Mortality Response Relationship of Irritant and Systemically Acting Vapours and Gases. J. Hazardous Materials, 13 pp 301 - 309.
9. Quarantelli, E. L., (1984) *Evacuation Behaviour and Problems: Findings and Implications from the Research Literature.* DRC Final Report No 2 University of Delaware.
10. Johnson, Jr.J and Zagler, D (1986) *Modelling Evacuation Behaviour During the Three Mile Island Reactor Crisis.* Socio-Economic Planning Science 20 pp 165 - 171
11. O'Riordan, T.R., Purdue, M. and Kemp (1985) *Sizewell B* MacMillan London.
12. IBC Conference on Disasters and Emergencies April 1988.
13. Perera, E. and Parkins, L. (1992) *Airtightness of UK Buildings, Status and Future Possibilities.* Environmental Policy and Practice 2 143 - 160.
14. Health and Safety Executive. *A Guide to the Genetically Modified Organisms (Contained Use) Regulations, 1992.* HMSO
15. UK Home Office, Fire Department. July 1989.

REPORT OF THE PLENARY SESSION

Introduction

The Plenary Session at the end of the conference was run as an open forum. Many of the delegates used the session as an opportunity to open up discussion on specific points arising out of papers that had been presented but, in addition, a general discussion on some fundamental issues developed as a result of the audience mix. Of the many interesting matters raised, the following is an almost verbatim report of the main points of general import which engaged the majority of the delegates, the speakers and the panel of the plenary session.

The discussions ranged over four main areas: the nature, and need, of risk assessment (and of operational risk assessment, in particular), the relationship between researchers and operational practitioners, education and research, and the qualifications of, or for, fire engineering.

Operational risk assessment

The idea of Operational Risk Assessment is not totally new because, in Germany, they already have the "feuerwehreinsatzplane" (the fire brigade access plan) which would identify the premises, the hazards in the premises, and the way to deal with the fire in the premises. Of course, as everything else, these are not perfect but they form one of the most important and major tools - providing primary information for the fire officers on how to deal with a fire on the premises. This plan is requested from the owner or occupier of a building, or a person running a plant, who is legally obliged to provide the information and these "plans" to the fire brigade. On the basis of these plans the fire brigade comes to the premises, checks the premises, puts additional questions and then either asks for additional hazard descriptions to be incorporated in these plans or accepts that the plan is sufficient for their purposes. The Risk Card system in the UK provides somewhat similar information.

However, what has to be recognised is that if you have a static system, or even if it is on computer, there is still an element of operational risk assessment. The officer in charge of the first attendance has to do their own risk assessment "on the ground" based the sort of basic information with which they are provided by the risk card or the "feuerwehreinsatzplane". Basically, the UK and German systems are the same but having formally to risk assess premises makes one much more aware that from time to time you inspect a property where maintenance activities are being conducted and it is found that things are not as they are on the plans. In such circumstances it falls to the officer in charge to do an operational risk assessment; most fire services work in that way.

There are, however, noticeable differences, both in philosophy and in legal requirements between the UK and Germany. It is only becoming so in the UK that employers or occupiers of premises unless they are registered, unless they have to register under certain regulations, would be expected to provide the fire service with plans. It's usually up to the service to get at the information by asking the correct questions. There is also one other thing that is actually quite different: the concept of "as far as reasonably practicable". This has to do with quite basic differences between the legal systems. Most European legal systems are Napoleonic in philosophy which means that, or it has the result that, many things are fairly prescriptive. They are described in great detail within the legislation. The British system is based on a double pronged approach: a.) statute law (with enforcing Authorities) and b.) case law in which the senior judges actually make the law by interpreting what has happened; and it may be that this interpretation evolves over very, very many years. By way of an example of this, it is only very recently that corporations have been able to be sued for manslaughter. This idea of immunity from suit was based on a judgement by the then Lord Chief Justice, Lord Holt (in 1702), that a corporation could not appear or plead in court because it was not a natural (or physical) person. This legal concept survived until very recently; however, the first corporation that was sued for manslaughter (but the case was dropped) was the company that owned the Herald of Free Enterprise. The second action, and the case was successful, occurred at Winchester Crown Court. This was the Lyme Bay canoe disaster in which the firm that owned the leisure centre was successfully prosecuted for manslaughter. Therefore, when talking about this subject in a European sense, we have to be aware not only of detailed differences in the legal background, or the way that things are done, but also philosophical differences of how it is perceived that things should be done.

Research relationships
What has come to the fore during this conference is that there is a great deal of work going on but that the communication in the broadest sense simply does not exist. It is vital for researchers, academics, engineers and the fire service members to have this exchange of information and, therefore, avoid on the one hand duplication and, on the other, a concentration on some exceedingly profound research which, when you come to translate it into action during an incident, is not of any practical use. This is a great pity. So, if this EuroFire Conference has done anything, it has opened the dialogue. The exchange of views, hopefully, has caused some researchers to think, at least, in terms not just of the static controlled condition that can be achieved under research conditions, but also in terms of the dynamic conditions that persist on the fire ground.

They should note that, however well-planned the building has been, (presuming that the client knew what they wanted, anyway) and if the architect can be persuaded to suppress his own desires and actually construct something that will be a purely practical enterprise and if, once it is brought into commission, it actually performs on the basis for which it was planned, and the continuing and on-going storage, usage and transportation conforms to the original concept over a period of days or weeks (never mind years) then, when ultimately and regrettably something goes awry, the complex mix that has then occurred in the interim period is the situation (the workplace) within which the operational fire service must operate effectively and safely in the United Kingdom.

However, while this argument found very strong support, it should be noted that fire science or fire research is carried out not only to provide more information on the working arena for the fire brigade, but must also address the issue of such things as early fire detection. It is also concerned with the issue of reducing losses in terms of economic losses. It is important to note the requirements of operational activities, and in this respect research or development has gone strongly in the direction of purely answering scientific questions. It is necessary to go back more to questions that are real world questions and not only scientific questions. That research should answer fire brigade questions.

Research and education

Experience in Sweden shows that education is a very good tool of communication. It is possibly the best one available. It takes away a large number of misunderstandings between academe and the operational emergency services. Lund University has been educating future senior fire brigade officers since 1986 on the basis of, now, three and a half years' education at university and one year at an operational school. Now, the academic researchers are being asked to do research for operational purposes. This would not have happened at, let's say, five years ago. This indicates that, having started an educational activity much of the other things follow automatically.

It was, however, considered that education should go in two ways. First, the fire fighters should be going to the universities and getting to know more about the background to fire and fire science. Secondly, the scientists should be invited to take part in fire service operations and experience fire fighting - not actually in the front line but maybe in the next row.

These exchanges should happen automatically once this academic/education process has started. It is not a question of who is learning what, or teaching what; it all falls into place automatically, more or less. There is both a synergy and a symbiosis which goes on as these educational and research facilities develop. For example, some of the academic staff at the University of Central Lancashire have had the experience of going through the smoke house and of using the facilities at the International Fire Training School at Washington Hall, at the Fire Research Centre in Warrington and have also had an involvement in research from an insurance point of view; time has been spent shadowing the Chief Fire Officer of the local brigade and quite a lot is being learned about the kind of problems being experienced by the operational service. So it would appear that many members of the research and academic community are already working to overcome this problem. In addition, it was the view that working with the emergency service makes life much more interesting and dangerous! It is something which is really worth striving for.

Risk

Discussants then referred to the earlier point about not forgetting property losses. Basically, with risk assessment one of the things that the U.K.'s Health and Safety Executive would also support is the idea that when looking at risk reduction or control and elimination procedures it is necessary to do a cost-benefit analysis. Cost-benefit analysis doesn't just include losing property. It is also a fairly major cost to lose a fire crew and, once that is acknowledged, it is then possible to see that there is a need for a unified approach to risk assessment. All the costs must be considered against all the benefits so that one does not find oneself in the position of just arguing for the service or just arguing for the insurance companies. It is a mixture that must recognise all of society's views and requirements, both public and private.

It is not only a mixture but the techniques need to be mixed as well. The tendency, at the moment, is to look at techniques in isolation. There would be more benefit in taking a slightly more eclectic view of the application of well-developed cost benefit techniques that are available both within the insurance industry, within the construction industry, within actuarial science and in many other areas.

With the continued developing of the complexity of emergency situations, I could be argued that risk analysis can also help in resolving two other questions in relation, specifically, to fire risk. The first one is that putting the fire out could make things worse. This is especially so if dealing with, say, a chemical warehouse. So how to teach people in not go for what they have been trained to do for the last forty years, or even longer - i.e., if you see a flame, put it out? How to tell them that, in those circumstances, where there is a flame, put petrol on it and make it worse so that, at least, the fumes will go up and away from the emergency crews. Another interesting conceptual discussion, that has been in progress in the Netherlands for a long time, is where we have a situation where, let us suppose, the National Museum is on fire. There is a night watch on the premises. How many firemen are we prepared to put at risk in order to get the night watch out? You can do a cost-benefit analysis for a long time without being able to resolve that matter because you always hit the point that all human life is sacrosanct and there then follows the question of how this can be balanced against economic cost or the cost in terms of other life losses or potential life losses.

Fire engineering qualifications
During the three days of the conference, the expression "fire engineer" or "fire safety engineering" has been used in a number of different contexts. In fact, it seems as though it is now appearing in draft Codes of Practice, in ISO Standards and certainly within the UK in the draft Fire Engineering Code itself. In the draft, it refers to "... this approach, if used by a competent fire safety expert ..." The difficulties we have, in the area of regulation, is putting a definition to those words. The reason for this is that, within the UK, the only requirement necessary to become a fire safety engineer is to have sufficient money to buy four screws and a brass plate with "fire protection engineer" written on it and you can set up business. On a wider basis, on a European basis, it is necessary to bring together a discipline or to provide a statement of what is meant by "fire safety engineer" and to provide for the control and development of the profession.

There is, of course, a problem in the UK, not just with the definition of a fire engineer or a fire safety engineer but with the very definition of what is an engineer. For example, in the UK, the person you take your car to - to have its service - is an "engineer" - people who repair sewing machines are "engineers" and there are many other examples. There is also a problem in Europe. It is necessary to note the way in which the European education systems and professional institutions operate. First of all, they differ from country to country. They are certainly different to the UK in that the professional bodies are seen, in the mainland European countries, more as associations for the exchange of views and knowledge and not as examining or disciplinary/standards setting organisations. It will be difficult, therefore, to resolve the question. There is an organisation FEANI that has developed the "Euro-Ingenieur" code; there is the Engineering Council in the UK but we have to again look at that in order to try to get some kind of cohesion

Certainly in Germany it would not be allowed for a person to claim to be, for example, a KFN or a KFZ Meister able to repair cars, unless certificated. This is a difference in philosophy. The British will allow the plumber, the electrician, the car mechanic without certification. That is certainly not true in Germany where the Craft Master or the Master Craftsman is certificated and cannot operate as a self employed individual without that certificate.

This move towards "certification of competence" is coming to the UK. It is demonstrated in, for example, the new degree courses by the difference between the fire engineering routes leading to BEng and BEng(Hons.) and the non-engineering routes which give BSc or BSc(Hons.). The main difference between the two is that the fire engineering BEng routes are following recognised Engineering Council prescribed curricula where there is a sufficient quantity of engineering, mechanics, physics, chemistry, etc. Other non-engineering routes are intended for people who do not wish to take up engineering, *per se*. Although they cover the same, or lot of the same, materials in many cases, the students often do not aspire to "Chartered Engineer" status. There is, in the UK, a difference between the Chartered Engineer and these other generic terms, Fire Engineer, the Fire Safety Engineer or the Fire Consultant, that are in general European use. The word "chartered" would appear to be one of the main causes of confusion alongside the many misuses of the term "engineer". In most of continental Europe, it is the academic degree which gives entitlement to the use of "engineer", not membership of a professional association.

It should be noted that it is possible to have all sorts of Engineers, just as it is possible to have academics with all sorts of doctorates,. There is a European Directive which requires that certain qualifications be recognised between member states. In the Netherlands, for example, it is not possible to call yourself an engineer unless you went to a university. However, there is no such thing as, say, a risk analysis doctorate. It is possible to be a doctor in chemistry and to know something about risk analysis, but there is no scheme by which you can say "I am a qualified risk analysis person". In order to do that you first have to define the qualifications that you need to tack on to the basic qualification (i.e., of being a scientist) to indicate also that you have a specialised field. In summary, the problem is twofold; it is firstly one of the general acceptability of the terminology, i.e., fire engineer, and secondly one of exclusive use of the term "engineer". The latter is one that is going to take some time to sort out. It appears to be that it would be easier to work in a profession where there are defined rules - but engineers, whether mechanical or civil, seem to have some overlap with other engineering disciplines. However, it is difficult to define a fire engineer, or even a civil or a mechanical engineer for that matter, on a Europe-wide basis.

Closing remarks

In the closing discussions, it was remarked that it had been encouraging to see that this conference has opened our eyes to a dynamic network of people working in academe and of interested practitioners as well. As opposed to there being a hierarchy, there seems at least to be an interaction of information flowing from the academicians to the practitioners but some of the feedback that has been received at the conference seems to imply that the academics are not specifying the possible applications of their research. It seems to be that there is a lot of research going on but it is alleged that there are no specific objectives to much of the research that is beginning carried out. It is generic in nature. The general feedback suggests that academicians should be more specific about the applications on which they are working.

It was also suggested by one of the delegates that it would be interesting to see this network working not just across borders in Europe but also intercontinentally and, in particular, involving some of the African nations. With that addition, it was generally felt that it would be interesting to see the network developing dynamically across Europe and that the continuation of research meetings of this nature would help to relate the generic research to the required applications and enable the function and purpose of fire engineering and emergency planning to become a lot more clear in the future.

APPENDICES

APPENDIX ONE

A FRAMEWORK FOR RESEARCH IN THE FIELD OF FIRE SAFETY IN BUILDINGS BY DESIGN

WRITTEN BY
THE EUROPEAN GROUP OF OFFICIAL LABORATORIES FOR FIRE TESTING
(EGOLF).

PUBLISHED BY

EGOLF
REGISTERED OFFICE
LABORATORIUM VOOR AANWENDING DER
BRANDSTOFFEN EN WARMTE-OVERDRACHT
OTTERGEMSESTEENWEG 711
B-9000 GENT
BELGIUM

Tel. int + 32 09 2 22 25 50
Fax int. + 32 09 2 20 20 61

PRINTED BY

DRUKKERIJ VANDEVELDE
A. DE KEYSERESTRAAT 6
B-9700 OUDENAARDE
BELGIUM

Tel. int + 32 055 31 38 72
Fax int. + 32 055 31 92 32

ISBN number: 90-801859-1-4
D/1994/6964/1

EGOLF

(European Group of Official Laboratories for Fire-testing)

EGOLF MEMBERS AND ACTIVITIES

* EGOLF has over 40 members, all nationally accredited/approved within EC/EFTA member states for the purpose of building regulatory control.

* EGOLF defines levels of quality, competence, expertise and independence to be followed by all its members.

* Members provide fire resistance testing and/or reaction to fire testing of building materials, components and elements of building construction.

* Members may also provide investigations, research and development, consultancy and information services on fire related matters.

* EGOLF is actively contributing to the development of European fire test standards.

* EGOLF is active in promotion of fire safety by application of fire safety engineering to the design of buildings, building products and their use.

* Members work for the building industry in particular but also to the transport, off-shore and other industries.

* Some members carry out product certification for national markets.

* Some members also carry out testing, research and certification of active fire protection products.

EGOLF OBJECTIVES

* to contribute to the removal of technical barriers to trade within Europe.

* to improve collaboration between fire test laboratories in Europe.

* to facilitate the mutual acceptance of fire test reports.

* to contribute to the standardisation of fire test methods, reports and assessment by providing a forum for discussion.

* to promote research and development within fire safety and fire testing.

Registered Office: EGOLF v.z.w., Ottergemsesteenweg 711, B - 9000 Gent

CONTENTS

FOREWORD

Fire safety considerations are an important aspect of the design of any system. Within the building construction sector, the desire to provide life safety within buildings and to reduce losses has resulted in many prescriptive requirements which influence and control the design of buildings. However, some building designs become more complex and new materials or existing materials used in new ways may mean that the prescriptive approaches are inadequate or outdated.

The move towards a Single European market has focused attention on the differences in regulatory requirements, design practices and performance standards that exist throughout Europe. The possibility for harmony of the different approaches will only be realised if fire safety by design is addressed from a more fundamental basis.

The scientific understanding of fire has developed considerably over the last decade and the possibilities for engineering for fire safety has improved. Major gaps, however, still exist in the knowledge and much research needs to be done, and the results of this interpreted into practice, if more efficient and economic fire safety design of buildings is to be achieved.

This document has been written by EGOLF (within its Technical Committee TC3 - Development and Research). It, therefore, represents primarily the views of EGOLF members.

The preparation of this document was mainly motivated by the desire to establish an overview of the requirements for fire research at a European level; the objective being to establish a framework which may be used both by CEC and by EFTA as background for selecting research activities which are relevant to the provision of adequate fire safety in buildings.

The publication of this Technical Report is also intended to stimulate reaction from others, who have similar or different interests in providing fire safety in buildings.

EGOLF TECHNICAL REPORT 93-1   NOVEMBER 1993

KEYWORDS :   Fire, buildings, research

A FRAMEWORK FOR RESEARCH IN THE
FIELD OF FIRE SAFETY IN BUILDINGS BY DESIGN

1.   **INTRODUCTION**

The 'Construction Products Directive', Council Directive 80/106/EEC, includes an essential requirement for buildings to have 'safety in case of fire' and expresses five component parts of this requirement. The interpretation of those requirements which are for buildings into requirements for products is detailed in the 'Interpretative Document' - CEC/DGIII document TC2/023, rev. 1.

The Interpretative Document, within its explanation of the essential requirement 'safety in case of fire', recognises that it is possible to take a more fundamental approach to the provision of fire safety in buildings, and in several Member States the national regulations are expressed in terms of functional requirements which facilitates such an approach. Those Member State regulations which are written in prescriptive terms, have been largely generated historically as a result of some specific fire event which may have illustrated weaknesses in building design; the detailed philosophy behind these regulations is often lost and the actual target or achieved level of safety is unknown. The regulations often contain redundant requirements. The intimate relationship between the tools which are used to measure or predict fire behaviour (e.g. test and calculation methods) and the prescriptive requirements written in the regulations, have created difficulties in progressive harmonisation within the construction products area (e.g. with reaction to fire testing and classification). The overall effect is that regulations increase costs and act as obstacles to innovation.

For these reasons and recognising the advancements which are being made in the application of fire safety engineering to building design it is necessary and essential to promote and begin the essential research work necessary to establish a scientific basis which will lead to rational and efficient fire safety requirements.

2.   **IMPORTANCE OF THE PROBLEM**

2.1   **Human aspects**

Fire safety in buildings concerns every person in the European Community. The annual number killed in building fires is approximately 5.000, with 10 to 15 times as many people injured.

2.2   **Economic aspects**

The overall cost of fire within the European Community has been estimated at between 0.5 and 1.0% of the Gross National Product (GNP).

3.   **SCOPE OF THE RESEARCH PROGRAMME**

After discussion amongst the fire safety research and testing institutions who are members of EGOLF, the following framework for fire research to achieve fire safety in buildings has been generated.   The scope of this research programme is not necessarily complete nor comprehensive, but represents the perceived needs as expressed by EGOLF members. It is, however, essential that there is an understanding of the total needs in order to place individual components in context. The detailed content of any proposal is not given in this document although some proposals are indicated in outline together with an expression of the priority.

The total need is in this document sub-divided into the following eight individual component parts, viz.

(1)   risk assessment
(2)   development and propagation of fire
(3)   smoke and toxicity
(4)   fire detection and extinguishment
(5)   structural fire resistance
(6)   evacuation
(7)   fire investigation
(8)   fire brigade needs

For each of the component parts the following format of presentation is used:

-       overview of the current situation
-       strategic direction
-       programme proposals

The programme proposals which are included below have been suggested by EGOLF members, and are examples of the type of work which is needed, but are not the only possible activities which may be necessary.

3.1   **Risk assessment**

It is essential to have a systematic approach in relation to the quantification of fire safety, and to be able to define one's objectives in terms of criteria which can be quantified. A risk assessment is a formalised procedure which may be used to estimate the expected risk of outbreak of fire, and/or the risk to the building occupants, and/or to the fabric or contents of the building, and/or to rescue teams.

The risk assessment process may be global, being applied to evaluate the overall fire safety provided within the building. It may also be more specific, dealing with only one or more of the individual components of the total system. This particular section considers the needs for providing the overall methodologies.

3.1.1   **Overview of the current situation**

There is currently much activity on a worldwide basis to provide a global approach to the quantification of fire safety in buildings under the umbrella heading of 'fire safety engineering'. These approaches allow the contributions made by the various components of the fire safety system to be considered in a rational way using deterministic methodologies and/or probabilistic techniques. There are ongoing activities in several countries to provide guidance on the use of such techniques in Standards or Codes, notably in the Nordic countries and in the U.K., and outside Europe in Japan, Australia and the USA. At the international level there is an activity within ISO/TC92/SC4 which is addressing the subject of 'Fire Safety Engineering'. There are also pre-normative research activities, and more fundamental research activities, being conducted on a worldwide basis and which is the subject of co-ordination within CIB/W14.

The ability to use formalised risk assessment methodologies in practice is currently very much limited by the lack of available data, both in relation to the individual components (see the following sections) and to the overall evaluation system. There are a number of different ad-hoc methodologies for quantifying fire safety levels currently in use, which may broadly be grouped as follows:

(a)   points system or schedule approaches - which relies primarily on the distillation of the judgement of a group of 'experts' as to the value of various factors that contribute to, or detract from the realisation of fire safety. These judgements are formulated into a scoring scheme on the basis that certain premises, usually those which follow existing regulations, are regarded as being 'safe enough'.

Examples of this approach are the Gretener scheme (from Switzerland), various schemes used to assess hospitals (used in the UK), and the tariff systems used by insurance companies. The judgement of the 'experts' can result from extensive personal experience, anecdotal experience, scientific or statistical knowledge. The skill of the method relies heavily on the way in which the agreed values for the various points are synthesized from what may be diverse perceptions of the 'experts'.

(b)   logic tree approaches - this is a more rational, and from an engineers perspective, a more acceptable approach. The problem is decomposed into elements which allow scientific and statistical data to be introduced. However, judgement is still required to fill in some of the gaps but these must be reduced to a minimum.

(c)   probabilistic simulation approaches - which generally requires a computer based solution. A number of methodologies have been developed incorporating different combinations of stochastic state transition models, network modelling techniques (e.g. Petri networks) and Monte Carlo simulation. Quantification of the level of safety relies heavily on statistical information and techniques. The models are computer based and consequently they are sometimes critizised by engineers as being too complicated and difficult to understand.

The various approaches need to be reviewed with a view to establishing their applicability and their usefulness in risk assessment methodologies used in the fire safety field. Recommendations should be given regarding their appropriateness to various situations, together with guidance on their limitation, validity, sensitivity etc.

3.1.2   **Strategic direction**

The future needs may be considered under three broad headings, the basic design data, the methodologies, the definition of acceptable levels of risk:

-    Basic assumptions need to be made regarding various characteristics of buildings, their contents and their occupants which are independent of any separate considerations within the different components of the fire safety system, e.g. fire loadings for different building types, anticipated number of fire starts, occupant efficiency etc. These data are fundamental inputs to any fire safety design and need to be determined and agreed.

-    Acceptable approaches to a global evaluation of fire safety on a performance basis need to be reviewed, validated and agreed.

-    The basis of defining acceptable levels of risk need to be common between all Member States of the Community.

To provide the basic design data the following areas are examples of those in need of activity - collection and collation of fire loadings and fire load distribution in different building types; an estimation of the number of fire starts that may be expected for any type of use of building and the distribution of the type of fire between slow, medium and fast development; reliability of different fire protection measures; assumed number, distribution and characteristics of the occupants of different building types, their mobility, dependency, efficiency, awareness, etc; environmental aspects. The results of these activities, together with information derived from other sections regarding fire performance of materials, components, constructions and systems should form part of a central European data base.

To provide for an agreement on acceptable approaches for assessment of risk and to ensure compatibility of the procedures used to derive information in support of these, methodologies need to be derived and agreed in relation to the following - gathering of statistical data from fire occurrences and how to bring this into the design system; the validation of modelling processes which may form part of the system; the structure of a global approach to fire safety and procedures for combining and evaluating the interactions between the various components of the fire safety system.

To provide a common basis for defining acceptable levels of risk, the following areas are examples of those which should be addressed - a comparison between the fire risk accepted in different Member States, using an agreed risk assessment approach, under their current regulatory requirements; tenability thresholds for building occupants in terms of their exposure to heat, smoke and toxic products; risk to life in terms of the 'acceptable' probability of single or multiple death fires; the risk of socially unacceptable occurrences, e.g. leading to pollution/environmental damage; the 'acceptable' level of economic loss resulting from fire.

3.1.3   **Programme proposals**

From the above, the following items may be listed as headings for specific research programme proposals. They are indicated here as short term or long term based on the urgency of the need and their ability to be achieved:

(i)   Basic design data

- establish a procedure for measurement of fire loading in buildings and for determining that proportion of it which is available for contribution to the fire, e.g. not contained within a cabinet which would protect it (short term)

- survey of different types of buildings, in various Member States, using an agreed detailed methodology to determine the fire loading and fire load distribution characteristics for the building type and rooms within building type (short term)

- critical review of statistical data, available within all Member States, supplemented by 'Delphe' approach if necessary, to provide information regarding the frequency of fire starts in different types of buildings, and the subsequent rate of fire development, i.e. slow, medium, fast (short term)

- from statistical data, or from testing if necessary, determine the reliability and availability of different fire protection measures, both active and passive (long term)

- establish a procedure for characterising occupants of buildings in relation to their responsiveness to fire and their efficiency in terms of their ability to escape (short term)

- review of the number, distribution and characteristics of the occupants that may be expected within different types of buildings (short term)

- establish a basis for establishing the internal and external environmental influences that may affect the development and spread of fire and its effluents (short term).

(ii)   The methodologies

- provide a basis for routinely gathering statistical data from fires which can be applied in a uniform manner throughout all Member States; this information needs to be related to the objectives which were the basis of the fire safety design (short term)

- develop procedures for the assessment and verification of models used as part of fire safety design or assessment including the establishment of 'bench mark' (reference) testing for validation of models (short term)

- development of a performance based design framework for achieving fire safety in buildings, and allowing for the contribution and interactions of the various components of the fire safety system to be quantified (short term)

(iii)   Acceptable levels of risk

- using an agreed assessment approach, for different types of buildings in different Member States, determine the fire risk currently acceptable under existing regulations as being safe enough. This will provide a comparison of safety levels existing in different Member States and the possibility for agreement on acceptable levels of risk (short term)

- to provide a basis for assessing the risk to life from a fire, determine maximum levels of exposure to heat, smoke and toxic products that are tenable to humans (short term)

- using the results of the first proposal listed above under (iii) and by reference to other statistical data, determine the acceptability of loss of life in fire (single and multiple fatality), the currently tolerated risk of major pollution or environmental damage, and the economic losses (short term).

3.2   <u>**Development and propagation of fire**</u>

Development and propagation of a fire (i.e. the pre-flashover period) covers the period from initial ignition until, if a fire is contained within a space, a broadly constant rate of burning occurs. In most cases, where ventilation is at a sufficient level this occurs after flashover and involves all of the combustibles within the space. If ventilation is sufficiently limited then growth may be inhibited so that flashover may not take place and restricted combustion may occur. In a very large or open space the fire may not involve all of the combustible material at one time and may spread in the form of a fire front.

The rate at which a fire develops is an essential component in assessing hazard associated with any chosen fire scenario. Initial fire development will depend upon the ignitability of single items of contents and surface materials and their ability to sustain combustion. Subsequent development depends upon the way the fire continues to spread between items of contents, between contents and surface linings and along surface linings. This in turn depends upon the geometry of the space, the nature, quantity and location of contents and the nature and distribution of wall linings. The complexity of this process means that most real fire scenarios have to be simplified both to provide a basis for assessing hazard and to allow appropriate methods to be derived for testing and classifying the characteristics of materials that contribute to the fire growth process.

In general, building regulatory authorities are concerned with surface materials, principally those lining internal spaces (such as rooms, escape routes) but also including external surfaces (facades, roofs). However, building surface materials are rarely the item first ignited in a fire and the ignitability of contents cannot be excluded, if only to indicate the level of potential ignition sources for linings. In consequence some authorities in member states regulate for the ignitability and post-ignition behaviour of contents (most notably upholstered furniture and bedding materials).

3.2.1   **Overview of the current situation**

(a) Existing test methods

A multiplicity of test methods for measuring the reaction to fire performance of construction and other materials are currently in use in CEC and other European countries. Since the middle 1980's efforts have been made, principally in connection with the need for harmonised test methods to reduce barriers to trade, to provide some form of equivalence between methods or to reduce their number to a few acceptable tests. Most notably the Blachere Report and other studies instituted by the Commission have contributed to the debate. While there may be a continuing short-term need to rationalise current test methods, there is general recognition that the future is likely to lie with measurement techniques which provide more fundamental information on the fire performance of a material than that required for classification alone.

(b) Second generation test methods

The recognition in the early 1980's of the potential of oxygen depletion calorimetry for the measurement of rate of heat release from a burning fuel has generated new approaches. ISO has played a leading role in these developments, in particular with the publication of standards for a bench-scale calorimeter (ISO 5660-1) and a room calorimeter (ISO 9705). The ISO methods are being increasingly used, although recognition in national regulations is limited. Their principal advantage lies in the ability to provide measurements of performance in terms which can be much more readily related to the development of real fires and, hence, to the level of hazard. However, the methods are worthy of further refinement and development.

(c) Theories of fire propagation

The development of the second generation test methods has provided the basis for improved models for the spread of flame across surfaces. In the last five years substantial progress has been made, particularly in Sweden, Finland and the USA, in the application of these models to the prediction of fire growth in relatively simple full-scale scenarios such as the ISO room/corner test (ISO 9705). Further work is required to improve the capability of these models and to extend their application to a much wider range of scenarios.

(d) Scenarios other than the small room

Most test methods and theoretical models are applicable to the simple scenario of a small, well-ventilated room. To date relatively little work has been undertaken to extend the understanding of fire growth in other spaces within buildings such as corridors, shafts (including staircases) and ducts.

Fire spread at building facades is also of concern, particularly with the development of new systems for weather-protection and thermal insulation. Although within individual member states full-scale test methods have been developed, there is no international agreement on a standard test procedure.

(e) Recent and current European research programmes

A comprehensive review of European reaction to fire tests for construction materials carried out in the late 1980's culminated in 1989, in the Blachere Report, mentioned above.

Between 1989 and 1992, a group of Nordic fire laboratories carried out a co-ordinated programme of research on European reaction to fire classification (the EUREFIC programme). This set out to show that the second generation reaction to fire test methods could be used to evaluate the fire behaviour of building products. This programme also included comparisons for a limited range of products between second generation test methods and a number of those currently in use in member states, as well as taking forward the development of theoretical models for relating bench-scale to full-scale test methods.

A number of European laboratories are currently carrying out a co-ordinated programme, known as the Charlemagne Project, to compare the performance of a range of construction products using national reaction to fire test methods and the cone calorimeter.

Considerable planning and discussion took place during 1992 to prepare a set of proposals for research to establish a harmonised approach to European reaction to fire testing of building products. However, no such research programme has yet been agreed which is acceptable to all member states and the Commission.

Work has, however, recently started on a programme, funded by the Commission, on the combustion behaviour of upholstered furniture. This programme, which is scheduled to be completed by early 1995, includes both bench and full-scale studies of the post-ignition behaviour of upholstered furniture, as well as the assessment of the level of hazard created within enclosed spaces (with different levels of ventilation) which contain burning items of furniture.

3.2.2   **Strategic direction**

The long-term goal is to fully understand the physical processes which determine the growth of fire in buildings and to use these in models, in conjunction with typical fire scenarios and appropriate test methods for construction materials and contents, to determine levels of hazard. In pursuit of this aim the following are proposed:

(i)   To develop further and validate models of fire growth to include, not only room surfaces but also surfaces where enhanced flow and re-radiation occur (such as ducts and shafts, including corridors and staircases).

(ii)   To develop further both existing, and possibly new, 'second generation' test methods, to overcome some of the problems, e.g. those associated with evaluating products in end use situations where fixing and jointing systems may significantly influence fire behaviour.

3.2.3   **Programme proposals**

While the strategic direction above sets out the long-term aims, reflected in some of the proposals set out below, others are included which are designed to satisfy shorter term needs related to the implementation of the Construction Products Directive.

(i) Inter-comparison of national test methods

Prior to a unique European testing and classification system, it will probably be necessary to provide a compromise approach to ensure that construction products can be freely traded across the borders. A programme is, therefore, proposed to compare the performance of a sufficiently wide selection of construction products when exposed to the principal national reaction to fire test methods. The programme could also include those ISO test methods which may be used as a basis for harmonisation in the future. It would have two objectives:

-   to underpin any arrangements made to provide for CE marking of construction products

-   to enable national regulators, and product manufacturers, to assess the implications of any such arrangements and to judge the impact on regulations and products of future harmonised European requirements.

The programme would include the following stages:

-   Review of existing data (from EUREFIC programme, ISO inter-laboratory comparisons, the Charlemagne Project etc.)

-   Identification, selection and programmed acquisition of a range of construction products.

-   Measurements using a variety of national and international standard test methods.

(ii)   Fire development in rooms

Whilst much is already understood about the influences of fire development in rooms, a better understanding is still required regarding the effect of various parameters. This proposal would be aimed at providing clearer understanding of various influences such as:

-   the influence of room size and configuration

-   the relative importance of materials used on walls, ceilings and/or floors

- the influence of ventilation, both restricted and well ventilated and the effect of cross ventilation etc

- comparison with performance of component materials in bench-scale tests using appropriate models.

(iii)   Fire spread - external walls and facades

This proposal would aim to provide the basis for an agreed method for testing the reaction to fire properties of external walls and facades and would include the following components:

- Review of scenarios; existing test methods; experimental data and theoretical models.

- Experimental programme to investigate test rig configuration and dimensions, possible ignition sources, etc.

- Co-ordinated programme on generic types of external walls and facades.

- Comparison, where appropriate, with performance of component materials in bench-scale tests using appropriate models.

Note: This programme may also address the additional problems associated with fire spread via the junction between a curtain wall facade and a floor slab of a building.

(iv)   Fire growth in corridors and staircases

In comparison with the well-ventilated small room there is relatively little information and data concerning the way in which fires develop in building circulation spaces such as corridors and stairways. Unlike what is normally considered as the case for a room, it is possible that the performance of floor coverings may need to be considered in addition to wall and ceiling linings.

The programme would be principally experimentally based to provide basic information but would include the development and validation of models. Amongst the factors to be considered are:

- Ignition sources

- Corridor/staircase configuration and dimensions

- Levels and methods of ventilation

- Range of materials for wall, ceiling and floor linings

(v)   Development of theoretical models

The EUREFIC programme and other studies have provided data which has been used to demonstrate that the current theoretical understanding of flame spread, taken together with properties of linings measured using the second generation test methods, can predict full-scale fire development (i.e. time to flashover) in the relatively simple situation of a small, well-ventilated room with well-defined ignition source. To date, these methods are insufficiently general to be extended to other scenarios. To achieve scenario-independence, research is required to develop flame spread models in combination with improved predictions of the gas phase conditions, using computational fluid dynamics. It will also be important to be able to predict both the early development of fire at surfaces, subject to different levels of ignition source, as well as the later stages approaching flashover.

(vi)   Further development of second generation test methods

Fire Calorimetry is accepted as a principal measurement technique. The current bench-scale test is the cone calorimeter, which attempts to simulate the exposure of a sample of material to constant levels of thermal radiation chosen to represent those which typically occur in a given fire scenario. Whilst some attempts have been made to simulate (e.g. in ISO 9705) other conditions which may have a significant effect on combustion, such as the vitiation of the air in the proximity of the combustion zone or the variability of heat flux, no standardised approach is yet available. The cone calorimeter is also constructed in its ability to evaluate the performance of some products in end use conditions, and as a consequence there may be a need for an intermediate scale fire calorimeter test (between the bench-scale, ISO 5660-1, and the ISO room/corner, ISO 9705) to provide the necessary additional data.

This proposal would include the following:

-   Full-scale experiments to identify conditions which occur in practice.

-   Measurements of ignitability and heat release rate using a suitably modified cone calorimeter under conditions of transient heat flux and reduced oxygen supply.

-   The evaluation of the need for, and the value of, additional data which may be provided from an intermediate fire calorimeter test.

-   The development of, if required, an intermediate scale fire calorimeter test.

3.3 **Smoke and Toxicity**

This section covers both the generation of smoke and toxic effluent by materials and the movement throughout a building as a consequence of the influence of the fire or the provision of systems within the building to limit their movement. The hazard represented by smoke and toxic gases is relevant to both the density and quantity of the smoke generated, in so far as it impedes visibility and the quantity and level of toxic potency of the gases generated, in so far as they represent a threat to human life. There are additional considerations with regard to the corrosivity of the fire effluent in so far as this may cause damage to building contents, equipment or services.

3.3.1 **Overview of the current situation**

The ability to predict the quantity and density of smoke, together with toxic products which may be produced from a fire, is not as advanced as the prediction of the thermal exposure and other consequences which result from a fire. These effects depend not only on thermal exposure but also on the chemistry of the fire processes. International agreement has not yet been reached on the appropriate method(s) for assessing the rates of production of smoke and toxic products, and given this information, the models which allow for the assessment of potential harm have yet to be successfully developed. Models are available which facilitate the smoke movement throughout a building to be predicted and the influence of smoke control systems to be evaluated. These, however, are based on rather crude assumptions. The potential for harm, either to people or to property, which arises from smoke or toxic gas production depends upon may other factors than the rate of smoke or toxic gas generation, e.g. the ignitability and flammability of the material/product, and the ventilation conditions to the fire and elsewhere in the building.

In this section the format of presentation used is:

- smoke quantity and density

- toxicity of fire effluents

- movement of fire effluents

(a) Smoke quantity and density

Much of the current activity in relation to smoke generation centres around a measurement of the optical density of smoke generated by burning materials in large and small scale experiments. In large scale experiments the smoke generated has always been measured in a dynamic manner, i.e. the optical density of the smoke generated is derived throughout the complete fire process, whilst in bench-scale smoke tests specifically designed only to measure smoke generation, integrated measurements have historically been the norm, i.e. the optical density of the smoke accumulated in a given volume during the fire process is measured.   However, in bench-scale experiments which are designed to provide more information than just smoke generation, dynamic measurements have nearly always been made.   Very little attention has been given to obtain rates of mass of smoke production.

The ISO/TC92 attitude regarding the measurement of smoke produced by burning materials is given in ISO TR 11696 (under preparation).

The report compiles the latest developments in methods of measurement of the optical density or light extinction coefficient.   The experimental methods developed by ISO/TC92 are described in ISO DP 5924 Dual Chamber Test (Static type), ISO 5660-1 Cone Calorimeter Test and ISO 9705 Room/corner Test (both dynamic type).   Variables affecting smoke development and obscuration as a result of the burning environment are known from experiments to be present.   Variables such as - radiant heat flux, oxygen concentration, ventilation, sample orientation and geometry, moisture content of sample.

The report concludes that predictive methods for large scale performance on the basis of small scale test data are in a developing stage.   Among the small scale test procedures, the dynamic cone calorimeter is most promising, even though its capability for smoke prediction in large scale has not yet been fully proven.   Correlations between small scale and large scale smoke test results may be improved by adding parameters describing the burning rate and fire spread behaviour.

(b)   Toxicity of fire effluent

The potential for harm from the toxicity of fire effluents is reported in ISO/CD 13344 "Determination of the lethal toxic potency of fire effluents" together with informative annexes ISO/TR9122 Part 1-2 and ISO/DTR9122 part 3-6.

The ISO technical committee TC92 has a well established work programme on the assessment of toxic hazard presented by fire effluents. Work continues on selecting the appropriate fire models and an analytical method for identifying toxic species in blood. A guidance document is being prepared which will give guidance to regulatory authorities on how data derived from toxic potency tests can be used. Work in connection with estimating potential toxic hazards and minimising the use of tests on animals is continuing.

(c)   Movement of fire effluents

Many countries have guidance or standards which indicate means by which the hazard presented by smoke may be mitigated as far as it impedes successful escape of building occupants, or presents a threat to fire fighters. These normally are prescriptive requirements in terms of what must be provided, or are simplified calculations based on having a fire of a given size and generating a known rate of hot gases; the interface between the smoke and 'no-smoke' is normally represented by a given temperature isotherm. There is no international agreements on the methodology employed.

Guidance may be found, e.g. in:

(i)   The SFPE Handbook of Fire Protection Engineering, SFPE/NFPA, 1988.

(ii)   Fire Protection Handbook - seventeenth edition, NFPA, 1991.

(iii)   Design principles for smoke ventilation in enclosed shopping centres, BRE, 1990.

3.3.2   **Strategic direction**

The importance of smoke toxicity has increased and will continue to increase, not only amongst those directly interested in providing safety in case of fire but also in the minds of the public at large. These factors are now clearly identified as the major cause of deaths.   Much greater attention must, therefore, be given to providing all the information relevant to the prediction of these factors and the threat which they present to life safety.

The future objectives must be to better understand the factors which influence the rate and density of smoke and toxic gas production from products, to identify evaluation methods for products and associated models which will allow the potential for harm to be assessed, to understand the individual and synergistic affects on life safety, to provide better methods for predicting smoke movement and management in buildings.

In determining the smoke and toxic gas production characteristics of construction products consistent conditions should be established with those used for evaluating other reaction to fire properties such that the data is comparable and compatible. However, the worst case for smoke and toxic gas production will not be that, for example, for maximum rate of heat release. Never-the-less, to allow proper application of the data, it is important that all reaction to fire data is relevant to the design fire scenario.

3.3.3   **Programme proposals**

The following headings for research proposals may be envisaged as a consequence of the observations above:

(a)   Smoke quantity and density

-    evaluate the appropriateness of the principle of smoke measurement (e.g. using dynamic or static methods, or both) to be used for a range of fire scenarios

-    provide appropriate small scale test method(s) which can derive the important smoke production characteristics of products (including the smoke mass generation rate).

-    evaluate the influence of those factors (e.g. ventilation, oxygen concentration, moisture content) which have a bearing on the generation of smoke and its characteristics (e.g. longevity)

-    provide validated models to predict the smoke potential in large scale fire scenarios on the basis of small scale test data.

(b)   Toxicity of fire effluents

-    selection of one or more fire models (small and/or large scale) to be used as the basis of a European system of testing and evaluation.

-    study of fire scenarios with special attention to scaling effects.

-    validated models between small scale tests and large scale experiments.

-    study of on-line measurement techniques.

(c)   Movement of fire effluents

-    validation of existing models considering especially their applicability to complex spaces

-    assessment of the influence of environmental effects and the containing structure on the fire effluent temperature and movement

-    assessment of the influences of dilution and coagulation on the density of the smoke within the hot gas layer

-    the effectiveness of smoke venting systems (natural vs powered) and associated calculation methods.

3.4   **Fire Detection and Extinguishment**

This section covers the requirements for providing a better understanding of the response and reaction of automatic fire detection systems and fixed fire extinguishing systems, i.e. those permanently installed in the building (excluding first aid-fire fighting facilities, e.g. extinguisher/hose reels). It considers the interaction of the extinguishing system with the developing fire and the consequence on the fire of the extinguishing media.

3.4.1   **Overview of the current situation**

The relative degree of emphasis placed upon active and passive fire protection measures in various buildings depends upon a wide variety of factors some of which include - national requirements (codes of practice, standards, etc), requirements of other financially involved bodies such as insurers, the proposed end use of the building, the relative costs of various methods or the contribution of the building structure, the current fashion in fire protection techniques.

The existing national regulations provide fire safety in buildings predominately by passive fire protection measures. Detection and extinguishment is, therefore, normally added to the passive fire safety provisions rather than integrated into a comprehensive, effective and economic fire safety design. They are normally provided to the requirements of a design code and are not 'tailored' to the specific fire hazard. Both for automatic detection and for extinguishment systems, it is a characteristic that developments are centred around development of products which will satisfy simple performance objectives. The relevance and relationship to real fires and the science governing their interaction with the fire are not significantly researched.

(a) Automatic fire detection

The current trend in developing automatic fire detection systems is to exploit computer technology in support of the management (i.e. the centralised control) of automatic detection systems, thereby reducing the number of false alarms. The various types of detector heads are still tested using principles established many years ago which provide little or no data which is helpful to assess response time in real fires. Whilst considerable reliability data is available, this does not relate to the ability to detect real fires, but is based predominantly on laboratory testing which may not relate to real fire conditions.

(b) Fixed extinguishing systems

The current methods of evaluating the performance of fixed fire extinguishing systems varies between the different types of media involved, i.e. there is no common concept.   None of the existing testing methodologies can be correlated to real fire scenarios and the information produced is of little value in determining the reaction time and the interaction between the extinguishment and the fire.

3.4.2   **Strategic direction**

The most important goal of any research in this area must be aimed at a better scientific understanding of the performance of detection and extinguishing systems in real fires.   This information is needed in support of fire engineering approaches to the provision of fire safety and will allow an integrated approach to the problem combining passive and active fire protection measures.

3.4.3   **Programme proposals**

(a)   Automatic Fire Detection

- develop testing procedures for fire detectors which will provide data for calculation of response times, e.g. regarding the influence of rate of heating, velocity of gas, density of smoke.

- develop procedures for evaluating the sensitivity of detectors with special regard to the problem of false alarms.

(b)   Fixed extinguishing systems

- in response to the commitment to phase out Halon gas, as agreed in the Montreal Protocol, urgent research is needed to evaluate the advantages of alternatives. A series of tests are required to look at the effectiveness of suppression, the aggression of the fire environment and the consequential damage to goods, electrical equipment etc

- the interaction between sprinkler systems and the developing fire for various rates of fire growth and the fire load storage configuration

- the interaction of sprinklers with the spread of smoke and hot gases and the consequent influence on smoke management systems e.g. smoke ventilators, curtains etc

- the performance of sprinklers relative to their height above the fire and their position relative to the floor soffit or suspended ceiling

3.5   **Structural Fire Resistance**

This section considers the detailed needs of fire resistance testing as well as the more general needs for calculation of fire resistance of building elements. It covers the relationship between performance in real fires and fire tests, and calculations and how those may be enhanced by the use of different heating rates.

3.5.1   **Overview of the current situation**

For the time being most of the work within the field of fire resistance tests is related to the CEN activities and performed in accordance with the CEN/TC127 work programme on fire resistance. This programme includes all the relevant types of construction elements, structures and components, as well as the time schedule.

The CEN work programme also indicates laboratory work which is needed, both round robin tests and proving tests to support the basis for further development and implementation of the CEN standards.

Work with respect to theoretical verification of fire resistance is related to the relevant activities dealt with by CEN TC250-Eurocodes. A separate part of the different Eurocodes is allocated to structural fire design, and this part gives both basic principles as well as calculation models at different levels.

In addition to the above mentioned laboratory work the following issues need to be addressed and resolved by research activities:

-   Establishment of a system and guidelines for further application of test results.

-   Thermal properties of construction materials.

-   Survey and evaluation of available computer codes for temperature analysis and analysis of mechanical behaviour of construction elements and structures, with the aim of selecting codes which are usable for structural design and classification purposes.

3.5.2   **Strategic direction**

The strategy within this field may be related to the following matrix

Model for structure	S_1 Element	S_2 Substructure	S_3 Complete structure
Model for thermal exposure			
H_1 Nominal Temperature/ Time Curves	Test Calculation Tabulated data	Calculation Exeptionally testing	
H_2 Parametric Fire Exposure	Calculation Experimental simulation	Calculation Exceptionally experimental simulation	Calculation

The aim is to provide a set of efficient and reliable tools for verification of fire resistance of both construction elements and complete structures exposed to different heating regimes. Having historically been working mostly in the upper left corner of this matrix, the tendency now is to move to the right and downwards. However, this requires further research work on both input data and model development.

The strategy must also include a system for combined verification of experimental and theoretical methods to make the results as reliable as possible.

The structural models in the matrix are more or less self explanatory. However, consideration must be taken with respect to controlled supporting conditions and mechanical load.

The models for thermal exposure range from nominal temperature/time curves such as the standard curve, the external fire curve and the hydrocarbon curve, to parametric fire exposure conditions, zone models and field models.

Application of the thermal actions according to ENV 1991 Part 2.2 and the design of structures according to the fire design parts of ENV 1992 to 1996 and ENV 1999 is illustrated in the following table.

Thermal actions given in ENV 1991, Part 2.2:	according to national specifications: for verifying	design by prescriptive rules/tabulated data given in ENV 1992-1996, 1999	design by calculation models given in ENV 1992-1996, 1999
standard temperature-time curve	standard fire resistance requirements	as relevant [1] or from fire resistance tests	as relevant [1]
other nominal temperature-time curves	other nominal fire resistance requirements	mainly from fire resistance tests	as relevant [1]
standard temperature-time curve	fire resistance - for equivalent time of fire exposure	as relevant [1]	as relevant [1]
parametric fire exposure	fire resistance - for specified period of time or - for entire fire duration	not applicable	as relevant [1]

[1] depending on the assessment methods included in the respective fire parts and the relevant scope of application.

A primary aim is to establish efficient and reliable tools for combined fire analysis, temperature analysis and response analysis.

3.5.3   **Programme proposals**

As mentioned in clause 3.5.1 much activity is going on within CEN/TC127 and CEN/TC250. Therefore different programme proposals for fire resistance must have objectives and scopes of work related to the outcome of the CEN activity.

From the above, the following items may be listed as headings for programme proposals for the additional issues which need to be addressed and resolved by research activities:

(i)   Establishment of a system and guidelines for further application of test results.

The aim of this work is to give harmonised, quantitative rules for both direct and extended application of test results. The description of the work to be undertaken will be based upon the chapters on direct application in the CEN standards for the different types of products as well as on the separate TC127 proposal for extended application.

An important input will also be the work performed by ISO/TC92/SC2/WG2/TG1.

(ii)   Thermal properties

A programme proposal should satisfy the needs for pre-normative research on the thermal properties of construction materials, which are to be identified by Ad-hoc 13 (Hot Data) of TC127. Prenormative research on the thermal properties of insulating materials is also needed.

(iii)   Survey and evaluation of available computer codes for temperature analysis and mechanical behaviour of construction elements and structures.

The aim of this work is to select codes which are usable for structural design and classification purposes.

A description of the work to be undertaken should be prepared based on the latest drafts of the fire design part of the relevant Eurocodes, and connected to the work performed by ISO/TC92/SC2/WG2/TG1 and TG2.

3.6   **Evacuation**

This section addresses the requirements for providing a rational approach for determining the ability and needs of human beings to escape and evacuate a building. It includes consideration of the models for predicting the rate of escape and evacuation in the event of fire, and considers the influence on these predictions of the ability of the occupants to respond in terms of their mobility and efficiency, etc.

3.6.1   **Overview of the current situation**

Currently safe evacuation from buildings is provided mainly by prescription, in regulations and codes, of adequate means of escape, e.g. by specifying requirements for protected routes for escape and maximum escape distances.

A more fundamental approach to evacuation of people from buildings is needed and in this context evacuation may be defined as:

"the process whereby people by themselves or assisted by other people inside the building are moved to a place where they are safe from the fire".

This does not include the situation where people are moved by means of external personnel and technology (e.g. smoke divers, helicopter, turntable ladder, etc). The latter situation is denoted *rescue operation*.

Evacuation is a process which can be described in terms of a number of different stages of human behaviour. These stages are as follows:

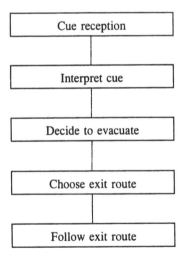

Each of these stages takes time and for an evacuation to be successful this sequence of behaviour has to be executed within the time between alerting and untenable conditions. The critical factor to consider in evacuation safety is therefore the period between the points of alert and escape limit. The relationship between different time factors is illustrated below. From this it can be seen that evacuation safety depends on the ratio of 'Available escape time' and 'Needed escape time'.

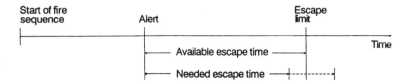

Assessment of evacuation safety is integrated in methods which assess the overall fire safety in a given building. There is currently much activity worldwide to provide such methods, notably in the Nordic countries, UK., Japan, Australia and USA. The common features of the evacuation part of these methods are tabulation of occupant loads, locations, activities and characteristics by occupancy. Thereafter they tabulate a series of delay times for alerting, decision making, investigation, fire fighting, movement speeds,' etc., all to be used in computing needed escape time. The occupants exposure to heat and toxic gas during evacuation are evaluated against limiting values. This approach is a great step forward compared to earlier approaches, but more empirical data are needed in order to model human evacuation behaviour. There is also a need to validate model predictions against empirical data.

3.6.2   **Strategic direction**

The future needs may be considered under three broad headings, the approach to functional requirements for evacuation, the methodologies and the need for data:

-   It is necessary to agree upon an approach to functional requirements to evacuation safety. This entails development of a conceptual model of evacuation safety. Functional requirements are useless unless there exist some way to confirm that the requirements are fulfilled. Consequently it is necessary to agree upon a way to do this control. Further, there is a need to reach an agreement on acceptable risk level, e.g. which fire scenarios should it be possible to evacuate from.

-   From the above follows that there is a need to develop methods to evaluate evacuation safety. This activity should be closely related to the development of methods for assessment of the overall fire safety.

- There is a need for data about human behaviour in a fire situation, and several of the basic design data needed for assessment of overall fire safety are also relevant here; e.g. distribution of fire types, distribution and characteristics of the occupants of different building types, their mobility etc.

3.6.3   **Programme proposals**

From the above, the following can be listed as headings for specific research programme proposals. They are indicated here as short term or long term based on the urgency of the need and their ability to be achieved:

(i)   Approach to functional requirements

- establish a conceptual model of evacuation safety that can be used to generate functional requirements (short term)

- define acceptable risk level with respect to evacuation (short term)

- development of a framework for evaluation of evacuation safety in existing and planned buildings, e.g. use of computer simulation in the design phase, walk through in existing buildings, standardized evacuation drills, etc (short term).

(ii)   The methodologies

- development of specific methods and tools to evaluate evacuation safety (long term)

- validation of methods and tools (long term).

(iii)   Data needs

- systematic collection of data about human behaviour in fires occurring in different countries (short term)

- the effect of a building's spatial complexity on human evacuation performance (long term)

- the effect of various information systems (e.g. signs, voice messages, alarms, etc) on human evacuation performance (long term)

- the effect of management on human evacuation performance (long term).

3.7   <u>**Fire Investigation**</u>

Experience from fires have, and have had, a major influence on national fire requirements, codes and standards. Both specific information on major fires and fire statistics is necessary information for authorities, fire protection industries and fire researchers.

To improve the overall understanding of fire and its consequences and to be able to justify the approaches taken in prediction of fire safety, and incorporated within building design, it is necessary to ensure that there is adequate feedback from real fires, especially those of exceptional or unusual behaviour. This section, therefore, envisages improvements in, and agreement on, the methods used for fire investigation and extension beyond normal fire cause/arson/insurance type issues.

3.7.1   **Overview of current situation**

Fire investigation is carried out mainly on major fires either involving multiple deaths or significant financial losses. In a number of countries, fire investigation is also carried out on a regular basis in specific areas, e.g. fire cause investigation, assessments of structural fire damage, role of furnitures, plastics, arson etc in fires.

Little systematic fire investigation, however, takes place which is aimed at providing feedback to fire research and fire safety design. The lack of common guidelines on the type of information needed from fires and the techniques available for fire investigations hinder a rational approach in this field and limit utilization of the large amount of valuable information gained from inspection fires.

In Europe, CEA collects information on fires with losses above a certain amount. Systematic studies on major fires is carried out in the USA with the aim to compare fire models against the real incident, on specific fire problems, and on peoples behaviour in fires. A CIB W14 guidance document reviewing the type of information which should be collected from fires is available.

3.7.2   **Strategic direction**

The need in this field can be covered by following main headings:

Collection of general information on the spread of fire and smoke in buildings, and the influence of building layout, contents, construction products, fire protection measures and fire fighting. Such information is of special interest in the overall fire safety design concept to satisfy the need for identification of realistic design fire scenarios and for validation of fire models.

Detailed information on the performance of structures, materials used for linings etc, fire detection and suppression systems etc, is needed for product specification and for the development of test procedures which give information on products in end use conditions.

Insufficient knowledge on peoples behaviour in a real fire situation is one of the main problems for design of warning systems and safe evacuations routes and procedures. There is a need for collection of peoples reaction pattern and behaviour in major fires, especially fires leading to multiple fire deaths [cf Section 3.6.3 (iii)].

A European system for gathering fire statistics with common terminology and definitions is required. In order to identify problems for detailed investigation and to evaluate the effect of actions taken, national fire statistics and the possible utilization of data from several countries is necessary.

3.7.3   **Programme proposals**

The following specific actions are proposed:

- Development of a guideline for the collection of information and setting up a European database on major fires.

- Setting up a common terminology and framework for European fire statistics, so that present national fire statistics can be used on a European level.

- Preparation of a guidance document on fire investigation, techniques and equipment available for detailed fire investigation.

- Training and education of fire investigators on a European wide basis.

3.8 **<u>Fire Brigade needs</u>**

This section considers the specific requirements of fire brigades in terms of their safety or their efficiency "in carrying out rescue operation and fire fighting" within buildings. These aspects are generally taken into consideration in the design of buildings and are specified in regulations and fire codes. This section does not deal with protective equipment for fire and rescue personnel, extinguishing facilities carried by the fire brigade, transportation and fire trucks, radio communication systems etc.

3.8.1 **Overview of current situation**

National regulations normally specify that buildings shall be so designed and laid out, that the safety of fire fighting and rescue teams are taken into consideration and rescue and fire fighting operations around and within the building are facilitated. The safety of rescue teams is normally ensured through adequate fire resistance of structures, limitations on combustibility of construction materials and the safe storage of flammable liquids and gases, toxic chemicals and compressed gas cylinders.

To facilitate easy access to buildings for the purposes of fire fighting and rescue, safe access routes may be established by providing one or more of the following - protected corridors and staircases, fire fighting lifts, protected areas, fire compartmentation and smoke control systems. Detailed requirements will depend upon the size, complexity and use of the building.

Despite the importance of efficient and rapid intervention both to protect occupants and to limit fire damage, as well as the potential restraints on building design implied by this, only relatively little research is currently carried out internationally in this area and the resources allocated have been limited.

3.8.2 **Strategic direction**

The future needs may be addressed under following headings:

(i) Building layout and fire and rescue intervention planning

The increasing problem of providing easy and safe access for fire fighting and rescue teams within especially large and complex smoke-filled buildings requires further research and development in relation to fire fighting operations, building design as well as to internal communication and guidance systems. Development of design criteria and concepts for means of escape, including corridors and stairways, fire lifts and internal protected areas and the ability to utilise fire and smoke control systems as part of fire fighting are all areas that need to be addressed in the future.

(ii)   Fire fighting capability

There is a need to ensure adequate water supply, e.g. in high rise buildings, for fire fighting as well as other measures and equipment necessary for increasing the capability of rescue and fire fighting teams when working inside buildings.

(iii)   Rescue capability

There is a need to develop methods of identifying the location of fires and persons trapped in the buildings as well as systems to guide rescue and fire fighting teams to such locations.

3.8.3   **Programme proposals**

The following headings indicate areas in need of consideration:

-    identification of benefits which can be provided to fire fighters by specific design features incorporated in the building which aid access, rescue and fire fighting

-    to identify tenability criteria for fire brigade search and rescue operations to enable design of buildings or to engineer systems to provide for this.

APPENDIX TWO

The Role of Public Perception in the Response Planning for Major Incidents:

Public Perception and Memory Retention Questionnaire Survey

Final Report

Steve Carver [1] and Alan Myers [2]

November 1993

[1] School of Geography
University of Leeds
Leeds LS2 9JT
(Formerly of University of
Newcastle upon Tyne)

[2] Tyne and Wear Emergency Planning Unit
Floor 2, Portman House
Portland Road
Newcastle upon Tyne
NE2 1AQ

Contents

Acknowledgements

The authors would like to acknowledge the assistance of the Home Office in providing funding enabling this work to proceed and would also like to acknowledge the help and co-operation of Sterling Organics and Newcastle Airport Authority.

List of figures

List of tables

1.0   Introduction

The following report presents the findings of a questionnaire survey into the public perception of major hazards. The survey was funded by the Home Office and carried out by the Department of Geography at the University of Newcastle upon Tyne in conjunction with the Tyne and Wear Emergency Planning Unit. The survey is part of a wider proposal written jointly by the University of Newcastle upon Tyne and the Tyne and Wear Emergency Planning Unit for a GIS (Geographic Information System) based approach to response planning for major incidents. This proposal is outlined in the document; The role of public perception in the response planning for major incidents: a proposal for a GIS based strategy (Carver, et al., 1992), included here as appendix C. At the time of writing, this work is well underway including the linking of existing dense gas dispersion models into the GIS.

2.0   Questionnaire outline

The public perception survey described here was initiated by a Home Office research grant to investigate public perception of emergency planning and major hazards together with memory retention after information campaigns regarding particular hazards. The questionnaire survey was carried out in selected areas of the Newcastle and North Tyneside districts on a door-to-door basis during November 1992. The survey yielded over 350 responses amounting to a 2% sample of the total population living in the targeted areas. This figure and all those used throughout this preliminary report are based on 1991 Census population figures unless stated otherwise. The following sections describe the specific and general aims of the survey in greater detail together with questionnaire design, sampling strategy, coding and analysis of survey returns.

2.1   **Survey aims and objectives**

The aims and objectives of the survey, as stated in the original project proposal, are to carry out a survey of the public perception of emergency planning, including variables to account for memory retention about instructions given in previous information campaigns around top tier (ie. those subject to regs. 7-12) CIMAH (Control of Industrial Major Accident Hazard) sites as well as variables accounting for hazard perception and response. In general terms the survey aims to investigate the extent to which the public are aware of the threats that exist to life and the environment and of the level of mitigation that might be possible through properly written MIPs (Major Incident Plans). In particular the survey aims to:

1.   ascertain the public's awareness of risks;

2.   ascertain the public's perception of the involvement of local authorities and emergency services in public protection;

3.   ascertain the public's perception of the importance of emergency planning; and

4.   investigate levels of memory retention in members of the public subject to information campaigns regarding a particular hazard.

It is intended that results from the analysis of questionnaire returns be included as one or more map data layers in the GIS described in the wider project proposal. This may pay dividends in improving the quality of risk maps and zones of consequence defined in the GIS analysis.

2.2   Questionnaire design

The questionnaire used in the door step survey was designed in two halves, the first containing questions relating to respondent profiling, and the second and main half containing questions relating to hazard perception, memory retention and emergency planning. These two halves are themselves split into seven separate sections relating to personal details, general hazard perception, perception of hazards from air crashes, perception of hazards from chemical releases, knowledge of emergency planning, response to risk maps, and prior involvement through occupation in related fields. The questionnaire form is included in appendix A.

One of the main actors affecting questionnaire results is the chemical company Sterling Organics. Sterling Organics have a factory at Dudley within the study area which in the past has used certain toxic chemicals (i.e. phosgene) in its manufacturing process thus designating it a top tier CIMAH site (see figure 1). Although the plant no longer uses these chemicals and so by law does not have to be a top tier CIMAH site, Sterling Organics still use other hazardous chemicals and so have kept their CIMAH status though on a voluntary basis. As a result of the site at Dudley being designated as a CIMAH site, Sterling Organics have carried out a number of public information campaigns and mailshots within the neighbouring community to inform residents what to do in the event of an accidental release of toxic chemicals and of the procedures implemented by Sterling Organics and the emergency services. This included the delivery of information cards to be retained by the public for reference.

Figure 1. Sterling Organics, Dudley Plant

Initially, only one questionnaire was to be used in the current survey, but later a second slightly reduced questionnaire was produced as it was clear that those questions relating specifically to the information cards delivered by Sterling Organics would only have any meaning to people living in the Dudley area (see appendix A, question's 4.3-4.7). Just before the survey was carried out, consultations with the Newcastle Airport Authority resulted in an additional question being included in the questionnaire used in the Woolsington area. This was added to find out what proportion of Woolsington residents knew that the take-off and landing flight paths for Newcastle Airport did not pass over residential areas of Woolsington (see Appendix A, question 3.4). As a result, three different questionnaire forms were used in the actual survey, the main body of each being identical.

2

The questions themselves were designed for both ease of understanding (from the point of view of the respondent) and ease of coding/interpretation whilst being meaningful within the aims and objectives of the survey. The actual wording of questions was greatly helped by reference to previous work by the Emergency Planning College (Kirkwood, ongoing research). Each questionnaire took between 5 and 15 minutes to complete, depending on the respondent. The three basic forms of question are used, those involving binary 'yes/no' type answers, those involving categorisation or multiple choice type answers and those involving an individual response. Coding of binary and categorical/multiple choice questions is achieved simply by assigning numbers to possible responses (e.g. yes = 1, no = 0). Individual responses to the more open questions (see appendix A, questions 2.1, 2.4 and 5.1) had to be previewed to enable categorisation and the assignment of a number code. An example coding sheet is shown in appendix B.

Great care was taken in the design and ordering of questions in the questionnaire so as not to bias the answers given by the respondent through information given in previous questions or the written categories in the multiple-choice type questions. For example, question 4.1 asks the respondent whether they know what to do in the event of a chemical release. The knowledge of the respondent is only tested later in question 4.8. Similarly question 3.1 asks the respondent whether they think that an air crash is a major hazard in the area with the subsequent question 3.3 asking them to clarify their answer by saying just how likely they think such a disaster actually is. With the multiple-choice type questions care was taken not to lead the respondent on to answering 'toxic gas' or 'air crash' (question 2.2) by providing a wide range of alternatives together with the option of specifying 'other' if the hazard they thought most important was not on the list.

The way questions were actually asked was also thought to be important in influencing how people were likely to answer. No pressure was put on the respondent to answer well or correctly and although the questions were asked, and the forms filled in by the survey operative the form was kept in full view of the respondent all the time in order to gain their attention and facilitate their understanding of the question. Throughout the completion of each questionnaire the survey operatives explained the questions and terminology used (e.g. what is meant by 'toxic chemical release' in question 4.1 and what is the distinction between the 'individual' and the 'general public' in questions 2.1 and 2.2), wherever appropriate so as to avoid any ambiguities in the response.

Whilst it is understood that there will always be some uncertainty in how people respond to questionnaire surveys the design of the questionnaire used here and the measures taken should have kept this to a minimum. For example, at first because the study was jointly undertaken by Newcastle University and the Tyne and Wear Emergency Planning Unit, both the logos were to have appeared on the questionnaire. It was clear however, that this could colour the responses of those surveyed. The Tyne and Wear Emergency Planning Unit's logo was therefore removed before the survey took place.

2.3   Sampling strategy

The areas sampled in the survey were chosen on the basis of their being representative of the types of hazard that are of interest (air crash and toxic gas release). Three areas were targeted. These were the residential areas around Dudley, Brunswick/Woolsington/Dinnington and Kingston Park The relative location of these areas is shown on the map in figure 2. The area around Dudley is of interest because of the proximity and association with Sterling Organics both in terms of their status as a voluntary CIMAH site and their history of public information campaigns in Dudley. The combined areas of Brunswick, Woolsington and Dinnington are of interest because of their location around the perimeter of Newcastle Airport and associated risk of a major air crash (see figure 3). The Kingston Park area was included as a control since although it is within the study area it is considered sufficiently far away from both the airport and Sterling Organics to make the influence of these hazards minimal.

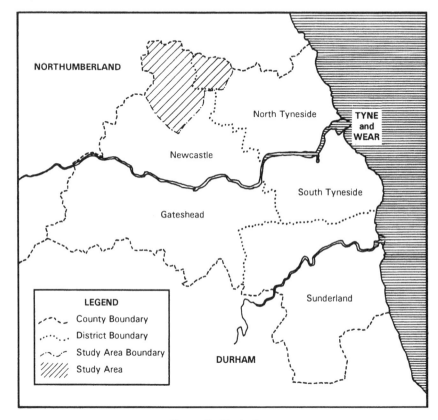

Figure 2. Study Area

Figure 2a. Dudley

Figure 2b. Brunswick/Wideopen

Figure 2c. Woolsington

Figure 2d. Dinnington

Figure 2e. Kingston Park

Figure 3. Newcastle Airport

Table 1. Statistical significance of sample size			
Sample size *(n)*	Standard error *(SE)*	Range % (Based on eg of 66%)	Reduction in *SE* with each extra 100 in n
100	4.79	61.28 - 70.74	-
200	3.75	62.65 - 69.35	1.39
300	2.73	63.27 - 68.73	0.62
400	2.37	63.63 - 68.37	0.36
500	2.12	63.88 - 68.12	0.25
800	1.67	64.33 - 67.67	0.15(av)
1000	1.50	64.50 - 67.50	0.12(av)

NB   In this example, if the surveyor wished to half the standard error (SE), and thus increase the accuracy of their extrapolations to the population, then the sample size must be increased two fold. Alternatively for this example, for each additional 100 questionnaire returns in the sample, the reduction in the *SE* would be by one quarter. It can therefore be seen that a law of diminishing returns operates, ie. doubling the sample size from 100 to 200 only results in a quartering of the *SE*.

SE calculated as..... $SE = \sqrt{\frac{pq}{n}}$

When p = % responding 'Yes' (is this example 66%)
           q = the balance (is 100% - p)
           n = sample size

(Adapted from Gardner, 1976)

It is suggested that a sample size of 1-2% of the total population be suitable for the purposes of the study. This is generally considered to be correct but what really matters in terms of the overall representativeness of the sample and statistical significance of the results is the absolute number of responses (almost regardless of the size of the population) and the variability within the data (Gardner, 1976). Population of the targeted areas was calculated at enumeration district (ED) level from the 1991 Census returns. The population distributions for ED's within the study area are shown in figure 4.

A stratified sampling approach was used to target individual groups of houses or streets and ensure a representative and comprehensive cross-section of the population. This was done on the basis of postcodes using the SuperProfiles™ demographic classification system developed at the University of Newcastle upon Tyne (Openshaw, 1989). Briefly, the SuperProfiles™ system defines clusters of similar groups of people on the basis of variables taken from the 1981 Census of Population. These include information on property characteristics, employment, geographical location, and family structure. Census returns from individual postcodes can then be used to classify its demographic and socio-economic make-up into particular lifestyle types and consumer groupings A selection of photographs showing the types of some of the areas sampled are included in figure 2. The location of the sampled housing within the study area is shown in figure 5.

All the individual unit postcodes within the chosen areas were classified according

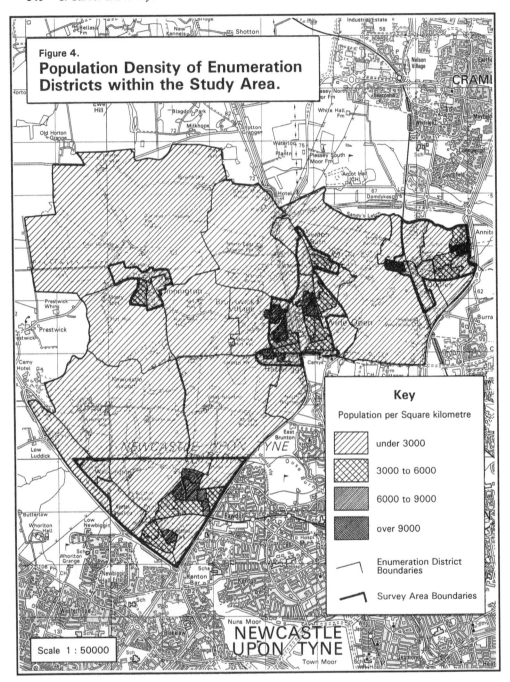

Figure 4.
Population Density of Enumeration Districts within the Study Area.

Key

Population per Square kilometre

under 3000

3000 to 6000

6000 to 9000

over 9000

Enumeration District Boundaries

Survey Area Boundaries

Scale 1 : 50000

8

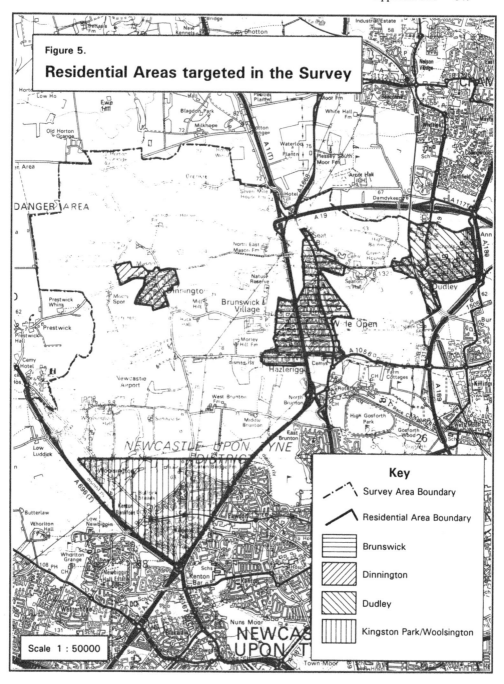

Figure 5.

Residential Areas targeted in the Survey

Key

Survey Area Boundary

Residential Area Boundary

Brunswick

Dinnington

Dudley

Kingston Park/Woolsington

Scale 1 : 50000

9

Figure 6.

Postcodes within the Target Areas

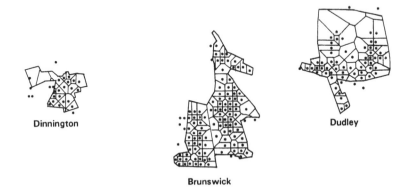

Dinnington

Dudley

Brunswick

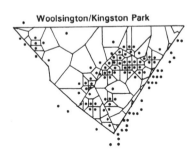

Woolsington/Kingston Park

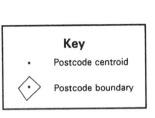

Key

· Postcode centroid

⬦ Postcode boundary

10

to their lifestyle and consumer type as determined by the SuperProfiles™ system. All postcodes within the targeted areas are shown in figure 6. Whilst most of the postcodes within areas of interest could be put into a relatively small number of groups, a number of less common, yet more varied, groups existed in some postcodes. It is considered important to try and get as wide a cross-section of the population as possible, so these rarer groups were positively targeted whilst a larger random sample was taken from the more common groups. The 2% sample taken from the targeted population is represented in figure 7 and a summary of the lifestyle/consumer types sampled is contained in table 2.

Table 2. Lifestyle/Consumer types sampled (from SuperProfiles™)		
Lifestyle	Type	Brief description
a Stockbroker Belt	a1	Middle age families in exclusive suburbs
b Metro Singles	b4	Young professionals in bedsits
c Young Married Suburbia	c10	Young well-to-do married in high turnover semis
d Rural Britain	d14	Affluent farming communities
e Older Suburbia	e11 e20	White collar family pensioners Middle aged white collar couples
f Lower Middle Class	f15 f18 f21	Lower middle class metropolitan semis Lower middle class in provincial semis Upper middle class in semis and terraced
g Multi-Ethnic Areas	None	No areas from this type within the study boundary
h Dark Satanic Mills	h24 h26	Skilled and semi-skilled in improved terraced Skilled and semi-skilled in poorer terraced
i Council Tenants	i27 i28 i29 i31 i34	Middle aged and older couples in council flats Blue collar workers in established council houses Blue collar workers with high unemployment Mature blue collar workers in mining areas Very low income council houses
j Under privileged Britain	j33 j35	High unemployment semi-skilled in council houses Highly unemployed in crowded council houses
k Unclassified	k37	All other areas unclassified

2.4   **Analysis**

Once coded into a computer readable format the questionnaire returns were read into a statistical analysis package (MINITAB) for analysis.

The first stage of any statistical analysis on a numerical dataset must be some form of preliminary analysis in order to identify basic parameters (size, shape, range, variation, etc.). More importantly, however, the figures in a categorical dataset such as on the one discussed here are often significant in themselves. Analysis presented here involves a detailed look at individual variables and at relationships between variables, In particular it is noted that certain responses vary spatially (influenced directly by demographic and environmental factors) and within local populations (again influenced by demographic and environmental factors). From

Figure 7.

Distribution of Responses by Postcode

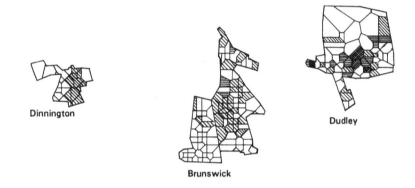

Dinnington

Brunswick

Dudley

Woolsington/Kingston Park

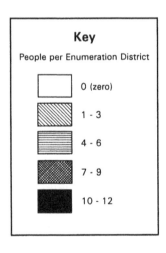

Key

People per Enumeration District

☐	0 (zero)
▨	1 - 3
▤	4 - 6
▨	7 - 9
■	10 - 12

12

these analyses it is possible to draw some reasonable relationships between demographic/environmental factors and public perceptions and likely responses to particular hazards. These will be included as further layers in the GIS database.

3.0 Analysis of questionnaire returns

Overall, 358 complete responses were obtained. The spatial distribution of these is shown in table 3 and figure 7. Most questionnaires were fully completed (i.e. all questions were answered), although a significant number of 'don't know' responses were obtained from some questions. A number of potentially interesting results have been identified from a preliminary analysis of the questionnaire returns and are discussed under each of the seven sections in the questionnaire below. Reference to specific questions from the questionnaire form are shown in brackets.

Table 3. Number of Responses by Target Area	
Responses	Target Area
130	Dudley
147	Brunswick/Dinnington/Woolsington
81	Kingston Park (control)

3.1 Section 1: Personal details

The age distribution of respondents (question 1.1) is positively skewed towards the older age groups. This is shown in figure 8. This is mostly to do with the fact that probability dictates that when sampling a population in a door-step survey it is more likely that those people found 'at home' will be of older age groups because of their habits and occupational status (i.e. more sedentary and no longer economically active). Most younger age groups being either at work or spending their leisure time away from home show up less in the questionnaire sample. It is noted that the survey was carried out on different days of the week including Wednesdays, Fridays and Sundays in an attempt to ensure a representative cross section of the population. An analysis of the 1991 Census Small Area Statistics (SAS) reveals a fairly even age distribution of respondents but with a tendency towards the older age groups in the areas sampled indicating a relatively 'top heavy' demographic structure of an ageing population (see figure 9).

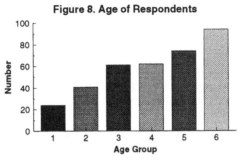

Figure 8. Age of Respondents

1 = 16-24, 2 = 21-30, 3 = 31-40, 4 = 41-50, 5 = 51-65, 6 = 65+

13

Figure 9. Age structure in survey area
(Based on Ward structure average)

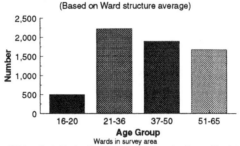

Wards in survey area
Blakelaw, Castle, Fawdon, Woolsington, Camperdown, Longbenton, Weetslade

The proportion of male and female respondents (question 1.2) is roughly equal but slightly biased towards a larger number of female respondents (i.e. male = 41%, female = 59%). Again, this may be due the underlying demographic and occupational characteristics of the population sampled (i.e. the 65 + age group tending to be dominated by women and the number of house-wives sampled, see figure 10), but could easily have occurred by chance.

As above, the proportion of households with or without children or other dependants (question 1.3) is roughly equal (i.e. dependants = 47%, no dependants = 53%).

The distribution of occupations recorded by the survey (question 1.4) reinforces the skewed age distribution and the slightly higher proportion of female respondents, with higher numbers of 'retired' (33%) and 'house-wife', (18%) responses, respectively. The remainder of responses are distributed more or less evenly throughout the other occupation groups. This is shown in figure 10.

Figure 10. Occupation of Respondents

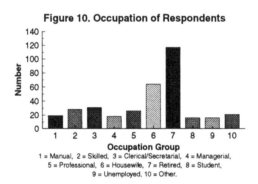

Occupation Group
1 = Manual, 2 = Skilled, 3 = Clerical/Secretarial, 4 = Managerial,
5 = Professional, 6 = Housewife, 7 = Retired, 8 = Student,
9 = Unemployed, 10 = Other.

The distribution of period of residence (question 1.5) is again skewed towards the longer period groups. This is shown in figure 11 and may again be a result of the underlying structure of the population or due to the higher proportion of older age groups sampled.

In none of the five questions in section 1 were any missing data values recorded.

14

Figure 11. Period of Residence

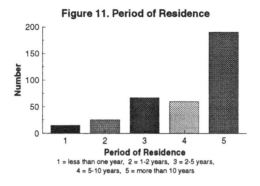

Period of Residence
1 = less than one year, 2 = 1-2 years, 3 = 2-5 years,
4 = 5-10 years, 5 = more than 10 years

3.2 **Section 2: General Hazard Perception**

Perception of greatest risk of injury/death to individuals (question 2.1) exhibits a low degree of variance and is dominated by road traffic accidents and related risks (i.e. responses stating 'crossing the road', 'speed of traffic', etc.) with over 57% of respondents stating this as the main risk. Other often stated risks include 'burglaries' (6%), 'chemical factories' (6%) (mainly due to the higher level of awareness about this hazard in Dudley - this will be discussed later) and 'illness' (13%). The full list of stated risks is shown in table 4 and the distribution of responses in figure 12.

Figure 12. Risk of Injury/Death to the Individual

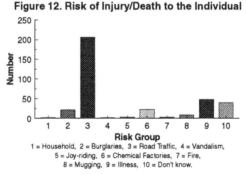

Risk Group
1 = Household, 2 = Burglaries, 3 = Road Traffic, 4 = Vandalism,
5 = Joy-riding, 6 = Chemical Factories, 7 = Fire,
8 = Mugging, 9 = Illness, 10 = Don't know.

Perception of greatest risk of injury/death to the general public involving larger numbers of people (question 2.2) is again dominated by road traffic accidents with this alone accounting for over 50% of the responses.

Table 4. Categories of risk of injury/death to the individual			
Code	Risk	Code	Risk
1	Household accidents	6	Chemical factories
2	Burglaries	7	Fire
3	Traffic accidents	8	Mugging
4	Vandalism	9	Illness
5	Joy-riding	0	Don't know

15

Other common responses include 'fire' (10%) and 'air crash' (16%). The distribution of responses across all 9 risks is shown in figure 13.

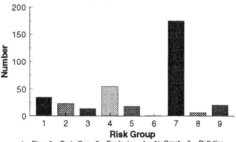

Figure 13. Risk of Injury/Death to the General Public

1 = Fire, 2 = Toxic Gas, 3 = Explosion, 4 = Air Crash, 5 = Pollution,
6 = Pesticide, 7 = Road Traffic, 8 = Military low flying, 9 = Other

The proportion of respondents being aware or unaware of living near to a potential hazard (question 2.3) is roughly equal (yes = 53%, no = 47%). Of those who were aware of living near to a potential hazard the list of stated hazards (question 2.4) was dominated by Sterling Organics in Dudley responses (54% overall and 75% of Dudley respondents alone). Other common responses included the 'airport' (28%) and 'road traffic' (11%). The full list of stated hazards is shown in table 5 and the distribution of responses in figure 14.

Table 5. Catagories of local hazards			
Code	Hazard	Code	Hazard
1	Winthrops	6	Metro
2	Chemical factories/Sterling Organics	7	Mining subsidence
3	Airport	8	Waste disposal
4	Road traffic	9	Pylons
5	Industrial estate	0	Other (mainly crime related)

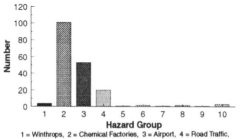

Figure 14. Stated Local Hazards

1 = Winthrops, 2 = Chemical Factories, 3 = Airport, 4 = Road Traffic,
5 = Industrial Estate, 6 = Metro, 7 = Mining Subsidence,,
8 = Waste Disposal, 9, Pylons, 10 = Other

16

3.3   **Section 3: Air Crash**
Although the proportion of respondents who thought that an air crash was or was not a major hazard (question 3.1) is roughly equal (yes = 55%, no = 45%), the proportion of respondents who thought there should be detailed plans for responding to the hazard (question 3.2) was biased towards a positive response (i.e. yes = 80%., no = 20%). In those target areas round the perimeter of Newcastle Airport (Brunswick, Dinnington and Woolsington) only 42% of respondents perceived an air crash to be a major hazard indicating that proximity to the airport does not affect peoples perception of the air crash hazard to a great degree. It is notable, however, that in Woolsington alone (the area nearest the airport) only 27% saw an air crash as being a major hazard perhaps indicating a negative relationship. In fact analysing individual feedback from respondents indicated that most people in Woolsington were more concerned about accidents happening as a result of air crash training exercises (especially speeding fire/emergency vehicles on local roads) rather than an actual air crash itself.

When asked how likely they thought such a disaster actually was (question 3.3), the response was negatively skewed towards the unlikely with 44% of respondents thinking such a disaster 'quite unlikely'. This is shown in figure 15. It is noted that this perception is recorded against the background of the recent Amsterdam air disaster and proximity to a major regional airport.

Figure 15. Perceived probability of Air Crash

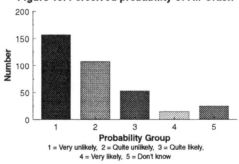

Probability Group
1 = Very unlikely, 2 = Quite unlikely, 3 = Quite likely,
4 = Very likely, 5 = Don't know

All but one of the 26 respondents in Woolsington understood that the Woolsington residential area was not on either the take-off and landing flight path for Newcastle Airport (question 3.4).

3.4   **Section 4: Chemical Hazard**
On the question of whether respondents knew what to do if there was a toxic chemical release (question 4.1) the majority (74%) said they did not. This is mainly the results of never having been told (when asked whether they had in question 4.2, 83% said 'no'). It is notable that of those who had been informed what to do at some time, most were in the Dudley area (i.e. having been informed by Sterling Organics' information campaigns, and of those outside the Dudley area, most had worked in related occupations (see section 3.7).

The following paragraphs refer to those questions about the Sterling Organics' information campaign unique to the Dudley questionnaire.

Of the 130 respondents from the Dudley area only 41% remembered the information card delivered by Sterling Organics (question 4.3). Only 28% still had possession of the card (question 4.4) but of those 64% knew where it was (question 4.5) and 57% said it was easy to get at (question 4.6). Of the 41% who remembered the card being delivered, 57% said they knew what it said on the card (question 4.7).

Of the total 358 responses collected the majority of people (68%) had little if no knowledge of what to do in the event of a toxic gas release (question 4.8), 18% knew enough to 'stay indoors' and 'close windows' etc. Only 9% knew about turning off fans, gas fires, etc. and staying in the room furthest away from the source of release and only 4% knew not to telephone the emergency services, etc. Only 2 respondents remembered everything that was on the card. The distribution of scores is shown in figure 16.

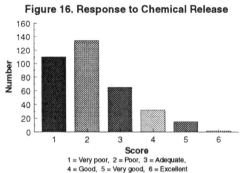

Figure 16. Response to Chemical Release

1 = Very poor, 2 = Poor, 3 = Adequate,
4 = Good, 5 = Very good, 6 = Excellent

By differentiating between responses from the Dudley area and those from other areas a better knowledge of what to do in the event of a toxic gas release can, on the whole, be seen in the responses of people living in the Dudley area. This can be seen in figure 17 but the relationship is by no means clear cut.

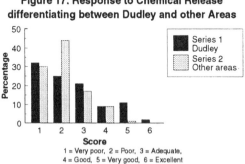

**Figure 17. Response to Chemical Release
differentiating between Dudley and other Areas**

Series 1 Dudley
Series 2 Other areas

1 = Very poor, 2 = Poor, 3 = Adequate,
4 = Good, 5 = Very good, 6 = Excellent

When asked how likely they thought a toxic gas release was (question 4.9) the response was negatively skewed towards the unlikely with 40% of respondents thinking such an incident 'quite unlikely'. However, the skewness of the distribution was not as marked as for the similar air crash question (question 3.3) since a significant proportion (24%) thought such an incident 'quite likely'. This is due also to the 'Dudley effect' with it's associated higher level of informed members of the public and proximity to the Sterling Organics site. The distribution of perceived probability is shown in figure 18.

Figure 18. Perceived probability of Chemical Release

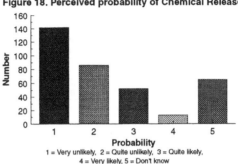

Probability
1 = Very unlikely, 2 = Quite unlikely, 3 = Quite likely,
4 = Very likely, 5 = Don't know

By differentiating between those households within and outside the Health and Safety Executive's public information zone defined around the Sterling Organics site at Dudley then a much clearer picture of the effect of information campaigns on public perception and memory retention can be seen. Sterling Organics only carried out the mailshot within the public information zone (defined by a radius of 900 metres from the former phosgene storage tank, see figure 19). This provides the opportunity to study the effect of an information campaign on two halves of a single population living in close proximity to the same hazard. By dividing the responses from the Dudley area into two groups, those inside the Sterling Organics public information zone and those outside the public information zone, a clear and significant difference can be seen (this is highlighted in section 3.8). Table 6 summarises this split in the responses from inside and outside the public information zone.

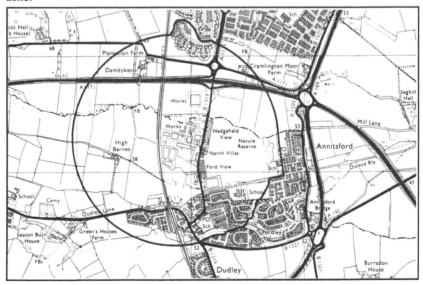

Figure 19. Sterling Organics' Emergency Planning Zone

19

Question	Inside (79 respondants)	Outside (52 respondants)
Table 6. Summary of other data relevant to chemical release and Dudley inside/outside public information zone split		
2.1 Individual risk	19% (15) = Chemicals	12% (6) = Chemicals
2.2 Public risk	16% (13) = Toxic gas 11% (9) = Explosion 13% (10) = Pollution	14% (7) = Toxic gas 4% (2) = Explosion 8% (4) = Pollution
2.3 Awareness	86% (68)	73% (37)
2.4 Named hazard	78% (62) = Chemicals	69% (37) = Chemicals
4.1 Know what to do	63% (50)	21% (11)
4.2 Ever been informed	56% (44)	20% (10)
4.3 Remember card	54% (43)	20% (10)
4.4 Still got it	42% (33)	8% (4)
4.5 Know where it is	28% (22)	4% (2)
4.6 Easy to get at	24% (19)	4% (2)
4.7 Know contents	34% (27)	6% (3)

Whilst general awareness of the risks associated with living near to a chemical works are fairly uniform across the whole range of Dudley respondents, the detailed knowledge of the individual on what to do in the event of a chemical release has been greatly enhanced by Sterling Organics' information campaign, This can be seen in the responses to questions in section 2 of the questionnaire with between 11% and 19% of respondents and between 4% and 12% of respondents stating chemicals related risks inside and outside the public information zone respectively. Perception of the probability of a chemical release from the Sterling Organics site was also fairly uniform across the range of Dudley respondents with 36% of respondents inside the public information zone thinking a chemical release 'quite likely' or 'very likely' as compared to 37% of respondents thinking the same outside the public information zone (see figure 20). Whether this is a function of living within the public information zone and having received an information card or just living closer to the Sterling Organics plant is not known.

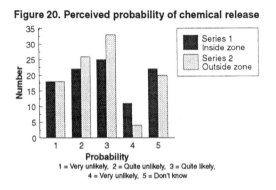

Figure 20. Perceived probability of chemical release

1 = Very unlikely, 2 = Quite unlikely, 3 = Quite likely,
4 = Very unlikely, 5 = Don't know

In relation to memory retention the responses of those people having received an information card indicates a quite marked drop in memory retention. Of those people questioned within the Dudley public information zone, only 54% remembered the card and only 34% said they knew what was on it.

In testing the knowledge of respondents on what to do in the event of a chemical release, respondents inside the public information zone faired much better than those outside it with more people having a better idea of what to do and fewer having no idea at all. Of those people questioned inside the public information zone, 53% said they knew what to do in the event of a chemical release as opposed to only 22% of those outside the public information zone. These figures are backed up by the test of people's actual knowledge (question 4.8) with 19% of respondents inside the public information zone scoring 4 or higher as opposed to only 2% scoring this high outside the public information zone (see figure 21).

Figure 21. Response to Chemical Release

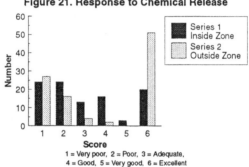

Score
1 = Very poor, 2 = Poor, 3 = Adequate,
4 = Good, 5 = Very good, 6 = Excellent

3.5 **Section 5: Emergency Planning**
The majority of people when asked which organisation co-ordinates the response to local emergencies and disasters (question 5.1) said they did not know (43%). The next most popular reply was the Police (21%), followed by the Fire Brigade (15%), the emergency services taken as a whole (11%) and the local authorities (6%). A complete list of answers given is shown in table 7 and the distribution of responses shown in figure 22. Seven respondents (2%) from the Dudley area thought that Sterling Organics co-ordinated the response, presumably as a result of their information campaigns.

Table 7. Organisations co-ordinating the response to local emergencies and disasters			
Code	Organisation	Code	Organisation
1	Police	6	Public Health Department
2	Fire Brigade	7	Community Group
3	Local Authorities (Council, etc.)	8	Ambulance Service
4	Emergency Services	9	Sterling Organics
5	Civil Defence		

Figure 22. Co-ordination of Local Emergencies

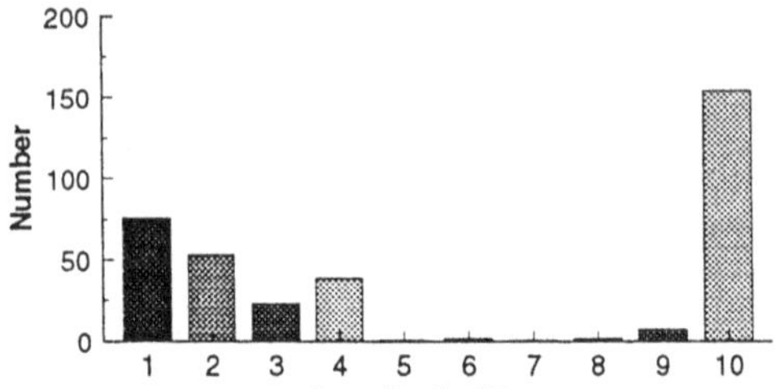

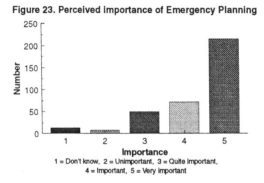

Organisation/Group

1 = Police, 2 = Fire Brigade, 3 = Local Authorities, 4 = Emergency Services,
5 = Civil Defence, 6 = Public Health, 7 = Community Group,
8 = Ambulance, 9 = Sterling Organics, 10 = Don't know

The distribution of responses to the question asking how important the respondents thought planning for emergencies and disasters was in reducing their overall impact (question 5.2) is positively skewed towards important with 60% of respondents answering 'very important'. This is shown in figure 23.

Figure 23. Perceived Importance of Emergency Planning

Importance

1 = Don't know, 2 = Unimportant, 3 = Quite important,
4 = Important, 5 = Very important

When asked which organisation writes local emergency plans (question 5.3) 30% of respondents thought it was the Police, 28% thought it was the Emergency Planning Unit (plus 15% Fire and Civil Defence Authority, since Emergency Planning Unit comes under the umbrella of Fire and Civil Defence Authority) and 21% thought it was the Fire Brigade. It is noted that although in total by combining the scores of the Emergency Planning Unit and the Fire and Civil Defence Authority (43%) many respondents only said one of these two options after reading it on the questionnaire form. The distribution of responses to the question asking the same question but for times of 'national crisis' (question 5.4) most people thought it was either the Fire and Civil Defence Authority (41%) or the military (32%) with those thinking it was the Emergency Planning Unit amounting to only 12%. The distribution of responses for these two questions is shown in figure 24.

Figure 24. Organisations writing Local Emergency Plans

1 =Police, 2 = Fire Brigade, 3 = Emergency Planning Unit,
4 = Fire and Civil Defence Authority, 5 = Military,
6 = All five, 7 = Other

3.6 **Section 6: Risk Map**

In section 6 of the questionnaire survey a hypothetical risk map of the study area is shown to the respondent. This shows areas purported to be at the most risk from toxic gas release and air crashes. The map is included here in appendix A. When asked after being shown the map whether it would affect any of their responses to previous questions (question 6.1) the majority of people said it would not (78%). Upon being asked whether they now thought emergency planning more or less important than before (question 6.2) 68% replied that it was more important with 26% saying that it made no difference.

3.7 **Section 7: Area of Work**

The final question in the questionnaire was aimed at gauging the likely prior knowledge of respondents of emergency planning. Most of the respondents (76%) did not work, or had not previously worked in the emergency services, chemical industry, health service, local government, or at the airport.

4.0 **Discussion**

More detailed analysis of the spatial variation in questionnaire responses reveals that for the bulk of responses geographical location in either real or relative terms does not exert a significant control. Neither has it been found that socio-economic or demographic factors have a significant influence on responses. The following discussion therefore concentrates on those areas where significant or probable relationships have been found to exist and are likely to have a bearing on future work.

Results from the differentiation of responses inside and outside the Dudley public information zone reveal that the information campaigns by Sterling Organics have had a marked effect in improving the knowledge of the public of what to do in the event of a toxic chemical release. This is supported by a chi squared (χ^2) statistic of 8.96 (at <0.01 level of significance and with degrees of freedom $= 1$) for high scoring responses (question 4.8). It may be therefore assumed that people having received information leaflets, etc. about local hazards (reinforced by the visible proximity of the hazard itself and past incidents at other locations, e.g. Seveso and Bhopal) will have a better knowledge of what to do should an emergency arise. In the case of other variables such as perceived hazards to the individual and the general public and the perceived risk of a chemical release then location relative to the public information zone would appear to no effect. It must be wise therefore to temper any conclusions about the predicted response of the public to a particular hazard with caution.

Comparing the response from the whole of the Dudley area (regardless of location relative to the public information zone) with those from all other areas, then it can be seen that Dudley respondents are generally more aware of the risks to both the individual and the general public posed by living in close proximity to a CIMAH site. This is supported by chi squared statistics of 47.18 and 21.59 (at <0.001 level of significance and with degrees of freedom $= 1$ in both cases) for responses to questions 2.1 and 2.2 stating 'chemical factories', respectively. It is suggested that this is due to perceived proximity to the hazard rather than a calculated linear distance due to the subjective nature of responses to questions of this kind.

In relation to the potential hazard posed by air crashes at or near Newcastle Airport a similar relationship exists. Although nobody listed the airport or air crashes as the greatest risk of injury/death to the individual, significantly more people listed 'air crash' as the main risk to the general public in question 2.2. Differentiating between those areas chosen for their proximity to the airport (i.e. Woolsington, Brunswick and Dinnington) and those some distance away (i.e. Dudley and the control group at Kingston Park) then a significant difference can be seen between the two groups. Of the respondents living near the airport 30% listed 'air crash' as the major hazard to the general public as opposed to only 5% of the respondents in other areas. This produces a chi squared statistic of 31.16 (at <0.001 level of significance and with degrees of freedom $= 1$). However, the same relationship is not repeated in peoples perception of the probability of an air crash associated with the airport. In this case only 16% of the respondents living near the airport thought that an air crash was 'quite likely' or 'very likely' as opposed to 21% of people living in other areas. This demonstrates that although proximity to the airport increases peoples awareness of the potential hazard it does not seem to make people think such a disaster is more likely.

5.0   Conclusions

From the above analysis and discussion some potentially interesting results can be seen. Firstly, it is obvious that information campaigns and mailshots by Sterling Organics have had a marked effect in increasing the public's knowledge of what to do if an accidental release should occur and may also have increased their awareness of the hazards associated with living in close proximity to a chemical plant (especially those within the public information zone). What is evident however, is that despite information campaigns by Sterling Organics, many people did not keep the information cards for future reference and memory retention was poor.

Responses given to questions under general hazard perception would indicate that most people seem more concerned about road traffic related hazards, i.e. car accidents, speeding traffic, crossing the road, etc. when considering risks to the individual and general public. Overall, the response regarding the hazard posed by air crashes showed that although people living near to the airport are more likely to be aware of the potential risk posed by air crashes they still perceive such an event as unlikely. This is especially interesting as the questionnaire survey was carried out just a few weeks after the Amsterdam air disaster when public awareness should have been at a higher level.

Despite most people appearing ignorant of the existence of the Emergency Planning Unit as an organisational body many respondents thought that 'Emergency Planning Unit' or 'Fire and Civil Defence Authority' were sensible responses to questions relating to who writes local emergency plans. Although it appears that the risk map did not have a significant impact on the views of respondents, most people thought that emergency planning was more important than they first thought. This is supported by the majority thinking emergency planning 'important' (20%) or 'very important' (60%) in reducing the overall impact of disasters and emergencies. Again in relation to air crashes the majority (80%) of people responded positively when asked if MIPs should exist for this hazard.

Further analysis of some of the more important variables relating to the current study shows few of the above results to be significant in a statistical sense. It may be that much of the less clear relationships are the result of chance but the more obvious and statistically significant relationships can be said to be the result of a combination of demographic, socio-economic and/or environmental factors (e.g. perceived proximity to particular hazards, being inside the Dudley public information zone, etc.).

The conclusions derived from the analysis of survey returns contained in this report suggest therefore, that:

1. most people regard emergency planning as being important though may not have heard of the Emergency Planning Unit directly;

2. public information campaigns and mailshots do have an effect of both increasing public awareness of hazards and improving response to those hazards;

3. memory retention after information campaigns is low indicating that more frequent information campaigns are required to remind the public of the risks and appropriate actions (this must be tempered with the possibility that over-exposure of the public may anaesthetise them to the risks);

4. perceived proximity (as opposed to calculated linear distance) is important in predicting public awareness of visible hazards (e.g. the airport and Sterling Organics' Dudley site);

6.0 Recommendations

Recommendations derived from the findings of the survey as described in this report and from the report's conclusions outlined above can only be that further and increased effort needs to be made to increase both the public's general awareness of the hazards associated with living near to particular hazards and their overall longer term memory retention. At present it seems that whilst proximity to the hazard in question increases awareness, memory retention, despite information campaigns, is still low. This is true at least in the case of the population targeted in this particular study. Whether it is possible to extrapolate from these particular results to other areas needs to be clarified by further research. The recommendations of this report, therefore, are as follows:

1. investigations need to be carried out as to how general awareness of hazards can be raised in local populations and how memory retention of key information can be increased in the longer term (i.e. the intervening periods

between information campaigns). More frequent information campaigns are a possibility, but care needs to be taken not to over familiarise the public so that they become blase;

2. greater emphasis needs to be placed on the role of the Emergency Planning Unit, working in liaison with the managers of CIMAH sites and other potential hazards, in increasing public awareness and memory retention;

3. information campaigns need to be instigated to establish the role of the Emergency Planning Unit in the public eye; and

4. further research is necessary to both widen the current results to a range of hazards and locations and to establish whether the findings of the current survey are representative of populations around CIMAH sites in general and other hazards as a whole. In particular, research effort needs to be concentrated on establishing a set of geographic indicators (e.g. distance from hazard, census variables, etc.) which can be used to predict likely levels of awareness and memory retention in unsurveyed populations.

Consultations

Cumbria Emergency Planning Unit
Emergency Planning College
Sterling Organics
Newcastle Airport authorities

Bibliography

Carver, S, Myers, A and Newson, M., (1992) Response planning for hazardous gas releases: a proposal for a GIS based strategy. Proposal to Home Office. (see appendix C)

Gardner, G., (1976) Social surveys for social planners. Open University Press

Kirkwood, A., (ongoing research) Public perception questionnaires and 'risk ranking' exercise. Home Office Planning College (York) and Bradford University

Openshaw, S., (1989) Making geodemographics more sophisticated, in Journal of the Market Research Society 31, p.111-131

Appendix A   Questionnaire design

1.   Personal Details:

1.1   Age
What age group do you fall into?
16-20   21-36   37-50   51-65   65+

1.2   Sex
What sex is the respondent?
Male   Female

1.3   Dependants
Are there any children or other dependants living in the house?   Yes   No

1.4   Occupation
What occupation group best describes your work?
Manual   Skilled   Clerical/secretarial   Managerial   Professional
Housewife   Retired   Student   Unemployed   Other

1.5   Period of residence
How long have you lived at this address or at another address in the immediate area?
under 1 year   1-2 years   2-5 years   5-10 years   10 +years

2.   General Hazard Perception:

2.1   What do you think is the greatest risk of injury/death to individuals in your area?

2.2   What do you think is the greatest risk of injury/death to the **general public** in your area?
Fire   Toxic gas   Explosion   Air crash   Pollution
Pesticides   Road traffic   Military low flying   Other

2.3   Are you aware of living near to any potential hazard?   Yes   No

2.4   If 'yes' to 2.3, what is that hazard?

3.   Air Crash

3.1   Do you think that an air crash is a major hazard in this area?   Yes   No

3.2   Should there be detailed plans for responding to this hazard?   Yes   No

3.3   How likely do you think such a disaster actually is?
Very unlikely   Quite unlikely   Quite likely   Very likely   Don't know

3.4   Did you know that this area is not on the take-off and landing flight path for the airport?   Yes   No

27

4.   **Chemical Hazard:**

4.1   Do you know what to do in the event of a toxic chemical release?   Yes   No

4.2   If 'no' to 4.1, Have you ever been informed what to do?   Yes   No

4.3   In July 1990 Sterling Organics delivered a card giving information
on what to do in the event of a release of toxic gas.
Do you remember?   Yes   No

4.4   If 'yes' to 4.3, have you still got the card?   Yes   No

4.5   If 'yes' to 4.4, do you know where it is?   Yes   No

4.6   If 'yes' to 4.5, is it easy to get at?   Yes   No

4.7   If 'yes' to 4.3, do you know what is says on the card?   Yes   No

4.8   If 'yes' to 4.7, what should you do if there is a release of toxic gas?
1   2   3   4   5

4.9   How likely do you think such an incident actually is?
Very unlikely   Quite unlikely   Quite likely   Very likely   Don't know

5.   **Emergency Planning:**

5.1   Which organisation co-ordinates the response to local emergencies and disasters?

5.2   How important do you think planning for emergencies and disasters is in reducing
their overall impact?
Unimportant   Quite important Important Very important Don't know

5.3   Which organisation writes local emergency plans?
Police   Fire Brigade   Emergency Planning Unit   Fire & Civil Defence
Military

5.4   Which organisation deals with emergency planning in times of 'national crisis' (i.e.
war)?
Police   Fire Brigade   Emergency Planning Unit   Fire & Civil Defence
Military

6.   **Risk Map**

6.1   This is an example of a map showing areas at most risk from toxic gas release and
air crashes. Would this affect any responses to previous questions?   Yes   No

6.2   This type of map is used in emergency planning. Do you think emergency planning
is more or less important than you first thought?   More   Less

7.   **Area of Work**

7.1.   Do you work in either of the following?
Emergency services   Chemical industry   Health service   Local government
Airport

N.B.   Question 3.4 applies to Woolsington questionnaire only
Questions 4.3-4.7 inclusive apply to Dudley questionnaire only

Appendix B.   Public Perception Survey Coding Form

1.   Personal Details

1.1   Age:

16-20	(1)
21-30	(2)
31-40	(3)
41-50	(4)
51-65	(5)
65+	(6)

1.2   Sex:

Male	(1)
Female	(2)

1.3   Dependants:

Yes	(1)
No	(0)

1.4   Occupation:

Manual	(1)
Skilled	(2)
Clerical/secretarial	(3)
Managerial	(4)
Professional	(5)
Housewife	(6)
Retired	(7)
Student	(8)
Unemployed	(9)
Other	(0)

1.5   Residence:

under 1	(1)
1-2	(2)
2-5	(3)
5-10	(4)
10+	(5)

2.   General Hazard Perception

2.1   Individual risk:

Household	(1)
Burglaries	(2)
Traffic	(3)
Vandalism	(4)
Joy-riding	(5)
Chemical factories	(6)
Fire	(7)
Mugging	(8)
Illness	(9)
None	(0)

2.2   Public risk:

Fire	(1)
Toxic gas	(2)
Explosion	(3)
Air crash	(4)
Pollution	(5)
Pesticides	(6)
Road traffic	(7)
Military low flying	(8)
Other	(9)

2.3   Awareness:

Yes	(1)
No	(0)

2.4   Named Hazard:

Winthrops	(1)
Chemical factories	(2)
Airport	(3)
Traffic	(4)
Industrial estate	(5)
Metro	(6)
Mining subsidence	(7)
Waste disposal	(8)
Pylons	(9)
Other	(0)

3.   **Air Crash** (question 3.4 on Woolsington responses only)

3.1   Major hazard:

Yes	(1)
No	(0)

3.2   Plans:

Yes	(1)
No	(0)

3.3   Likelihood:

Very unlikely	(1)
Quite unlikely	(2)
Quite likely	(3)
Very likely	(4)
Don't know	(0)

3.4   Flight path:

Yes	(1)
No	(0)

4.   **Chemical Hazard** (questions 4.3 - 4.7 on Dudley responses only)

4.1   Know what do:
      Yes   (1)
      No   (0)

4.2   Informed:
      Yes   (1)
      No   (0)

4.3   Remember card:
      Yes   (1)
      No   (0)

4.4   Still got it:
      Yes   (1)
      No   (0)

4.5   Know where:
      Yes   (1)
      No   (0)

4.6   Easy to get at:
      Yes   (1)
      No   (0)

4.7   Know contents:
      Yes   (1)
      No   (0)

4.8   What should do:
      1-5   (1)
      Don't know   (0)

4.9   Likelihood:
      Very unlikely   (1)
      Quite unlikely   (2)
      Quite likely   (3)
      Very likely   (4)
      Don't know   (0)

5   **Emergency Planning**

5.1   Organisation:
      Police   (1)
      Fire Brigade   (2)
      Local authorities   (3)
      Emergency services   (4)
      Civil defence   (5)
      Public health   (6)
      Community group   (7)
      Ambulance   (8)
      Sterling Organics   (9)
      Don't know   (0)

31

5.2   Importance:

Unimportant	(1)
Quite important	(2)
Important	(3)
Very important	(4)
Don't know	(0)

5.3   Local plans:

Police	(1)
Fire Brigade	(2)
Emergency Planning Unit	(3)
Fire & Civil Defence	(4)
Military	(5)

5.4   War plans:

Police	(1)
Fire Brigade	(2)
Emergency Planning Unit	(3)
Fire & Civil Defence	(4)
Military	(5)

6.   Risk Map

6.1   Affect response:

Yes	(1)
No	(0)

6.2   Importance:

More	(1)
Less	(2)

7.   Area of Work

7.1   Related work:

Emergency services	(1)
Chemical industry	(2)
Health Service	(3)
Local Government	(4)
Airport	(5)
None of these	(0)

Appendix C   **The Role of Public Perception in the Response Planning for Major Incidents - A Proposal for a GIS based strategy**

Steve Carver (1), Alan Myers (2) and Malcolm Newson 1

1   Department of Geography
     University of Newcastle upon Tyne
     Newcastle upon Tyne
     NE1 7RU

2   Tyne and Wear Emergency Planning Unit
     Floor 2, Portman House
     Portland Road
     Newcastle upon Tyne
     NE2 1AQ

1   **Introduction**

The following research proposal outlines a combined GIS (Geographic Information Systems) and public perception survey approach to assist the Emergency Planner in improving current plans for coping with for example hazardous gas releases from industrial chemical plants designated as CIMAH (Control of Industrial Major Accident Hazard) sites and tanker collision/derailment along transport routes. GIS provide the user with the capability to assimilate large volumes of spatial data from a variety of sources in a highly efficient and accurate manner. One variable which may vary considerably within an area of interest is the Public's perception of emergency planning. This is a seldom thought of, but crucially important aspect of planning for major incidents which pose a threat to the general public. In many cases public perception may determine and/or correlate to the public response to instructions given, either previously or during in incident, and hence affect the severity of the outcome. Results from public perception surveys would form an important information layer in any GIS used for emergency planning for hazardous gas releases.

With the above discussion in mind, then the combined GIS and public perception survey approach proposed would include:

a)   defining more realistic zones of consequence based on likely dispersal scenarios used in conjunction with relevant ground information (i.e. census - based population data, local infrastructure and land use);

b)   carrying out a survey of public perception of emergency planning in the study area (this may include variables to account for memory retention about instructions given in previous mailshots around CIMAH sites as well as variables accounting for hazard perception and response);

c)   aiding and development of better response plans for the emergency services (i.e. local authority, police, fire and ambulance) and for the evacuation of the population at risk based on analysis in (a) and feedback from surveys in (b);

d)   simulating chemical releases for use in training exercises; and

e)   developing field-based decision support tools for use in making better informed 'real-time' decisions.

It is proposed to investigate the potential of GIS and survey returns as a development tool through a detailed case study of one or more areas containing CIMAH sites. a break down of estimate cost is given in annex A of this appendix

2   **Developing improved zones of consequence**

By modelling the likely dispersal patterns of gases under various conditions specific to the study site and combining these in the GIS with local information on population infrastructure (e.g. road and rail networks) and other relevant factors it is suggested that more realistic zones of consequence than the concentric ring and sector models used hitherto can be derived. Risk maps may then be defined detailing those areas most at risk under difference dispersal scenarios. In turn, using these may then lead to more informed decisions on emergency response planning.

2.1.   **Modelling gas dispersions**

Problems associated with this approach are mainly concerned with the ability to model the dispersal of gases under various physical conditions. If accurate models could be developed, combining outputs with local information to derive the 'definitive' risk map would be a relatively simple matter. As it is, the dispersal of gases into the atmosphere is, at best, extremely complex and governed to a large extent by chaos theory. Without resorting to a very complex models, then gas dispersal cannot be modelled with a high degree of accuracy. Reasonable approximations may, however, be derived in a GIS using basic dispersion models. Although these models simplify reality by making a number of assumptions regarding turbulence and diffusion etc., they may represent a considerable improvement over the concentric ring and sector model. The major advantage of GIS-based approach is that the model used can be modified using information from the GIS to take both spatial and temporal variations of controlling factors into account and so provide more realistic predictions.

It should be stressed at this point that the spatial level of dispersing modelling required makes highly detailed predictions of plume dispersal impossible within the confines of the current proposal. Whilst it has been shown that detailed plume modelling is possible on micro scale (i.e. around individual obstructions typically within a few hundred metres of the point of release) using for example, finite element analysis (Taylor 1989), model predictions on a meso scale (i.e. up to several kilometres from the point of release) are less accurate and require a more generalised approach.

It is proposed here to use modified cell-based and/or Gaussian plume and puff models (Miller and Holzworth, 1967) to simulate gas dispersal in a GIS framework. Inputs to the model will be provided from sources both internal and external to the GIS. Internal inputs will include spatial information on surface roughness, topography and the location of physical barriers. External inputs will include such basic information necessary for simulation as type of release, initial gas concentration, wind speed, wind direction and atmospheric stability.

A number of different outputs are envisaged. At the most basic level the maximum likely extent of the gas plume before it is diluted and dispersed beyond dangerous levels of concentration can be calculated and mapped. A more detailed picture may be obtained by looking at maximum and/or average gas concentrations within the plume. Alternatively, outputs may be temporal, providing a break down of plume development with time, again showing either maximum plume extent or in-plume gas concentrations. A logical extension of the time-based model is its use as an aid to real-time decision support in the field. Model predictions could be updated as and when controlling factors such a wind speed and direction change during a release incident.

From these outputs a number of dispersal scenarios may be calculated before an incident occurs which can be used in the definition of zones of consequence for use by the authorities and emergency services. A range of dispersal scenarios may be identified on the basis of likely combinations of the different controlling factors (i.e. wind speed and direction, type of release, initial gas concentration, etc.). By combining the set of possible release scenarios (i.e. instantaneous bulk release, continuous release with fire) with the set of average meteorological scenarios, then a reasonable range of likely dispersal scenarios can be calculated.

Because of errors arising from uncertainty in model inputs, temporal variations in controlling factors (i.e. wind shift, release rates, etc.) and the assumptions and simplification made by the model itself, model outputs will contain varying amounts of error. Allowances for model inaccuracies therefore need to be made in the presentation of results. In this respect previous work on error handling and accuracy in GIS operations may prove useful. An epsilon model (Brunsdon et al., 1990) or a Monte Carlo simulation approach (Openshaw et al., 1991) could be used in providing confidence regions for plume simulations.

2.2 Defining zones of consequence and risk mapping

It has been suggested that improved zones of consequence may be derived from combining dispersal scenarios with local information on population, infrastructure and land use. This can be easily achieved in a GIS framework by simple overlaying of dispersal scenarios on to relevant digital map information describing the distribution and characteristics of local population, transport routes, etc. within a specified distance of the CIMAH site, for example, the 1000 metre consultation distance specified by the Health and Safety Executive for chlorine and phosgene gas (Smallwood, 1990).

Obviously when considering accidental releases of hazardous gases such as chlorine and phosgene, then it is people (and livestock) which are at risk. It follows therefore, that it is possible to identify areas potentially at risk from a gas release (i.e. residential, retail, service, office and industrial areas where people live and work) and also linear features (i.e. railway lines, roads and other rights of way along which people travel in the vicinity of the release site). By overlaying the range of identified dispersal scenarios on to these areas and linear features then areas of highest potential risk from gas releases can be identified. Both areas potentially at risk and areas of highest potential risk can then be used in the definition of improved zones of consequence and an overall risk map for the area around the chemical site.

It is recognised that population patterns change with the season, day of the week and time of day and that this needs to be taken into account when defining areas at risk. All temporal information influencing population distribution could be considered if taking this aspect of the problem to its logical extent. It is envisaged that at its most basic level at least three risk maps may be necessary; one showing areas at risk during work hours (i.e. 0700-1700 hours); one showing areas at risk during non-working hours (ie. 1700-0700 hours) and holiday/weekends; and one showing areas at risk during those hours when people are moving around (i.e. lunch time and morning and evening rush hours).

3.   **Public perception of emergency planning**

To ascertain the extent to which public perception of emergency planning affects the outcome of a hazardous gas release, it is necessary to collect information from a sample of the population within the study area using a questionnaire survey. The overall aim of such a survey would be to investigate the extent to which the public are aware of the threats which exist to life and the environment and the level of mitigation which might be possible through properly written MIPs (Major Incident Plans). Specifically, questions in the survey might aim to:

a)   ascertain the public's awareness of risks;

b)   ascertain the public's perception of local authority involvement in protection;

c)   ascertain the public's perception of the involvement of Emergency Planning, Police, Fire Brigade and Ambulance Service and;

d)   ascertain the public's perception of the importance of emergency planning.

It is proposed that the questionnaire be devised by the University with assistance from the Emergency Planning Unit. Questionnaires devised at the Emergency Planning College (York) and Bradford University already exist (Kirkwood, ongoing research), and ideas and structures from these may be used where appropriate. Further, it is proposed that the questionnaire should incorporate a risk map and associated zones of consequence calculated prior to carrying out the public perception survey by the methods outlined above in section 2.

Results from the questionnaire survey could be coded according to levels of perception found and used as the basis of a further data layer in the GIS in order to assess any effects on risk maps and zones of consequence. In this way feed back in the form of questionnaire results could be used to produce modified risk maps of the study area and provide base data for use in further studies to aid emergency planning on a broader scale within the county.

4.   **Memory retention around CIMAH sites**

A further factor to be taken into consideration is how much the public remembers after being instructed what to do in case of a hazardous gas release. This may be useful in deciding how to educate the population around CIMAH sites in the future and may also be important when compiling plans to mitigate the effects of gas releases. It is suggested that data on memory retention may be collected within the questionnaire survey proposed above in Section 3, thereby reducing the time required to gather full information. Results on memory retention may then be coded to form a further layer in the GIS and be used alongside information on public perception to improve risk maps and zones of consequence. Again, this data may also be used to aid emergency planning on a broader scale within the county.

5.   **Response planning**

Response plans for use by the authorities and emergency services may be re-defined in reference to the dispersal scenarios, zones of consequence, risk maps and public perception data described above.

Dispersal scenario maps may be used in deciding which areas need evacuating and in

36

which areas evacuation should be a priority concern given the conditions of release. Similarly, areas where warnings may be issued to stay indoors with windows closed, etc., may also be defined using dispersal scenario and risk maps. Evacuation routes may be reviewed and altered by making reference to the local road network and especially how it lies relative to high risk areas and dispersal maps. Where road traffic and right of way control points are set up to prevent people entering the area may also be reviewed using this information. Again, the location of mobile incident control points may be reviewed in the same manner.

Problems of evacuation and traffic control aside, the information accessible by the GIS may provide valuable information and insights into wider response planning issues. These may include location of temporary accommodation for evacuees, indicators of likely public response (depending on patterns of public perception and memory retention), access to transport for immediate evacuation, etc.

The ability to simulate chemical releases and integrate dispersal maps with information on local population, transport, etc., all within a computer environment creates opportunities for advanced training facilities. Training exercises may be based around a decision support system such as that outlined in the next section.

6. **Field-based decision support**

The discussion entered into above has been based mainly around pre-processing information to simulate likely dispersal scenarios, identify zones of consequence, process information on public perception, draw risk maps and using these for reviewing response plans and developing training exercises. Fifth on the list of objectives in the introductory paragraphs is a reference to the possible use of a GIS-based decision support system as an aid to making real-time decisions in the field. It is suggested that some ability to modify plume model outputs if changes in wind speed and direction, release conditions, etc., occur would be extremely useful. This would allow the emergency services to respond appropriately to changing conditions affecting plume dispersal and new areas of greater risk.

Although quite complex analyses in the field (akin to those described above under defining zones of consequence and response planning) would be possible, the complexity of a field-based decision support system would need to be kept to an absolute minimum. With this in mind, functionality should be limited to:

a)   displaying the relevant dispersal scenario for the prevailing conditions;

b)   making modifications by re-running the plume model if changes in controlling factors occur; and

c)   displaying plume information (i.e. extent, gas concentrations and error boundaries) on a backcloth map of relevant strategic information for visualisation purposes.

Any more information/functionality than that outlined above would only serve to confuse.

7. **Data sources**

Data from various different sources are required for the implementation of the above methodology. These include meteorological records, large scale plans and maps, census statistics, aerial photographs, ground surveys and questionnaires. These are considered briefly in turn below.

Meteorological records are essential for providing information on wind speed and direction which are primary controlling factors in modelling plume dispersal. Although it is unlikely that detailed wind speed and direction records will be available for the site in question, records from the nearest weather station may provide a general picture of prevailing conditions. In addition, field measurements may prove useful in supplementing basic wind data to provide more detailed information on wind movements in and around the buildings of the site itself.

Large scale plans and maps are invaluable sources of information regarding land use, transport networks, topography, location of physical barriers to plume dispersal and indices of surface roughness. This can be digitised and stored in the GIS and used as model inputs in defining dispersal scenarios, identifying zones of consequence, drawing risk maps and developing response plans.

Information about population, both in terms of numbers of people and their characteristics, can be obtained from census statistics. Population density information is necessary for mapping populations at risk and defining zones of consequence whilst information provided on age structure, occupation, prosperity indices and housing may also be useful in assessing public response and daily variations in population density.

Information not readily available from maps and statistics may be gained from aerial photographs and ground surveys. These may assist in collating more detailed information on land use, estimates of surface roughness indices, physical barriers and possible collecting points for heavy gases such as surface depressions and the uphill sides of walls and embankments.

Information about levels of and variations in public perception and memory retention as eluded to in section 3 and 4, will need to be gathered from a sample population by questionnaire survey. It is proposed that this may be done by the University and helped by undergraduate and postgraduate students.

8.   Conclusions

This document outlines a proposal for a GIS-based approach to planning for and managing releases of hazardous gases from industrial chemical plants. An integrated GIS and modelling approach is forwarded as a means of gaining reasonable predictions of plume development from which a range of likely dispersal scenarios may be created. These together with local information concerning population, land use, infrastructure and public perception could be used to define zones of consequence, draw maps showing population at risk and review response plans. Attention is drawn to the possible use of GIS-based systems as a means of real-time decision support for emergency operations during a release incident. A structured approach to the collecting of information for the public perception surveys, which involves risk based maps being compiled both before and after the survey results are collected and added to the GIS, is recommended.

References

Brunsdon, C., Carver, S., Charlton, M. & Openshaw, S., (1990). A review of methods for handling error propagation in GIS. In Proc. 1st European Conf. on Geographical Information Systems. Vol 1, p.106-116, Amsterdam, The Netherlands, April 1990.

Kirkwood, A (ongoing research) Public perception questionnaires and 'risk ranking' exercise. Home Office Emergency Planning College (York) and Bradford University.

Miller, M. & Holzworth, G., (1967). An atmospheric diffusion model for metropolitan areas in JAPCA, Vol 17, p.46-50.

Openshaw, S., Charlton, M. & Carver, S., (1991). Error propagation: a Monte Carlo simulation in Handling Geographic Information: methodology and potential applications. I. Masser and M. Blakemore (eds.), Longman, London.

Smallwood, R., (1990). The UK experience in communicating risk advice to land use planning authorities in communicating with the public about major accident hazards. H. Gow & H. Otway (eds.), Elsevier Applied Science, London.

Taylor, C., (1989) 'CFD; what is it, what does it cost, and what does it do?' in Process Engineering, March 1989.

Index